Operator Theory: Advances and Applications
Vol. 8

Edited by
I. Gohberg

Editorial Office

School of Mathematical Sciences
Tel-Aviv University
Ramat-Aviv (Israel)

Birkhäuser Verlag
Basel · Boston · Stuttgart

I. Gohberg, P. Lancaster, and
L. Rodman

Matrices and Indefinite Scalar Products

1983

Birkhäuser Verlag
Basel · Boston · Stuttgart

Volume Editorial Office

School of Mathematical Sciences
Tel-Aviv University
Ramat-Aviv (Israel)

Library of Congress Cataloging in Publication Data

Gohberg, I. (Israel), 1928–
Matrices and indefinite scalar products.
(Operator theory, advances and applications ; v. 8)
Includes bibliographical references and index.
1. Linear operators. 2. Matrices. 3. Inner product spaces, Indefinite. I. Lancaster, Peter, 1929– II. Rodman, L. III. Title. IV. Series.
QA329.2.G63 1983 512.9′434 83-11887
ISBN 3-7643-1527-X

CIP-Kurztitelaufnahme der Deutschen Bibliothek

Gohberg, Izrail':
Matrices and indefinite scalar products / I. Gohberg ; P. Lancaster ; L. Rodman. - Basel ; Boston ; Stuttgart : Birkhäuser, 1983.
(Operator theory ; Vol. 8)
ISBN 3-7643-1527-X
NE: Lancaster, Peter:; Rodman, L.:; GT

Printed in Switzerland by Birkhäuser AG, Graphisches Unternehmen, Basel
ISBN 3-7643-1527-X

To M. G. Krein in appreciation of his pioneering and profound work in operator theory in indefinite scalar product spaces

TABLE OF CONTENTS

PREFACE

This book provides a comprehensive treatment of the theory and application of matrices in the presence of an indefinite scalar product. The theory is a natural extension of the classical theory of hermitian and unitary matrices in linear algebra, and detailed accounts of application to problems arising in differential equations and systems theory are included. The mathematical tools employed are mostly accessible even for undergraduate students, and where more advanced techniques are required, they are included in order to provide as self-contained a presentation as possible. The accessibility of the mathematics and the importance of the applications should make this a widely useful work for engineers, scientists, and mathematicians, alike.

The authors became actively engaged in this subject matter while working together on a comprehensive theory for matrix polynomials. In order to resolve certain questions concerning matrix polynomials with hermitian coefficients, it was necessary to understand fully, and to develop further, a theory of matrices which are selfadjoint, or unitary, in an indefinite scalar product. Once this was done it was realised that we had at our disposal some widely useful tools, and we soon became aware of other problem areas where these tools could be put to work. We also realised that the theory we had put together fills a gap in the mathematical literature, even though some of it already existed in the case of infinite dimensional spaces.

This volume is the result of three years of work by the authors in this direction. In it, the comprehensive theory referred to will be found, as well as detailed accounts of the applications to four different problem areas; namely, linear periodic Hamiltonian systems of differential equations, hermitian matrix polynomials, hermitian rational matrix functions, and symmetric algebraic Riccati equations.

Many colleagues have assisted us with discussion, criticism, and comments, to all of whom we are duly grateful. Especially we would like to thank our friends M.A. Kaashoek and A. Ran of the Free University, Amsterdam, for their help. We are also grateful to our employers, Tel-Aviv University and the University of Calgary, for the provision of supporting facilities and financial support as the authors travelled from one institution to the other. Support from the Natural Sciences and Engineering Research Council of Canada, as well as from the Nathan and Lily Silver

Chair for Mathematical Analysis and Operator Theory at the Tel Aviv University, is also gratefully acknowledged.

Last, but not least, we are glad to express our indebtedness to Sara Fliegelmann. She worked diligently, and with extraordinary speed and accuracy to produce the final typescript.

THE AUTHORS

Tel-Aviv
January, 1983

INTRODUCTION

The theory of hermitian matrices forms a fundamental part of linear algebra and it is well-known that this theory admits the analysis of pencils of matrices of the form $A_1\lambda + A_2$ where $\lambda \in \mathbb{C}$, provided A_1 and A_2 are hermitian and one of them, say A_1, is positive definite. One instructive way of making this generalization is to introduce a new scalar product $[.,.]$ on $\mathbb{C}^n$ defined by A_1 and the standard scalar product through the relation $[x,y] = (x,A_1y)$ for all $x,y \in \mathbb{C}^n$. Then it is easily seen that $A_1^{-1}A_2$ is selfadjoint with respect to the new scalar product and in this way analysis of the pencil $A_1\lambda + A_2 = A_1(\lambda I+A_1^{-1}A_2)$ is reduced to the classical hermitian case.

One of the objects of study of this book is matrix pencils of the form $A_1\lambda + A_2$, where A_1 and A_2 are hermitian but now, A_1 is merely nonsingular. The methods to be developed for this problem begin with a generalization of the remarks made in the first paragraph. A form $[.,.]$ is defined on $\mathbb{C}^n$ in what is formally the same way, but now determines an *indefinite* scalar product on $\mathbb{C}^n$. Again, $A_1^{-1}A_2$ is selfadjoint with respect to this indefinite scalar product. However, this simple generalization hides some dramatic differences in the algebraic and geometric properties of the pencils $A_1\lambda + A_2$. But before proceeding with discussion of the mathematical implications, consider first an important and mathematically interesting example arising in the analysis of second (or higher) order systems of differential equations with constant coefficients.

Let L_0, L_1, L_2 be hermitian matrices with L_2 nonsingular and consider the differential equation

$$L_2 \frac{d^2y}{dt^2} + L_1 \frac{dy}{dt} + L_0y = 0$$

where y is a $\mathbb{C}^n$-valued function of the real variable t. The characteristic equation of this system is

$$(L_2\lambda^2 + L_1\lambda + L_0)x = 0$$

where $\lambda \in \mathbb{C}$ and $x \in \mathbb{C}^n$. This is easily seen to be equivalent to the equation $(A_1\lambda+A_2)z = 0$ where

$$A_1 = \begin{bmatrix} L_1 & L_2 \\ L_2 & 0 \end{bmatrix}, \quad A_2 = \begin{bmatrix} -L_0 & 0 \\ 0 & L_2 \end{bmatrix}, \quad z = \begin{bmatrix} x \\ \lambda x \end{bmatrix}. \tag{1}$$

Clearly, A_1 and A_2 are hermitian and A_1 is nonsingular but, without further strong hypotheses, neither A_1 nor A_2 is a definite matrix. Note, in particular, that for the problem of vibrating systems where such equations arise, the coefficient matrices L_0 and L_1 are positive semidefinite while L_2 is positive definite. Analysis of this important problem therefore demands the introduction of an indefinite scalar product.

In general, if H is a nonsingular hermitian matrix, we define an associated indefinite scalar product on $\mathbb{C}^n$ by $[x,y] = (x,Hy)$ for all $x,y \in \mathbb{C}^n$ and say that matrix A is H-selfadjoint if $[x,Ay] = [Ax,y]$ for all $x,y \in \mathbb{C}^n$. This is easily seen to be equivalent to the condition $HA = A^*H$. Also, a matrix U is said to be H-unitary if $[Ux,Uy] = [x,y]$ for all $x,y \in \mathbb{C}^n$ and this is found to be equivalent to the condition that $U^*HU = H$.

For the analysis of hermitian matrices the standard scalar product on $\mathbb{C}^n$ leads to several natural geometric notions. These notions have analogues for the space $\mathbb{C}^n$ with an indefinite scalar product, but there are significant and far-reaching differences, such as the possibility of positive, negative and zero lengths.

The first aim of this book is to develop a systematic theory of linear algebra in the presence of an indefinite scalar product. The ideas to be developed are parallel to those of a standard linear algebra course, including canonical forms and their invariants for H-selfadjoint and H-unitary matrices, invariant subspaces of different kinds, and perturbation problems.

The second aim is to develop four different areas of application and to show how the theory works in each case. These are, briefly, linear periodic Hamiltonian systems of differential equations, the study of polynomial and rational hermitian matrix valued functions, and investigation of the symmetric algebraic Riccati equation. Each of these problems will be seen to originate from either systems of differential equations with hermitian matrix coefficients or matrix-valued functions taking hermitian values on the real line or unit circle.

Both aims of the book are achieved within a reasonably self-contained presentation. Necessary background material is provided in the text or in the form of appendices.

The four problem areas will now be introduced and it will be indicated how an indefinite scalar product makes its appearance in each case. First, let H be a nonsingular hermitian matrix, and $K(t)$ be a matrix-valued function taking hermitian values for all real t, and let $K(t)$ be periodic with period ω. Then an important class of differential equations, described here as Hamiltonian, has the form

$$H\frac{dy}{dt} = iK(t)y \tag{2}$$

where y is a $\mathbb{C}^n$-valued function of t. The set of all solutions is essentially determined by the unique solution of the initial value problem for the matrix variable $X(t)$:

$$H\frac{dX}{dt} = iK(t)X\ , \qquad X(0) = I\ . \tag{3}$$

This function $X(t)$ is known as the matrizant of the system *and is found to be* H-*unitary* for each t.

Now let $L_0, L_1, \cdots, L_{\ell-1}$ be $n \times n$ hermitian matrices and consider the (monic) matrix polynomial $L(\lambda)$ defined by $L(\lambda) = \lambda^{\ell} I + \sum_{j=0}^{\ell-1} \lambda^j L_j$. It is well-known that the spectrum of $L(\lambda)$, i.e. the zeros of $\det L(\lambda)$, coincide with the eigenvalues of C_L where

$$C_L = \begin{bmatrix} 0 & I & 0 & \cdot & \cdot & \cdot & 0 \\ 0 & 0 & I & & & & \cdot \\ \cdot & & & \cdot & & & \cdot \\ \cdot & & & & \cdot & & \cdot \\ \cdot & & & & & \cdot & \cdot \\ 0 & 0 & \cdot & \cdot & \cdot & \cdot & I \\ -L_0 & -L_1 & \cdot & \cdot & \cdot & \cdot & -L_{\ell-1} \end{bmatrix} . \tag{4}$$

Now C_L is generally not hermitian but, on introducing the "symmetrizer" B_L for $L(\lambda)$ where

$$B_L = \begin{bmatrix} L_1 & L_2 & \cdot & \cdot & \cdot & L_{\ell-1} & I \\ L_2 & & & & \cdot & I & 0 \\ \cdot & & & \cdot & \cdot & & \cdot \\ \cdot & & \cdot & \cdot & & & \cdot \\ \cdot & & \cdot & & & & \cdot \\ L_{\ell-1} & I & & & & & \cdot \\ I & 0 & \cdot & \cdot & \cdot & \cdot & 0 \end{bmatrix} \tag{5}$$

it is easily verified that $B_L C_L = C_L^* B_L$. Thus, since $B_L^* = B_L$ and B_L is nonsingular we may say that C_L *is selfadjoint in the indefinite scalar product on* $\mathbb{C}^{\ell n}$ *determined by* B_L.

Although matrix polynomials are a special class of rational matrix-valued functions, analysis of the latter requires the introduction of new ideas and techniques. Note that when matrix polynomials are viewed as rational functions we must acknowledge the presence of poles at infinity. In contrast, and for the purpose of illustration, consider now rational matrix valued functions with neither poles nor zeros at infinity. Thus, $W(\lambda)$ denotes a rational matrix valued function for which $W(\lambda) = W(\bar{\lambda})^*$ for all $\lambda \in \mathbb{C}$ at which $W(\lambda)$ is defined and, in particular, $W(\lambda)$ is hermitian whenever $W(\lambda)$ is defined and $\lambda \in \mathbb{R}$. Suppose that $W(\lambda)$ has a representation

$$W(\lambda) = I + C(\lambda I-A)^{-1}B \tag{6}$$

for all $\lambda \notin \sigma(A)$. If $HA = A^*H$ and $C^* = HB$ for some nonsingular hermitian matrix H, it is easily verified that $W(\lambda) = W(\bar{\lambda})^*$ as required. It turns out that there is a converse statement: Every rational matrix-valued function $W(\lambda)$ for which $W(\lambda) = W(\bar{\lambda})^*$ and $W(\lambda) \to I$ as $|\lambda| \to \infty$ has a representation (6) and

$$HA = A^*H\ , \quad C^* = HB \tag{7}$$

for some H with $\det H \neq 0$ and $H^* = H$. Thus, *an* H-*selfadjoint matrix* A enters the analysis once more.

Finally, let A, D, C be $n \times n$ matrices with D positive semi-definite and C hermitian. It is also assumed that

$$\mathrm{rank}[D,AD,A^2D,\cdots,A^{n-1}D] = n ,$$

i.e. that the pair (A,D) is controllable. The equation

$$XDX + XA + A^*X - C = 0 \tag{8}$$

for the unknown $n \times n$ matrix X is known as the symmetric algebraic Riccati equation. Here, the theory of solutions of (8) is based on analysis of the matrix

$$M = i\begin{bmatrix} A & D \\ C & -A^* \end{bmatrix} . \tag{9}$$

Clearly, M is not hermitian, but it is easily seen to be *selfadjoint with respect to either* H_1 *or* H_2 where

$$H_1 = i\begin{bmatrix} 0 & -I \\ I & 0 \end{bmatrix}, \quad H_2 = \begin{bmatrix} -C & A^* \\ A & D \end{bmatrix} . \tag{10}$$

(It can be assumed, without loss of generality, that H_2 is nonsingular. Also, $H_2 = H_1M$.) These remarks are the point of departure for investigation of the solutions of (8).

The contents of the book are divided into four parts. In the first part we develop the basic geometrical ideas concerning finite dimensional spaces with an indefinite scalar product, the canonical forms for matrices which are either selfadjoint or unitary in the indefinite scalar product, and some related questions of functional calculus.

Of course, the study of matrix pencils $A_1\lambda + A_2$, or of H-selfadjoint matrices, in the generality proposed here is not new. Detailed analysis and important results go back to the works of Kronecker, Frobenius and Weierstrass in the last century, for example. It should also be noted that problems of this kind arise frequently in other parts of mathematics. However, the present systematic development yields new results as well as unsolved problems. For example, the construction of canonical forms for H-normal matrices, remains an open question.

It is also true that, in spite of the long history and breadth of interest, there are only quite restricted expositions of this subject matter from the point of view of linear algebra to be found in the literature. The only ones known to the authors are due to Gantmakher [14], Glazman and Ljubich

[16] (in a sequence of exercises) and Mal'cev [35]. We should also mention a volume by Yacubovitch and Starzhinski [51] in which the theory of Hamiltonian systems is the main concern, and some necessary parts of the theory of matrices with indefinite scalar products also appear.

The second part of the book consists of a detailed introduction to the four main areas of application to be analysed subsequently, and also contains the first results obtained by applying the theory of Part I. The first of these areas concerns Hamiltonian equations of the form (2) (or more general systems with similar properties). In particular, the intimate relationship between the periodic coefficient $K(t)$ of (2) and the matrizant $X(t)$ of (3) is examined, and the first results concerning boundedness of solutions are established. Then an account of the basic facts concerning hermitian matrix polynomials and their factorizations is presented, together with applications to differential and difference equations. This topic is followed by an account of the theory of hermitian rational matrix functions. Special attention is paid to the questions of factorization and, for example, the description of divisors in terms of H-neutral subspaces, with H as in equation (7). Finally, the Riccati equation (8) is studied, with emphasis on the role of M-invariant subspaces (ref. equation (9)) in the description of the solutions.

The main ideas in the analysis of Hamiltonian systems in Part II are already well-established (see [51,17b,26,15]; only the exposition is original and the development in the context of a well-developed theory of H-selfadjoint and H-unitary matrices. The chapters on hermitian matrix valued functions are relatively new in origin. The theory of matrix polynomials with hermitian coefficients presented here is based on [19b] where earlier references can also be found. The development of the theory of hermitian rational matrix functions is based on the recent papers of Ran [38] and the authors [19d], as well as background material to be gleaned from the book of Bart, Gohberg and Kaashoek [4]. The importance of M-invariant subspaces in the analysis of the Riccati equation has been recognized for some time, but the treatment of Chapter 4 is based mainly on the paper of Lancaster and Rodman [31].

Part III is concerned with the ideas of perturbation and stability and begins with a chapter on general perturbations of pairs (A,H) in which A is H-selfadjoint to pairs (B,G) in which B is G-selfadjoint. Results are also obtained in the format of analytic perturbation theory. Perturbations of more special form are also considered in subsequent chapters of Part III, and applications are included to the problem areas developed in Part II. For example, results are obtained describing the dependence of divisors of matrix valued functions on the parent function and the dependence of solutions of the Riccati equation (8) on the coefficient matrices.

This part of the book contains well-known results of M.G. Krein [26] which were later extended by Gelfand and Lidskii [15], and Coppel and Howe [11], as well as generalizations by Levinson [32] to selfadjoint systems of differential equations. In this context, all of their results appear as applications of the general theory.

The first topic of Part IV concerns diagonable H-selfadjoint matrices with only real eigenvalues which have the property that they retain these properties after small perturbations which are also H-selfadjoint, for example. In particular, the connected components in the set of all such matrices is studied. It is shown how these connected components can be identified and applications are made to differential and difference equations and, finally to Hamiltonian systems of differential equations. These results are mainly due to Gelfand and Lidskii with subsequent extensions by other authors.

This completes our review of the book's contents, but our comments on the relevant research literature would not be complete without a comment on the interesting fact that indefinite scalar products and related operators have received a comprehensive treatment in the context of infinite dimensional spaces. Indeed, this has been an active area of research for 30 or 40 years and problems of this degree of generality also arise in many situations. An important role in the development of this theory is played by M.G. Krein; working together with colleagues and students. We do not go into the exciting developments in this area, but would mention the monographs by M.G. Krein [26b], Bognar [6], and Iohvidov, Krein and Langer [23], and the survey by Azizov and Iohvidov [2], for reviews of developments in the theory of spaces with indefinite scalar products and operators acting on them. Extensive lists of references can also be found in these sources. We would like also to mention the latest results of Ball and Helton [3] on Beurling-Lax invariant subspace representations (in the context of an indefinite scalar product space) and their applications to factorization and interpolation problems.

PART I

BASIC THEORY

The purpose of Part I is to develop the basic concepts of a traditional course in linear algebra in the context of finite dimensional spaces with an indefinite scalar product. Replacing the definite scalar product of a traditional approach by an indefinite one requires reconsideration of the basic concepts of length and orthogonality. When this is done the special classes of matrices (or operators) generated by the indefinite scalar product are studied, i.e. the natural generalizations of hermitian, unitary, and normal matrices. Special emphasis is placed on the development of canonical forms for these classes of matrices in the cases of both complex and real linear spaces. Also, some natural questions concerning functions of matrices in these classes are developed.

In general, the material presented in this Part has a long history going back to the nineteenth century. Since that time much has been added and rediscovered by several authors, including the present ones.

CHAPTER 1

INDEFINITE SCALAR PRODUCTS

In traditional linear algebra the concepts of length, angle, and orthogonality are defined by a definite scalar product. Here, the definite scalar product is replaced by an *indefinite* one, and this produces substantial changes in the geometry which are to be studied in this chapter.

1.1 Definition

Let $\mathbb{C}^n$ be the n-dimensional complex Hilbert space consisting of all column vectors x with complex coordinates $x^{(j)}$, $j = 1,2,\cdots,n$. The typical column vector x will be written in the form $x = \langle x^{(1)},x^{(2)},\cdots, x^{(n)}\rangle$. The standard scalar product in $\mathbb{C}^n$ is denoted by $(.,.)$. Thus,

$$(x,y) = \sum_{j=1}^{n} x^{(j)}\bar{y}^{(j)}$$

where $x = \langle x^{(1)},\cdots,x^{(n)}\rangle$, $y = \langle y^{(1)},\cdots,y^{(n)}\rangle$ and the bar denotes complex conjugation.

A function $[.,.]$ from $\mathbb{C}^n \times \mathbb{C}^n$ to $\mathbb{C}$ is called an *indefinite scalar product* in $\mathbb{C}^n$ if the following axioms are satisfied:

(i) Linearity in the first argument;

$$[\alpha x_1+\beta x_2,y] = \alpha[x_1,y] + \beta[x_2,y]$$

for all $x_1,x_2,y \in \mathbb{C}^n$ and all complex numbers α, β;

(ii) antisymmetry;

$$[x,y] = \overline{[y,x]}$$

for all $x,y \in \mathbb{C}^n$;

(iii) non-degeneracy; if $[x,y] = 0$ for all $y \in \mathbb{C}^n$, then $x = 0$.

Thus, the function $[.,.]$ satisfies all the properties of the standard scalar product with the possible exception that $[x,x]$ may be non-positive for $x \neq 0$.

It is easily checked that for every $n \times n$ invertible hermitian matrix H the formula

$$[x,y] = (Hx,y) \ , \quad x,y \in \mathbb{C}^n \tag{1.1}$$

determines an indefinite scalar product on $\mathbb{C}^n$. Conversely, for every indefinite scalar product $[.,.]$ on $\mathbb{C}^n$ there exists an $n \times n$ invertible and hermitian matrix H such that (1.1) holds. Indeed, for each fixed $y \in \mathbb{C}^n$ the function $x \to [x,y]$ $(x \in \mathbb{C}^n)$ is a linear form on $\mathbb{C}^n$. It is well-known that such a form can be represented as $[x,y] = (x,z)$ for some fixed $z \in \mathbb{C}^n$. Putting $z = Hy$ we obtain a linear transformation $H : \mathbb{C}^n \to \mathbb{C}^n$. Now antisymmetry and non-degeneracy of $[.,.]$ ensure that H is hermitian and invertible.

Note that here, and whenever it is convenient, an $n \times n$ complex matrix is identified with a linear transformation acting on $\mathbb{C}^n$ in the usual way.

The correspondence $[.,.] \leftrightarrow H$ established above is obviously a bijection between the set of all indefinite scalar products on $\mathbb{C}^n$ and the set of all $n \times n$ invertible hermitian matrices. This correspondence will be widely used throughout this work. Thus, the notions of the indefinite scalar product $[.,.]$ and the corresponding matrix H will be used interchangeably.

The following example of an indefinite scalar product will be important.

EXAMPLE 1.1 Put $[x,y] = \sum_{i=1}^{n} x_i \bar{y}_{n+1-i}$, where $x = \langle x_1,\cdots,x_n \rangle$, $y = \langle y_1,\cdots,y_n \rangle \in \mathbb{C}^n$. Clearly, $[.,.]$ is an indefinite scalar product. The corresponding $n \times n$ invertible hermitian matrix is

$$\begin{bmatrix} 0 & \cdot & \cdot & \cdot & 0 & 1 \\ \cdot & & & & 1 & 0 \\ \cdot & & & \cdot & & \cdot \\ \cdot & & \cdot & & & \cdot \\ 0 & 1 & \cdot & & & \cdot \\ 1 & 0 & \cdot & \cdot & \cdot & 0 \end{bmatrix}$$

This matrix will be called the *sip matrix* of size n (the standard involutary permutation). □

1.2 Orthogonality and Orthogonal Bases

Let $[.,.]$ be an indefinite scalar product on $\mathbb{C}^n$ and M be any subset of $\mathbb{C}^n$. Define the *orthogonal companion* of M in $\mathbb{C}^n$ by

$$M^{[\perp]} = \{x \in \mathbb{C}^n \mid [x,y] = 0 \quad \text{for all} \quad y \in M\}.$$

Note that the symbol $M^{[\perp]}$ will be reserved for the orthogonal companion with respect to the indefinite scalar product, while the symbol $M^{\perp}$ will denote the orthogonal companion in the original scalar product $(.,.)$ in $\mathbb{C}^n$, i.e.

$$M^{\perp} = \{x \in \mathbb{C}^n \mid (x,y) = 0 \quad \text{for all} \quad y \in M\}.$$

Clearly, $M^{[\perp]}$ is a subspace in $\mathbb{C}^n$, and we will be particularly interested in the case when M is itself a subspace of $\mathbb{C}^n$. In the latter case, it is not generally true (as experience with the euclidean scalar product might suggest) that $M^{[\perp]}$ is a direct complement for M. The next example illustrates this point.

EXAMPLE 1.2 Let $[x,y] = (Hx,y)$, $x,y \in \mathbb{C}^n$, where H is the sip matrix of size n introduced in Example 1.1. Let M be spanned by the first *unit vector*, e_1, in $\mathbb{C}^n$ (i.e. $e_1 = \langle 1,0,\cdots,0\rangle$). It is easily seen that $M^{[\perp]}$ is spanned by $e_1,e_2,\cdots,e_{n-1}$ and is not a direct complement to M in $\mathbb{C}^n$. □

In contrast, it is true that, for any subspace M,

$$\dim M + \dim M^{[\perp]} = n\ . \tag{1.2}$$

To see this observe first that

$$M^{[\perp]} = H^{-1}(M^{\perp})\ . \tag{1.3}$$

For, if $x \in M^{\perp}$ and $y \in M$ we have

$$[H^{-1}x,y] = (HH^{-1}x,y) = (x,y) = 0 \tag{1.4}$$

so that $H^{-1}(M^{\perp}) \subset M^{[\perp]}$. Conversely, if $x \in M^{[\perp]}$ and $z = Hx$ then, for any $y \in M$,

$$0 = [x,y] = [H^{-1}z,y] = (z,y) .$$

Thus, $z \in M^{\perp}$ and $x = H^{-1}z$ so that $M^{[\perp]} \subset H^{-1}(M^{\perp})$ and (1.3) is established. Then (1.2) follows immediately.

It follows from Equation (1.2) that, for any subspace $M \subset \mathbb{C}^n$,

$$(M^{[\perp]})^{[\perp]} = M .$$

Indeed, the inclusion $(M^{[\perp]})^{[\perp]} \supset M$ is evident from the definition of $M^{[\perp]}$. But (1.2) implies that these two subspaces have the same dimension, and so (1.5) follows.

A subspace M is said to be *non-degenerate* (with respect to the indefinite scalar product $[.,.]$) if $x \in M$ and $[x,y] = 0$ for all $y \in M$ imply that $x = 0$. Otherwise M is *degenerate*. For example, the defining property (iii) for $[.,.]$ ensures that $\mathbb{C}^n$ itself is always non-degenerate. In Example 1.2 the subspace M is degenerate because $\langle 1,0,\cdots,0\rangle \in M$ and $[\langle 1,0,\cdots,0\rangle,y] = 0$ for all $y \in M$.

The non-degenerate subspaces can be characterized in another way:

PROPOSITION 1.1 *$M^{[\perp]}$ is a direct complement to M in $\mathbb{C}^n$ if and only if M is non-degenerate.*

PROOF. By definition, M is non-degenerate if and only if $M \cap M^{[\perp]} = \{0\}$. In view of (1.2) this means that $M^{[\perp]}$ is a direct complement to M. □

In particular, the orthogonal companion of a non-degenerate subspace is again non-degenerate.

In terms of the hermitian matrix H corresponding to $[.,.]$ (so that $[x,y] = (Hx,y)$, non-degenerate subspaces can also be characterized as follows: M is non-degenerate if and only if $PH|_M : M \to M$ is an invertible linear transformation, where $P : \mathbb{C}^n \to M$ is the orthogonal (with respect to the underlying scalar product $(.,.)$) projector on M. Note that the linear transformation $PH|_M : M \to M$ is always hermitian (whether M is degenerate or non-degenerate).

If M is any non-degenerate non-zero subspace, Proposition 1.1 can be used to construct a basis in M which is *orthonormal* with respect to the indefinite scalar product $[.,.]$, i.e. a basis $x_1,\cdots,x_k$ satisfying

$$[x_i,x_j] = \begin{cases} \pm 1 & \text{for} \quad i = j \\ 0 & \text{for} \quad i \neq j \end{cases}.$$

To start the construction observe that there exists a vector $x \in M$ such that $[x,x] \neq 0$. Indeed, if this were not true, then $[x,x] = 0$ for all $x \in M$. Then the easily verified identity

$$[x,y] = \frac{1}{4}\{[x+y,x+y] + i[x+iy,x+iy] - [x-y,x-y] - i[x-iy,x-iy]\} \tag{1.5}$$

shows that $[x,y] = 0$ for all $x,y \in M$; a contradiction.

So it is possible to choose $x \in M$ with $[x,x] \neq 0$ and write $x_1 = x/\sqrt{|[x,x]|}$ so that $[x_1,x_1] = \pm 1$. By Proposition 1.1 (applied in M), the orthogonal companion $(\mathrm{Span}\{x_1\})^{[\perp]}$ of $\mathrm{Span}\{x_1\}$ in M is a direct complement to $\mathrm{Span}\{x_1\}$ in M and is also non-degenerate. Now take a vector $x_2 \in (\mathrm{Span}\{x_1\})^{[\perp]}$ such that $[x_2,x_2] = 1$, and so on, until M is exhausted.

In the next proposition we describe an important property of bases for a non-degenerate subspace which are orthonormal in the above sense. By sig Q, where $Q : M \to M$ is an invertible hermitian linear transformation from M to M, we denote the *signature* of Q, i.e. the difference between the number of positive eigenvalues of Q and the number of negative eigenvalues of Q (in both cases counting with multiplicities).

PROPOSITION 1.2 *Let* $[.,.] = (H.,.)$ *be an indefinite scalar product with corresponding hermitian invertible matrix* H, *and let* $x_1,\cdots,x_k$ *be an orthonormal (with respect to* $[.,.]$*) basis in a non-degenerate subspace* $M \subset \mathbb{C}^n$. *Then the sum* $\sum_{i=1}^{k} [x_i,x_i]$ *coincides with the signature of the hermitian linear transformation* $PH|_M : M \to M$, *where* P *is the orthogonal projector (with respect to* $(.,.)$*) of* $\mathbb{C}^n$ *onto* M.

PROOF. Since M is non-degenerate, the transformation $PH|_M$ is invertible. Now for $y = \sum_{i=1}^{k} \alpha_i x_i$ we have

$$(PH|_M y,y) = \sum_{i=1}^{k} |\alpha_i|[x_i,x_i] .$$

Thus, the quadratic form defined on M by $(PH|_M x,x)$ reduces to a sum of squares in the basis $x_1,\cdots,x_k$. Since $[x_i,x_i] = \pm 1$ it follows that $\sum_{i=1}^{k} [x_i,x_i]$ is just the signature of $PH|_M$. □

Note, in particular, that the sum $\sum_{i=1}^{k} [x_i,x_i]$ does not depend on the choice of the orthonormal basis.

1.3 Classification of Subspaces

Let $[.,.]$ be an indefinite scalar product on $\mathbb{C}^n$. A subspace M of $\mathbb{C}^n$ is called *positive* (with respect to $[.,.]$) if $[x,x] > 0$ for all non-zero x in M, and *nonnegative* if $[x,x] \geq 0$ for all x in M. Clearly, every positive subspace is also nonnegative but the converse is not necessarily true (see Example 1.3 below). Observe that a positive subspace is non-degenerate. If the invertible hermitian matrix H is such that $[x,y] = (Hx,y)$, $x,y \in \mathbb{C}^n$, we say that a positive (resp. nonnegative) subspace is *H-positive* (resp. *H-nonnegative*).

EXAMPLE 1.3 Let $[x,y] = (Hx,y)$, $x,y \in \mathbb{C}^n$, where H is the sip matrix of size n, and assume n is odd. Then the subspace spanned by the first $\frac{1}{2}(n+1)$ unit vectors is nonnegative, but not positive. The subspace spanned by the unit vector with 1 in the $\frac{1}{2}(n+1)$-th position is positive. □

We are to investigate the constraints on the dimensions of positive and nonnegative subspaces. But first a general observation is necessary.

Let $[.,.]_1$ and $[.,.]_2$ be two indefinite scalar products on $\mathbb{C}^n$ with corresponding invertible hermitian matrices H_1, H_2 respectively. Suppose, in addition, that H_1 and H_2 are congruent, i.e. $H_1 = S^*H_2S$ for some invertible matrix S. (Here, and subsequently, the adjoint S^* of S is taken with respect to $(.,.)$.) In this case, a subspace M is H_1-positive if and only if SM is H_2-positive, with a similar statement replacing "positive" by "nonnegative". The proof is direct: take $x \in M$ and

$$[Sx,Sx]_2 = (H_2Sx,Sx) = (S^*H_2Sx,x) = (H_1x,x) = [x,x]_1 .$$

Thus, $[x,x]_1 > 0$ for all non-zero $x \in M$ if and only if $[y,y]_2 > 0$ for all non-zero y in SM.

THEOREM 1.3 *The maximal dimension of a positive, or of a nonnegative subspace with respect to the indefinite scalar product* $[x,y] = (Hx,y)$ *coincides with the number of positive eigenvalues of* H *(counting multiplicities).*

Note that the maximal possible dimensions of nonnegative and positive subspaces coincide.

PROOF. We prove only the nonnegative case (the positive case is analogous). So let M be a nonnegative subspace, and let $p = \dim M$. Then

$$\min_{\substack{(x,x)=1 \\ x \in M}} (Hx,x) \geq 0 . \tag{1.6}$$

Write all the eigenvalues of H in the non-increasing order: $\lambda_1 \geq \lambda_2 \geq \cdots \geq \lambda_n$. By the max-min characterization of the eigenvalues of H, we have

$$\lambda_p = \max_{L} \min_{\substack{(x,x)=1 \\ x \in L}} (Hx,x) ,$$

where the maximum is taken over all the subspaces $L \subset \mathbb{C}^n$ of dimension p. Then (1.6) implies $\lambda_p \geq 0$ and, since H is invertible, $\lambda_p > 0$. So $p \leq k$ where k is the number of positive eigenvalues of H.

To find a nonnegative subspace of dimension k, appeal to the observation preceding Theorem 1.3. There exists an invertible matrix S such that S^*HS is a diagonal matrix of +1's and -1's:

$$H_0 \overset{\text{def}}{=} S^*HS = \text{diag}[1,\cdots,1,-1,-1] , \tag{1.7}$$

where the number of +1's is k. Hence, it is sufficient to find a k-dimensional subspace which is nonnegative with respect to H_0. One such subspace (which is even positive) is spanned by the first k unit vectors in $\mathbb{C}^n$. □

A subspace $M \subset \mathbb{C}^n$ is called H-*negative* (where H is such that $[x,y] = (Hx,y)$, $x,y \in \mathbb{C}^n$), if $[x,x] < 0$ for all non-zero x in M. Replacing this condition by the requirement that $[x,x] \leq 0$ for all $x \in M$, we obtain the definition of a *nonpositive* (with respect to $[.,.]$), or H-*nonpositive* subspace. As in Theorem 1.3 it can be proved that the maximal possible dimension of an H-negative or of an H-nonpositive subspace is equal to the number of negative eigenvalues of H (counting multiplicities).

Note also the following inequality: Let M be an H-nonnegative or H-nonpositive subspace, then

$$|(Hy,z)| \leq (Hy,y)(Hz,z) \tag{1.8}$$

for every $y,z \in M$. The proof of (1.8) is completely analogous to the standard proof of Schwarz inequality.

We pass now to the class of subspaces which are peculiar to indefinite scalar product spaces and have no analogues in the spaces with the usual scalar product. A subspace $M \subset \mathbb{C}^n$ is called *neutral* (with respect to $[.,.]$), or H-*neutral* (where H is such that $[x,y] = (Hx,y)$, $x,y \in \mathbb{C}^n$) if $[x,x] = 0$ for all $x \in M$. Sometimes, such subspaces are called *isotropic*. In Example 1.3 the subspaces spanned by the first k unit vectors, for $k = 1,\cdots,\frac{n-1}{2}$, are all neutral.

In view of the identity (1.5) a subspace M is neutral if and only if $[x,y] = 0$ for all $x,y \in M$. Observe also that a neutral subspace is both nonpositive and nonnegative, and is necessarily degenerate.

We have seen in Example 1.3 that the nonnegative subspace spanned by the first $\frac{1}{2}(n+1)$ unit vectors is a direct sum of a neutral subspace (spanned by the first $\frac{1}{2}(n-1)$ unit vectors) and a positive definite subspace (spanned by the $\frac{1}{2}(n+1)$-th unit vector). This is a general property, as the following theorem shows.

THEOREM 1.4 *A nonnegative (resp. nonpositive) subspace is a direct sum of a positve (resp. negative) subspace and a neutral subspace.*

PROOF. Let $M \subset \mathbb{C}^n$ be a nonnegative subspace, and let M_0 be a maximal positive subspace in M (since $\dim M$ is finite, such an M_0 always exists). Since M_0 is non-degenerate, Proposition 1.1 implies that $M_0 \dotplus M_0^{[\perp]} = \mathbb{C}^n$, and hence

$$M_0 \dotplus (M_0^{[\perp]} \cap M) = M .$$

It remains to show that $M_0^{[\perp]} \cap M$ is neutral. Suppose not, so there is an $x \in M_0^{[\perp]} \cap M$ such that $[x,x] \neq 0$. Since M is nonnegative it follows that $[x,x] > 0$. Now for each $y \in M_0$ we have $[x+y,x+y] = [x,x] + [y,y] > 0$, in view of the fact that M_0 is positive. So $\mathrm{Span}\{x,M_0\}$ is also a positive subspace; a contradiction with the maximality of M_0.

For a nonpositive subspace the proof is similar. □

The decomposition of a nonnegative subspace M into a direct sum $M_0 + M_1$, where M_0 is positive and M_1 is neutral, is not unique. However, $\dim M_0$ is uniquely determined by M. Indeed, let P be the orthogonal (with respect to $(.,.)$) projection on M; then it is easily seen that $\dim M_0 = \mathrm{rank}\, PH|_M$, where $PH|_M : M \to M$ is a selfadjoint linear transformation.

One can easily compute the maximal possible dimension of a neutral subspace.

THEOREM 1.5 *The maximal possible dimension of an* H-*neutral subspace is* $\min(k,\ell)$, *where* k *(resp.* ℓ*) is the number of positive (resp. negative) eigenvalues of* H, *counting mulciplicities.*

PROOF. In view of the remark preceding Theorem 1.3 it may be assumed that $H = H_0$ is given by (1.7). The existence of a neutral subspace of dimension $\min(k,\ell)$ is easily seen. A basis for one such subspace can be formed from the unit vectors $e_1,e_2,\cdots$ as follows: $e_1+e_{k+1},e_2+e_{k+2},\cdots$.

Now let M be a neutral subspace of dimension p. Since M is also nonnegative it follows from Theorem 1.3 that $p \leqslant k$. But M is also nonpositive and so the inequality $p \leqslant \ell$ also applies. Thus, $p \leqslant \min(k,\ell)$.□

CHAPTER 2

CLASSES OF LINEAR TRANSFORMATIONS

When a scalar product is introduced on $\mathbb{C}^n$, then certain $n \times n$ matrices (seen as linear transformations of $\mathbb{C}^n$) have symmetries defined by the scalar product. If the scalar product is definite this leads to the usual classes of hermitian, unitary, and normal matrices. If the scalar product is indefinite then analogous classes of matrices are defined.

2.1 Adjoint Matrices

Let $[.,.]$ be an indefinite scalar product on $\mathbb{C}^n$ and let H be the associated invertible hermitian matrix as in Equation (1.1). Let A be an $n \times n$ complex matrix considered as a linear transformation on $\mathbb{C}^n$. The H-*adjoint* of A is the unique $n \times n$ matrix, written $A^{[*]}$, which satisfies

$$[Ax,y] = [x,A^{[*]}y] \tag{2.1}$$

for all $x,y \in \mathbb{C}^n$. The H-adjoint of A may also be described as the adjoint of A with respect to $[.,.]$. Expressing (2.1) in terms of H we have

$$(HAx,y) = (Hx,A^{[*]}y) \tag{2.2}$$

for all $x,y \in \mathbb{C}^n$, and, hence

$$(x,A^*Hy) = (x,HA^{[*]}y)$$

where (here and elsewhere) $A^* : \mathbb{C}^n \to \mathbb{C}^n$ is the usual adjoint of A (i.e. $(x,A^*y) = (Ax,y)$ for all $x,y \in \mathbb{C}^n$). It follows that

$$A^{[*]} = H^{-1}A^*H\ . \tag{2.3}$$

In particular, this representation confirms the existence of $A^{[*]}$ and shows that it is uniquely determined by A. It also follows from (2.3), or from the definition of $A^{[*]}$, that $(A^{[*]})^{[*]} = A$.

There are important and well-known connections between the images and kernels of A and A^*, its usual adjoint. The next proposition describes the extension of these results to the H-adjoint of A.

PROPOSITION 2.1 *Let* $A : \mathbb{C}^n \to \mathbb{C}^n$ *and let* $A^{[*]}$ *be its H-adjoint. Then*

$$\operatorname{Im} A^{[*]} = (\operatorname{Ker} A)^{[\perp]} \; ; \quad \operatorname{Ker} A^{[*]} = (\operatorname{Im} A)^{[\perp]} \; . \tag{2.4}$$

PROOF. Let $x \in \operatorname{Im} A^{[*]}$ so that $x = A^{[*]}y$ for some $y \in \mathbb{C}^n$. Then for every $z \in \operatorname{Ker} A$:

$$[x,z] = [A^{[*]}y,z] = [y,Az] = 0 \; ,$$

and it follows that

$$\operatorname{Im} A^{[*]} \subset (\operatorname{Ker} A)^{[\perp]} \; . \tag{2.5}$$

However, using (2.3) and (1.2),

$$\dim(\operatorname{Im} A^{[*]}) = \dim(\operatorname{Im} A^*) = n - \dim(\operatorname{Ker} A) = \dim(\operatorname{Ker} A)^{[\perp]} \; ,$$

so that equality must obtain in (2.5).

The proof of the second relation in (2.4) is similar. □

A subspace $M \subset \mathbb{C}^n$ is said to be *invariant* for an $n \times n$ matrix A (considered as a linear transformation from $\mathbb{C}^n$ to $\mathbb{C}^n$), or to be A-*invariant*, if $x \in M$ implies $Ax \in M$.

PROPOSITION 2.2 *Let* $A : \mathbb{C}^n \to \mathbb{C}^n$ *and let* $[.,.]$ *be an indefinite scalar product in* $\mathbb{C}^n$. *Then a subspace* M *is* A-*invariant if and only if its orthogonal companion* $M^{[\perp]}$ *is* $A^{[*]}$-*invariant.*

PROOF. Let M be A-invariant and let $x \in M$, $y \in M^{[\perp]}$. Then

$$[A^{[*]}y,x] = [y,Ax] = 0$$

since Ax is again in M. So $A^{[*]}y \in M^{[\perp]}$ and $M^{[\perp]}$ is $A^{[*]}$-invariant.

To prove the converse statement apply what is already proved to $A^{[*]}$ and $(A^{[*]})^{[*]} = A$, taking into account the fact that $(M^{[\perp]})^{[\perp]} = M$

(see Equation (1.5)). □

Now it is natural to describe a matrix as H-*selfadjoint* (or selfadjoint with respect to $[.,.]$) if $A = A^{[*]}$. In a similar way, a matrix U is said to be H-*unitary* if it is invertible and $U^{-1} = U^{[*]}$, and matrix N is H-*normal* if $NN^{[*]} = N^{[*]}N$. Clearly, H-selfadjoint matrices and H-unitary matrices are also H-normal. The rest of this part of the book is devoted mainly to the study of these classes of matrices.

The next result shows that if matrices used in defining indefinite scalar product are congruent, then matrices of these three types are transformed in a natural way.

PROPOSITION 2.3 *Let* H_1, H_2 *define indefinite scalar products on* $\mathbb{C}^n$ *and* $H_2 = SH_1S^*$ *for some invertible* $n \times n$ *matrix* S. *Then* A_1 *is* H_1-*selfadjoint (or* H_1-*unitary, or* H_1-*normal) if and only if* $A_2 = S^{*-1}A_1S^*$ *is* H_2-*selfadjoint (or* H_2-*unitary, or* H_2-*normal, respectively).*

PROOF. We consider the "only if" part of the statement. Proof of the converse statement is analogous. Suppose first that A_1 is H_1-selfadjoint so that, (see (2.3)), $H_1A_1 = A_1^*H_1$. Then

$$H_2A_2 = (SH_1S^*)(S^{*-1}A_1S^*) = SH_1A_1S^* = SA_1^*H_1S^* = (SA_1^*S^{-1})(SH_1S^*) = A_2^*H_2 ,$$

which implies that A_2 is H_2-selfadjoint.

If A_1 is H_1-unitary then $A_1^{-1} = A_1^{[*]}$, and it follows from (2.3) that $H_1A_1^{-1} = A_1^*H$. So we have

$$H_2A_2^{-1} = (SH_1S^*)(S^{*-1}A_1^{-1}S^*) = SH_1A_1^{-1}S^* = SH_1^*H_1S^* = (SA_1^*S^{-1})(SH_1S^*) = A_2^*H_2 .$$

Thus, A_2 is H_2-unitary.

In much the same way, it is easily seen that if A_1 is H_1-normal then $A_1H_1^{-1}A_1^*H_1 = H_1^{-1}A_1^*H_1A_1$ and that this implies a similar relation between A_2 and H_2. □

Proposition 2.3 allows us to study the properties of these selfadjoint, unitary and normal matrices in the context of a canonical indefinite scalar product of the form $[.,.] = (P.,.)$ where $P^* = P$ and $P^2 = I$. Indeed, given an invertible hermitian H there is an invertible S such that P is a diagonal matrix: $P = \mathrm{diag}[1,1,\cdots,1,-1,\cdots,-1]$. However, this reduction is achieved at the expense of replacing A by $S^{-1}AS$. This suggests that S may be chosen in such a way that *both* H and A are reduced to some simplest possible forms; an idea that will be developed in the next chapter.

2.2 H-selfadjoint Matrices; Examples and Simplest Properties

Let $[.,.] = (H.,.)$ be an indefinite scalar product in $\mathbb{C}^n$. As we have seen in the last section, an $n \times n$ matrix A is said to be H-selfadjoint (or selfadjoint with respect to $[.,.]$) if $A = A^{[*]}$ or, in other words, (see (2.3)) if

$$A = H^{-1}A^*H . \tag{2.6}$$

Thus, any H-selfadjoint matrix A is similar to A^*. We shall see later that the converse is also true: if a matrix A is similar to its adjoint (i.e. $A = S^{-1}A^*S$ for some S) then A is H-selfadjoint for some H. In other words, the similarity between A and A^* can be carried out by means of an invertible *hermitian* matrix. Observe also that the set of all $n \times n$ H-selfadjoint matrices form a real linear space; i.e. if A and B are H-selfadjoint then so is $\alpha A + \beta B$ where α, β are any real numbers.

For the case when $H^2 = I$ (see concluding remarks of the preceding section) it is easily seen that A is H-selfadjoint if and only if A^* is H-selfadjoint. This leads to the following observation: if $H^2 = I$ and H is hermitian, then A is H-selfadjoint if and only if both of

$$\mathit{Re}\, A = \frac{1}{2}(A+A^*) , \quad i\, \mathit{Im}\, A = \frac{1}{2}(A-A^*) \tag{2.7}$$

are H-selfadjoint.

The following examples of H-selfadjoint matrices are fundamental.

EXAMPLE 2.1 Let $[x,y] = (\varepsilon Px,y)$, $x,y \in \mathbb{C}^n$, where P is the $n \times n$ sip matrix introduced in Example 1.1, and ε is either 1 or -1. Further, let

$$J = \begin{bmatrix} \alpha & 1 & 0 & \cdot & \cdot & \cdot & 0 \\ 0 & \alpha & 1 & & & & \cdot \\ \cdot & & \ddots & & & & \cdot \\ \cdot & & & \ddots & \ddots & & \cdot \\ \cdot & & & & \cdot & \alpha & 1 \\ 0 & \cdot & \cdot & \cdot & \cdot & 0 & \alpha \end{bmatrix}$$

be the $n \times n$ Jordan block with *real* eigenvalue α. The equality $(\varepsilon P)J = J^T(\varepsilon P)$ is easily checked and, since $J^T = J^*$, it means that J is εP-

selfadjoint. □

EXAMPLE 2.2 Let $[x,y] = (Qx,y)$, $x,y \in ℂ^{2n}$ where

$$Q = \begin{bmatrix} 0 & P \\ P & 0 \end{bmatrix},$$

and P is again the $n \times n$ sip matrix. Let

$$K = \begin{bmatrix} J & 0 \\ 0 & \bar{J} \end{bmatrix},$$

where J is the $n \times n$ Jordan block with *non-real* eigenvalue α (so that $\bar{J}$ is the $n \times n$ Jordan block with the eigenvalue $\bar{\alpha}$). Again, one checks easily that $QK = K^*Q$, i.e. K is Q-selfadjoint. □

The H-selfadjoint matrices from Examples 2.1 and 2.2 will appear in the next chapter as elements of a canonical form for selfadjoint matrices in indefinite scalar product spaces.

We describe now some simple properties of H-selfadjoint matrices.

PROPOSITION 2.4 *The spectrum* $\sigma(A)$ *of an H-selfadjoint matrix* A *is symmetric relative to the real axis, i.e.* $\lambda_0 \in \sigma(A)$ *implies* $\bar{\lambda}_0 \in \sigma(A)$. *Moreover, in the Jordan normal form of* A, *the sizes of the Jordan blocks with eigenvalue* λ_0 *are equal to the sizes of Jordan blocks with eigenvalue* $\bar{\lambda}_0$.

PROOF. Using (2.6), write

$$\lambda I - A = H^{-1}(\bar{\lambda}I-A)^*H .$$

So $\lambda I - A$ is singular if and only if $\bar{\lambda}I - A$ is singular; i.e. $\lambda_0 \in \sigma(A)$ implies $\bar{\lambda}_0 \in \sigma(A)$. Further, let J be the Jordan form of A with reducing matrix $T : A = T^{-1}JT$.

Observe also that J^* is similar to $\bar{J}$ (as usual, $\bar{J}$ denotes the matrix whose entries are complex conjugate to the corresponding entries of J) and write $J^* = K^{-1}\bar{J}K$. Then, using the equation above,

$$\lambda I - A = T^{-1}(\lambda I-J)T = H^{-1}\{T^{-1}(\bar{\lambda}I-J)T\}^*H$$

so that

$$\lambda I - J = (TH^{-1}T^*)(\lambda I-J^*)(T^{*-1}HT^{-1}) = S(\lambda I-\bar{J})S^{-1}$$

where $S = TH^{-1}T^*K^{-1}$. Thus, J and $\bar{J}$ are similar and $\bar{J}$ can be obtained from J by permutation of some of its Jordan blocks. The proposition follows. □

Note, in particular, that non-real eigenvalues of an H-selfadjoint matrix can only occur in conjugate pairs and, consequently, their total number, whether counted as distinct eigenvalues or according to multiplicities, must be even.

For an $n \times n$ matrix A, the *root subspace* $\mathcal{E}_A(\lambda_0)$ corresponding to an eigenvalue λ_0 is defined as follows:

$$\mathcal{E}_A(\lambda_0) = \{x \in \mathbb{C}^n \mid (A-\lambda_0 I)^s x = 0 \text{ for some positive integer } s\} . \tag{2.8}$$

It is well known that $\mathcal{E}_A(\lambda_0)$ is indeed a subspace and that $\mathbb{C}^n$ is a direct sum of the root subspaces $\mathcal{E}_A(\lambda_0)$, $\lambda_0 \in \sigma(A)$. If A is hermitian, i.e. selfadjoint with respect to the ordinary scalar product $(.,.)$, then its root subspaces corresponding to different eigenvalues are orthogonal (with respect to $(.,.)$). It turns out that, with proper modification, this result extends to the case of H-selfadjoint matrices as well.

THEOREM 2.5 *Let* A *be an* H-*selfadjoint matrix and* $\lambda,\mu \in \sigma(A)$ *with* $\lambda \neq \bar{\mu}$. *Then*

$$\mathcal{E}_A(\lambda) \subset (\mathcal{E}_A(\mu))^{[\perp]}$$

i.e. the root subspaces $\mathcal{E}_A(\lambda)$ *and* $\mathcal{E}_A(\mu)$ *are orthogonal with respect to* $[.,.] = (H.,.)$.

PROOF. Let $x \in \mathcal{E}_A(\lambda)$ and $y \in \mathcal{E}_A(\mu)$ so that $(A-\lambda I)^s x = 0$ and $(A-\mu I)^t y = 0$ for some s and t. We are to prove that

$$[x,y] = 0 . \tag{2.9}$$

Proceed by induction on $s + t$. For $s = t = 1$ we have $Ax = \lambda x$, $Ay = \mu y$. Then

$$\lambda[x,y] = [Ax,y] = [x,Ay] = [x,\mu y] = \bar{\mu}[x,y] \tag{2.10}$$

and since $\lambda \neq \bar{\mu}$, we obtain (2.9).

Suppose now that (2.9) is proved for all $x' \in \mathcal{E}_A(\lambda)$, $y' \in \mathcal{E}_A(\mu)$ such that $(A-\lambda I)^{s'}x' = (A-\lambda I)^{t'}y' = 0$ for some s' and t' satisfying

$s'+t'<s+t$. Given x and y as above, put $x'=(A-\lambda I)x$, $y'=(A-\mu I)y$. Then, by the induction assumption $[x',y]=[x,y']=0$, which means that $\lambda[x,y]=[Ax,y]$; $\bar{\mu}[x,y]=[x,Ay]$. Now use the relations of (2.10) once more to complete the proof.

In particular, taking $\lambda=\mu$ non-real in Theorem 2.5 we obtain the following corollary.

COROLLARY 2.6 *Let* A *be an H-selfadjoint matrix and let* $\lambda_0\in\sigma(A)$ *be non-real. Then the root subspace* $E_A(\lambda_0)$ *is H-neutral.*

2.3 H-Unitary Matrices; Examples and Simplest Properties

As introduced in Section 2.1, an $n\times n$ matrix A is called H-unitary (or unitary with respect to $[.,.]$) if A is invertible and $A^{-1}=A^{[*]}$. In other words, A is H-unitary if and only if $[Ax,y]=[x,A^{-1}y]$ for all $x,y\in \mathbb{C}^n$, or

$$A=H^{-1}A^{*-1}H\ ;\quad A^*HA=H\ . \tag{2.11}$$

In particular, A is similar to A^{*-1}. The converse statement is also true (as in the case of H-selfadjoint matrices): if A is similar to A^{*-1}, then the matrix achieving this similarity can be chosen to be hermitian. This fact will be proved later.

Note that for a fixed H, the set of all H-unitary matrices form a group, i.e. if A, B are H-unitary then so are A^{-1}, B^{-1} and AB.

If H is reduced, as described at the end of Section 2.1, so that $H^2=I$, then it is easily seen that A is H-unitary if and only if A^* is H-unitary.

The following examples of H-unitary matrices are related to the canonical forms of H-unitary matrices (see Chapter 4).

EXAMPLE 2.3 Let $[x,y]=(\varepsilon Px,y)$, $x,y\in\mathbb{C}^n$, where P is the $n\times n$ sip matrix and $\varepsilon=\pm1$. Suppose that $\lambda\in\mathbb{C}$ with $|\lambda|=1$ and

$$A=\begin{bmatrix} \lambda & 2i\lambda & 2i^2\lambda & \cdot & \cdot & \cdot & 2i^{n-1}\lambda \\ 0 & \lambda & 2i\lambda & & & & \cdot \\ \cdot & & \lambda & \cdot & & & \cdot \\ \cdot & & & & \cdot & \cdot & \cdot \\ \cdot & & & & & \lambda & 2i\lambda \\ 0 & \cdot & \cdot & & & 0 & \lambda \end{bmatrix}.$$

It is easily verified that $A^*(\varepsilon P)A = \varepsilon P$ so that A is εP-unitary. □

EXAMPLE 2.4 Let $[x,y] = (Qx,y)$ for all $x,y \in ¢^n$ where

$$Q = \begin{bmatrix} 0 & P \\ P & 0 \end{bmatrix}$$

and P is the $n \times n$ sip matrix. For a non-zero $\lambda \in ¢$ such that $|\lambda| \neq 1$ put

$$A = \begin{bmatrix} K_1 & 0 \\ 0 & K_2 \end{bmatrix}$$

where

$$K_1 = \begin{bmatrix} \lambda & k_1 & k_2 & \cdots & k_{n-1} \\ 0 & \lambda & k_1 & & \vdots \\ \vdots & & \ddots & \ddots & \vdots \\ \vdots & & & \lambda & k_1 \\ 0 & \cdots & \cdots & 0 & \lambda \end{bmatrix}, \quad K_2 = \begin{bmatrix} \bar{\lambda}^{-1} & \kappa_1 & \kappa_2 & \cdots & \kappa_{n-1} \\ 0 & \bar{\lambda}^{-1} & \kappa_1 & & \vdots \\ \vdots & & \ddots & \ddots & \vdots \\ \vdots & & & \bar{\lambda}^{-1} & \kappa_1 \\ 0 & \cdots & \cdots & 0 & \bar{\lambda}^{-1} \end{bmatrix}$$

and $k_r = \lambda q_1^{r-1}(q_1-\bar{q}_2)$, $\kappa_r = \bar{\lambda}^{-1}q_2^{r-1}(q_2-\bar{q}_1)$, $r = 1,2,\cdots,n-1$, and $q_1 = \frac{1}{2}i(1+\lambda)$, $q_2 = \frac{1}{2}i(1+\bar{\lambda}^{-1})$.

A direct computation shows that $A^*QA = Q$, so A is Q-unitary. Note also that $K_2 = \bar{K}_1^{-1}$. □

Using the similarity of A and A^{*-1} for an H-unitary matrix A, one can prove the following analogue of Proposition 2.4 (its proof is similar to that of Proposition 2.4 and therefore is omitted).

PROPOSITION 2.7 *Let* A *be an H-unitary matrix. Then* $\sigma(A)$ *is symmetric relative to the unit circle, i.e.* $\lambda_0 \in \sigma(A)$ *implies* $\bar{\lambda}_0^{-1} \in \sigma(A)$. *Moreover, in the Jordan normal form of* A, *the sizes of Jordan blocks with eigenvalue* λ_0, *and the sizes of Jordan blocks with eigenvalue* $\bar{\lambda}_0^{-1}$, *are the same.*

There is a strong connection between H-unitary and H-selfadjoint matrices. As in the case of the usual scalar product one way to describe this connection is via Cayley transforms.

Recall that if $|\alpha| = 1$ and $w \neq \bar{w}$ then the map f defined by

$$f(z) = \alpha(z-\bar{w}) / (z-w) \tag{2.12}$$

maps the real line in the z-plane onto the unit circle in the ζ-plane, where $\zeta = f(z)$. The inverse transformation is

$$z = (w\zeta - \bar{w}\alpha) / (\zeta - \alpha) . \tag{2.13}$$

Then, if $w \notin \sigma(A)$ the function f is defined on $\sigma(A)$ and if A is H-selfadjoint one anticipates that $U = f(A)$ is H-unitary. This idea is developed in:

PROPOSITION 2.8 *Let A be an H-selfadjoint matrix. Let w be a non-real complex number with $w \notin \sigma(A)$ and let α be any unimodular complex number. Then*

$$U = \alpha(A-\bar{w}I)(A-wI)^{-1} \tag{2.14}$$

is H-unitary and $\alpha \notin \sigma(U)$.

Conversely, if U is H-unitary, $|\alpha| = 1$ and $\alpha \notin \sigma(U)$, then for any $w \neq \bar{w}$ the matrix

$$A = (wU-\bar{w}\alpha I)(U-\alpha I)^{-1} \tag{2.15}$$

is H-selfadjoint and $w \notin \sigma(A)$. Furthermore, formulas (2.14) and (2.15) are inverse to one another.

PROOF. If A is H-selfadjoint and $|\alpha| = 1$ it is easily seen that

$$(A^*-\bar{w}I)H(A-wI) = (\bar{\alpha}A^*-\bar{\alpha}wI)H(\alpha A-\alpha\bar{w}I) .$$

Premultiplying by $(A^*-\bar{w}I)^{-1}$ and postmultiplying by $(\alpha A-\alpha\bar{w}I)^{-1}$ it is found that $HU^{-1} = U^*H$ where U is defined by (2.14), and this means that U is H-unitary. Furthermore, it follows from (2.14) that

$$(U-\alpha I)(A-wI) = \alpha(w-\bar{w})I . \tag{2.16}$$

Thus, the hypothesis that w is not real implies that $U - \alpha I$ is invertible and so $\alpha \notin \sigma(U)$.

The relation (2.16) also gives

$$A = wI + \alpha(w-\bar{w})(U-\alpha I)^{-1} = [w(U-\alpha I) + \alpha(w-\bar{w})I](U-\alpha I)^{-1}$$
$$= (wU-\bar{w}\alpha I)(U-\alpha I)^{-1} ,$$

so that (2.15) and (2.14) are, indeed, inverse to each other.

Proof of the converse statement is left to the reader. □

Suppose that U is H-unitary and, as in the Proposition 2.8, A is the H-selfadjoint matrix given by (2.15). Then the root subspace of U corresponding to $\lambda_0 \in \sigma(U)$ is also the root subspace of A corresponding to its eigenvalue

$$\mu_0 = (w\lambda_0 - \bar{w}\alpha)(\lambda_0 - \alpha)^{-1} . \tag{2.17}$$

Thus,

$$E_U(\lambda_0) = E_A(\mu_0) . \tag{2.18}$$

This fact can be verified directly by using (2.15) but it is also a consequence of the following more general lemma. Although this is relatively well-known, a proof is included in the interests of a self-contained presentation.

LEMMA 2.9 *Let* S, T *be* $n \times n$ *matrices with the property that* $S = f(T)$, $T = g(S)$ *for some complex functions* f *and* g *which are analytic in neighbourhoods of* $\sigma(T)$, $\sigma(S)$ *respectively. Then for every* $\lambda_0 \in \sigma(S)$

$$E_S(\lambda_0) = E_T(g(\lambda_0)) . \tag{2.19}$$

PROOF. It is well known that $\tilde{g}(s) = g(s)$ where $\tilde{g}$ is a polynomial which, in particular, has the property that $g(\lambda_0) = \tilde{g}(\lambda_0)$ for $\lambda_0 \in \sigma(S)$. Let $g(\lambda) = \sum_{j=0}^{p} \alpha_j (\lambda - \lambda_0)^j$, so that

$$T - g(\lambda_0 I) = \tilde{g}(S) - \tilde{g}(\lambda_0)I = \sum_{j=0}^{p} \alpha_j (S - \lambda_0 I)^{j-1}(S - \lambda_0 I) = W(S - \lambda_0 I) ,$$

where $W = \sum_{j=0}^{p} \alpha_j (S - \lambda_0 I)^{j-1}$ is a matrix commuting with S. Now

$$(T - g(\lambda_0)I)^s = W^s (S - \lambda_0 I)^s \quad \text{for} \quad s = 0,1,\cdots$$

so that, using the definition (2.8), we have $E_S(\lambda_0) \subset E_T(g(\lambda_0))$. However, the same inclusion applies on replacing S, λ_0 by $T, g(\lambda_0)$ yielding

$$E_T(g(\lambda_0)) \subset E_{f(T)}(f(g(\lambda_0))) = E_S(\lambda_0),$$

so (2.19) follows. □

Now (2.18) is established by an application of the lemma and implies, in view of Theorem 2.5, that root subspaces $E_U(\lambda)$ and $E_U(\mu)$ are H-orthogonal provided $\lambda \neq \bar{\mu}^{-1}$. In particular (cf. Corollary 2.6), every root subspace of U corresponding to an eigenvalue not on the unit circle is H-neutral.

It is clear that these orthogonality properties of root subspaces for H-unitary matrices could also be proved directly using arguments analogous to the proof of Theorem 2.5.

A more general class of unitary matrices will subsequently be useful. Suppose that *two* indefinite scalar products are defined on $\mathbb{C}^n$ with associated invertible hermitian matrices H_1 and H_2. An $n \times n$ matrix A is said to be (H_1,H_2)-*unitary* if $[Ax,Ay]_{H_2} = [x,y]_{H_1}$ for all $x,y \in \mathbb{C}^n$. It is easily seen that this is equivalent to the relation

$$A^* H_2 A = H_1 , \tag{2.20}$$

which could be compared with (2.11). Note that an (H_1,H_2)-unitary matrix is necessarily invertible. Also, the relation (2.20) indicates that the notion of (H_1,H_2)-unitary matrices is meaningful only when H_1 and H_2 are congruent. Thus, the assertion that A is (H_1,H_2)-unitary implies the existence of an invertible S such that $H_2 = SH_1S^*$ and (2.20) is equivalent to the statement that S^*A is H_1-unitary or, alternatively, AS^* is H_2-unitary.

2.4 A Second Characterization of H-Unitary Matrices

It has been remarked in Section 2.2 that a matrix A is H-selfadjoint for some H if and only if the spectrum of A is symmetric with respect to the real line. With the aid of Cayley transformations it is easy to obtain the analogous result: A matrix U is H-unitary for some H if and only if the spectrum of U is symmetric with respect to the unit circle.

Now if A is nonsingular it is not difficult to verify that $\sigma(A^{-1}A^*)$ is symmetric with respect to the unit circle and hence $A^{-1}A^*$ is H-unitary for some H. It turns out that this property characterizes H-unitary matrices. In proving this result, however, we follow a quite different line of argument which makes less demands on spectral theory.

Note that in the lemma A is not necessarily hermitian.

LEMMA 2.10 *If* $U^*AU = A$, $\det A \neq 0$, *then there is an* H, *with* $H^* = H$ *and* $\det H \neq 0$ *such that* $U^*HU = H$.

PROOF. Let $H = \bar{z}A + zA^*$ for some z with $|z| = 1$. Then $U^*HU = U^*(\bar{z}A+zA^*)U = \bar{z}A + zA^* = H$ and $H^* = H$. To ensure $\det H \neq 0$ observe

$$H = \bar{z}A + zA^* = zA(z^{-1}\bar{z}I+A^{-1}A^*)$$

and so we have only to choose z (with $|z| = 1$) so that

$$-z^{-1}\bar{z} = -\bar{z}^2 \notin \sigma(A^{-1}A^*) \ . \ \square$$

THEOREM 2.11 *A matrix* U *is H-unitary for some* H ($H^* = H$, $\det H \neq 0$) *if and only if* $U = A^{-1}A^*$ *for some non-singular* A.

PROOF. If $U = A^{-1}A^*$ then

$$U^*AU = AA^{*-1}AA^{-1}A^* = A$$

and, from the lemma, U is H-unitary for some H.

Conversely, let U be H-unitary and let

$$A = i\beta(I-\alpha U^*)H$$

where $|\alpha| = 1$, $\alpha \notin \sigma(U)$ and $\bar{\beta} / \beta = \alpha$. Then

$$\begin{aligned} AU &= i\beta(I-\alpha U^*)HU = i\beta H(U-\alpha I) = i\alpha\beta H(\bar{\alpha}U-I) = -i\alpha\beta H(I-\bar{\alpha}U) \\ &= i\bar{\beta}H(I-\bar{\alpha}U) = A^* , \end{aligned}$$

so $U = A^{-1}A^*$. $\square$

CHAPTER 3

CANONICAL FORMS OF H-SELFADJOINT MATRICES

In this chapter we consider matrix pairs (A,H) in which A is H-selfadjoint. We develop a canonical form for such pairs under transformations of the form $A \to T^{-1}AT$, $H \to T^*HT$. This analysis requires the introduction of an invariant of the transformations known as the *sign characteristic* of the pair (A,H). This is a phenomenon which is insignificant in the more familiar case of a positive definite H. Once the canonical form has been fully developed, the remainder of the chapter is largely devoted to analysis of properties which are dependent on the sign characteristic for their description; for example, the analysis of A-invariant subspaces which are definite with respect to H, and matrices which are definite with respect to H.

3.1 Unitary Similarity

Let A_1 and A_2 be $n \times n$ matrices which are H_1-selfadjoint and H_2-selfadjoint, respectively. A notion of equivalence of such matrices appears naturally. Namely, A_1 and A_2 are unitarily similar if $A_1 = T^{-1}A_2T$, where the matrix T is invertible and (H_1,H_2)-unitary (i.e. $[Tx,Ty]_{H_2} = [x,y]_{H_1}$ for all $x,y \in \mathbb{C}^n$ or $H_1 = T^*H_2T$). In other words, A_1 and A_2 are unitarily similar if they are similar, and the similarity matrix is unitary with respect to the indefinite scalar products involved.

It will be convenient to study this equivalence in the framework of the set $\mathcal{U}$ of all pairs of $n \times n$ matrices (A,H) where A is an arbitrary complex matrix, and H defines an indefinite scalar product on $\mathbb{C}^n$, i.e. $H^* = H$ and $\det H \neq 0$. The pairs $(A_1,H_1),(A_2,H_2) \in \mathcal{U}$ are said to be *unitarily similar* if, for some invertible matrix T, we have

$$A_1 = T^{-1}A_2T \quad \text{and} \quad H_1 = T^*H_2T .$$

Thus, for unitarily similar pairs, A_1 and A_2 are similar, and H_1 and H_2 are congruent. In general, however, similarity of A_1 and A_2 and congruency of H_1 and H_2 do not guarantee that (A_1,H_1) and (A_2,H_2) are unitarily similar. To ensure this, the similarity and the congruence must be determined simultaneously by the same transforming matrix T.

It is easily verified that unitary similarity defines a relation on $\mathcal{U}$ which is reflexive, symmetric, and transitive, i.e. defines an equivalence relation on $\mathcal{U}$. The corresponding equivalence classes will be called the *unitary similarity classes* of $\mathcal{U}$. The observation that each such class is arcwise connected will be useful later.

THEOREM 3.1 *A unitary similarity class of pairs of matrices from* $\mathcal{U}$ *is arcwise connected.*

PROOF. Let (A,H) and (B,G) be in $\mathcal{U}$, and be unitarily similar. Thus, $A = S^{-1}BS$, $H = S^*GS$, for some invertible S.

Let $S(t)$, $t \in [0,1]$ be a continuous path of invertible matrices with $S(0) = I$, $S(1) = S$. To establish the existence of such a path, let J be a Jordan form for S and for each Jordan block $J_p = \lambda_p I + K$, $\lambda_p \neq 0$, (where K is the nilpotent matrix with ones on the super-diagonal and zeros elsewhere) define $J_p(t) = \lambda_p(t)I + tK$, where $\lambda_p(t)$ is a continuous path of non-zero complex numbers with $\lambda_p(0) = 1$, $\lambda_p(1) = \lambda_p$. Then let $J(t)$ be the block diagonal matrix made up of blocks $J_p(t)$ in just the way that J is made up from the blocks J_p. Then the construction implies $J(0) = I$, $J(1) = J_p$.

Now define $S(t) = TJ(t)T^{-1}$, where T is a matrix for which $S = TJT^{-1}$, and the construction is complete.

Using the path $S(t)$ construct a path of pairs of matrices $(B(t), G(t))$ in the unitary similarity class of (B,G) by

$$B(t) = S(t)^{-1}BS(t) , \quad G(t) = S(t)^*GS(t) ,$$

for $t \in [0,1]$. Then $(B(0),G(0)) = (B,G)$ and $(B(1),G(1)) = (A,H)$, as required. □

Note also the following property of unitary similarity (which is just another formulation of Proposition 2.3).

PROPOSITION 3.2 *Let* (A_1,H_1) *and* (A_2,H_2) *be unitarily similar. Then* A_1 *is selfadjoint (unitary, or normal) with respect to the indefinite scalar product defined by* H_1 *if and only if* A_2 *is selfadjoint (resp. unitary, or normal) with respect to the scalar product defined by* H_2.

In the sequel we shall generally apply the notion of unitary similarity to classes of pairs (A,H) in which A is H-selfadjoint, or A is H-unitary.

3.2 Description of a Canonical Form

Using the concepts of the previous section, we consider now a unitary similarity class of pairs of matrices $(A,H) \in \mathcal{U}$ *in which* A *is H-selfadjoint.* We seek a canonical form in such a class, i.e. a standard simple pair (J,P) in the class with the property that for each pair (A,H) in the class, there is a T such that

$$A = T^{-1}JT , \quad H = T^{*}PT .$$

In fact, the canonical pair (J,P) will consist of a matrix J in Jordan normal form, and a hermitian, invertible P of simple structure (in particular, it will transpire that $P^2 = I$).

Since A is H-selfadjoint we know (Proposition 2.4) that the number and the sizes of Jordan blocks in J corresponding to an eigenvalue λ_0 $(\neq \bar{\lambda}_0)$ and those for $\bar{\lambda}_0$ are the same. So we can assume that J is a direct sum of Jordan blocks with *real* eigenvalues and blocks of the type $\mathrm{diag}[J_k,\bar{J}_k]$, where J_k is a Jordan block with non-real eigenvalue.

It will be convenient to introduce the following notation. By $J_k(\lambda)$ we denote the Jordan block with eigenvalue λ of size k if λ is real, and the direct sum of two Jordan blocks of size $\frac{k}{2}$, the first with eigenvalue λ and the second with eigenvalue $\bar{\lambda}$, if λ is non-real. Often we shall write $J(\lambda)$ omitting the subscript k.

We have seen in Example 2.1 that $J_k(\lambda)$ is $\pm P$-selfadjoint, where P is the $k \times k$ sip matrix. The following theorem shows that the pair (A,H) is unitarily similar to a direct sum of blocks of types $(J(\lambda),\pm P)$ when λ is real and $(J(\lambda),P)$ when λ is non-real.

THEOREM 3.3 *Let* A *be H-selfadjoint and let*

$$J = J(\lambda_1) \oplus \cdots \oplus J(\lambda_\alpha) \oplus J(\lambda_{\alpha+1}) \oplus \cdots \oplus J(\lambda_\beta) \tag{3.1}$$

be a Jordan normal form for A, *where* $\lambda_1,\cdots,\lambda_\alpha$ *are real and* $\lambda_{\alpha+1},\cdots,\lambda_\beta$

are non-real eigenvalues of A. *Then* (A,H) *is unitarily similar to a pair* $(J, P_{\varepsilon,J})$, *where*

$$P_{\varepsilon,J} = \varepsilon_1 P_1 \oplus \cdots \oplus \varepsilon_\alpha P_\alpha \oplus P_{\alpha+1} \oplus \cdots \oplus P_\beta \ , \tag{3.2}$$

$P_1, P_2, \cdots, P_\beta$ *are sip matrices with the sizes of* $J(\lambda_1), \cdots, J(\lambda_\beta)$ *respectively, and* $\varepsilon = \{\varepsilon_1, \cdots, \varepsilon_\alpha\}$ *is an ordered set of signs* ± 1. *The set* ε *is uniquely determined by* (A,H) *up to permutation of signs corresponding to equal Jordan blocks.*

Conversely, if for some set of signs ε, *the pairs* (A,H) *and* $(J, P_{\varepsilon,J})$ *are unitarily similar, then* A *is H-selfadjoint.*

The set of signs ε appearing in this theorem will be called the *sign characteristic* of the pair (A,H), and recall that it consists of a +1 or -1 factor applied to each *real* Jordan block of the Jordan form J of A.

An alternative description of the sign characteristic can be made in terms of the elementary divisors of $I\lambda - A$: one sign (+1 or -1) is attached to each divisor $(\lambda-\lambda_0)^\alpha$ with a real λ_0. (Recall that there is a one-to-one correspondence between the elementary divisors of $I\lambda - A$ and the Jordan blocks in the Jordan normal form of A. In fact, each divisor $(\lambda-\lambda_0)^\alpha$ corresponds to a Jordan block of size α with eigenvalue λ_0.)

It is clear from the statement of Theorem 3.3 that the sign characteristic of (A,H) is uniquely defined if we apply the following normalization rule: in every subset of signs corresponding to Jordan blocks with the same size and the same real eigenvalue, +1's (if any) precede -1's (if any). Subsequently, it will always be assumed (unless stated otherwise) that the sign characteristic is normalized in this way.

The proof of Theorem 3.3 will be given in the next section. Here, we note two immediate corollaries. The first of which gives a complete description of all the invertible hermitian matrices H for which a given matrix A is H-selfadjoint.

COROLLARY 3.4 *Let* A *be an* $n \times n$ *matrix which is similar to* A^* *and let* J *be a Jordan form for* A *arranged as in* (3.1). *Then* A *is H-selfadjoint if and only if* H *has the form* $H = T^* P_{\varepsilon,J} T$ *where* $P_{\varepsilon,J}$ *is given by* (3.2) *for some set of signs* ε, *and* T *is an invertible matrix for which* $A = T^{-1}JT$.

In particular the following result, promised in Section 2.2, follows immediately.

COROLLARY 3.5 *If an* $n \times n$ *matrix* A *is similar to* A^*, *then there exists an invertible hermitian* H *such that* $A^* = H^{-1}AH$. *Such a matrix is given by* $H = T^*P_{\varepsilon,J}T$ *where* J *is a Jordan normal form for* A *(arranged as in* (3.1)), T *is any invertible matrix for which* $A = T^{-1}JT$, ε *is an arbitrarily chosen set of signs, and* $P_{\varepsilon,J}$ *is given by* (3.2).

The following example is a simple illustration of Corollary 3.4.

EXAMPLE 3.1 Let

$$J = \operatorname{diag}\left\{ \begin{bmatrix} 0 & 1 \\ 0 & 0 \end{bmatrix}, \begin{bmatrix} 1 & 1 \\ 0 & 1 \end{bmatrix} \right\}.$$

Then

$$P_{\varepsilon,J} = \operatorname{diag}\left\{ \begin{bmatrix} 0 & \varepsilon_1 \\ \varepsilon_1 & 0 \end{bmatrix}, \begin{bmatrix} 0 & \varepsilon_2 \\ \varepsilon_2 & 0 \end{bmatrix} \right\} \quad \text{for} \quad \varepsilon = (\varepsilon_1,\varepsilon_2), \quad \varepsilon_i = \pm 1 .$$

According to Theorem 3.2, the set $\Omega = \Omega_J$ of all invertible hermitian matrices H such that J is H-selfadjoint, splits into 4 disjoint sets Ω_1, Ω_2, Ω_3, Ω_4 corresponding to the sets of signs $(+1,+1)$, $(+1,-1)$, $(-1,+1)$ aand $(-1,-1)$ respectively: $\Omega = \Omega_1 \cup \Omega_2 \cup \Omega_3 \cup \Omega_4$. An easy computation shows that each set Ω_i consists of all matrices H of the form

$$H = \operatorname{diag}\left\{ \begin{bmatrix} 0 & a_1\varepsilon_1 \\ a_1\varepsilon_1 & b_1 \end{bmatrix}, \begin{bmatrix} 0 & a_2\varepsilon_2 \\ a_2\varepsilon_2 & b_2 \end{bmatrix} \right\},$$

where a_1, a_2, are positive and b_1, b_2 are real parameters; ε_1, ε_2 are ± 1 depending on the set Ω_i. Note also that each set Ω_i $(i = 1,2,3,4)$ is (arcwise) connected. □

3.3 First Applications of the Canonical Form

In this section we consider some conclusions that can be drawn readily from Theorem 3.3. The first concerns the structure of $\operatorname{Ker}(\lambda_0 I-A)$ when $\lambda_0 \in \sigma(A)$ and A is H-selfadjoint. Note that for any eigenvalue λ_0 the eigenspace $\operatorname{Ker}(\lambda_0 I-A)$ can be written as a direct sum

$$\operatorname{Ker}(\lambda_0 I-A) = L_1 \dot{+} L_2 \tag{3.3}$$

where L_1, L_2 are generated by all eigenvectors associated with linear, and nonlinear, elementary divisors of λ_0, respectively. These subspaces could be defined in other ways. For example

$$L_1 = \{x \in \mathrm{Ker}(\lambda_0 I-A) : x \notin \mathrm{Im}(\lambda_0 I-A)\} \cup \{0\}$$

$$L_2 = \mathrm{Ker}(\lambda_0 I-A) \cap \mathrm{Im}(\lambda_0 I-A) .$$

Clearly, $\dim L_1$ ($\dim L_2$) is just the number of linear (respectively non-linear) elementary divisors associated with λ_0.

Our first observation is that, *if* A *is* H-*selfadjoint and* $\lambda_0 \in \sigma(A)$ *is real then, in the decomposition* (3.3), L_1 *is* H-*nondegenerate and* L_2 *is* H-*neutral*. To see this, suppose that λ_0 has p linear, and q nonlinear elementary divisors. Then the corresponding submatrices of J and $P_{\varepsilon,J}$ from (3.1) and (3.2) have the form

$$J^{(0)} = \mathrm{diag}[\lambda_0,\cdots,\lambda_0,J_1,J_2,\cdots,J_q]$$

where λ_0 appears p times and J_j is a Jordan block of size $m_j \geq 2$ for $j = 1,2,\cdots,q$,

$$P_{\varepsilon,J} = \mathrm{diag}[\varepsilon_1,\cdots,\varepsilon_p,\varepsilon_{p+1}P_{p+1},\cdots,\varepsilon_q P_q] .$$

For these matrices the eigenspace associated with *linear* divisors, say $L_1^{(0)}$ is spanned by unit coordinate vectors $e_1,\cdots,e_p$ and, in the $P_{\varepsilon,J}^{(0)}$ scalar product, $[e_j,e_k] = \varepsilon_j\delta_{jk}$ for $j,k = 1,2,\cdots,p$. It follows readily that $L_1^{(0)}$ is $P_{\varepsilon,J}^{(0)}$-nondegenerate, and hence that L_1 is H-nondegenerate.

The eigenspace of $J^{(0)}$ associated with *nonlinear* divisors, say $L_2^{(0)}$ is spanned by unit vectors e_j wity $j = p+1,p+m_1+1,\cdots,p+\sum_{r=1}^{q-1} m_r+1$, and for indices j, k taking these values we clearly have $[e_j,e_k] = 0$ in the $P_{\varepsilon,J}^{(0)}$ scalar product. It follows that L_2 is H-neutral.

If $\lambda_0 \in \sigma(A)$ *and* $\bar{\lambda}_0 \neq \lambda_0$ *then the whole eigenspace* $L_1 \dotplus L_2$ *is* H-*neutral*. This can be deduced from the presence of zeros in certain strategic positions of $P_{\varepsilon,J}$, but this statement can be strengthened to the observation that the whole root subspace $X_{\lambda_0} = \mathrm{Ker}(\lambda_0 I-A)^n$, where n is the size of A, is H-neutral. This can be "seen" from the canonical pair $J,P_{\varepsilon,J}$, but has also been proved in Corollary 2.6.

Another useful observation can be made in the case that $\lambda_0 \in \sigma(A)$ and $\bar{\lambda}_0 \neq \lambda_0$; namely, that *the direct sum of root subspaces* $X_{\lambda_0} \dotplus X_{\bar{\lambda}_0}$ *is non-degenerate with respect to* H. This is easily verified for a canonical pair $(J(\lambda_0),P)$ and hence, in full generality.

Our next deduction from the canonical forms of Theorem 3.3 concerns the possibility of counting the negative (or the positive) eigenvalues of H, once the sign characteristic is known. To be precise about this, let the real Jordan blocks $J(\lambda_1),J(\lambda_2),\cdots,J(\lambda_\alpha)$ of (3.1) have sizes $m_1,\cdots,m_\alpha$ and note (from (3.2)) that the associated signs in the sign characteristic of (A,H) are $\varepsilon_1,\cdots,\varepsilon_\alpha$. If N is the number of *negative* eigenvalues of H (and hence of $P_{\varepsilon,J}$) it follows first from (3.2) that

$$\operatorname{sig} P_{\varepsilon,J} = \frac{1}{2}\sum_{i=1}^{\alpha}[1-(-1)^{m_j}]\varepsilon_i$$

and, hence, that

$$N = \frac{1}{2}n - \frac{1}{4}\sum_{i=1}^{\alpha}[1-(-1)^{m_j}]\varepsilon_i \ . \tag{3.4}$$

Note also that (3.4) implies immediately

$$N \geqq \frac{1}{2}n - \frac{1}{2}\sum_{i=1}^{\alpha} m_i$$

with equality when the m_i are all equal to one and $\varepsilon_1 = \cdots = \varepsilon_\alpha = 1$. Now the lower bound for N is just half the number of non-real eigenvalues of A. H *has* $\frac{1}{2}n$ *positive eigenvalues (and so* $\frac{1}{2}n$ *negative eigenvalues) if and only if the signs associated with real elementary divisors of odd degree (if any) are equally divided between* +1'*s and* -1'*s*.

3.4 Proof of Theorem 3.3

It is easily verified that, in the statement of Theorem 3.3, J is $P_{\varepsilon,J}$-selfadjoint. The converse statement of the theorem is then an immediate application of Proposition 3.2.

Now let A be H-selfadjoint, let $\lambda_1,\cdots,\lambda_\alpha$ be all the different real eigenvalues of A and $\lambda_{\alpha+1},\cdots,\lambda_\beta$ be all the different eigenvalues in the upper half of the complex plane. Decompose $\mathbb{C}^n$ into a direct sum:

$$\mathbb{C}^n = X_1 \dotplus \cdots \dotplus X_\alpha \dotplus X_{\alpha+1} \dotplus \cdots \dotplus X_\beta$$

where $X_1,\cdots,X_\alpha$ are the root subspaces corresponding to $\lambda_1,\cdots,\lambda_\alpha$ respectively, and for $j = \alpha+1,\cdots,\beta$, X_j is the sum of the root subspaces corresponding to λ_j and $\bar{\lambda}_j$. It follows immediately from Theorem 2.5 that for $j,k = 1,2,\cdots,\beta$ the subspaces X_j and X_k are orthogonal with respect to H, i.e. $[x,y] = (Hx,y) = 0$ for every $x \in X_j$, $y \in X_k$. From (1.2) it follows that for $i = 1,2,\cdots,\beta$

$$X_i^{[\perp]} = X_1 \dotplus \cdots \dotplus X_{i-1} \dotplus X_{i+1} \dotplus \cdots \dotplus X_\beta \tag{3.5}$$

and then Proposition 1.1 implies that each X_i is non-degenerate.

Consider a fixed X_i with $i \in \{\alpha+1,\cdots,\beta\}$, so that $X_i = X_i' \dotplus X_i''$ and X_i', X_i'' are the root subspaces corresponding to λ_i, $\bar{\lambda}_i$, respectively. Corollary 2.6 asserts that both X_i' and X_i'' are H-neutral.

There exists an integer m with the properties that $(A-\lambda_i I)^m|_{X_i'} = 0$, but $(A-\lambda_i I)^{m-1}a_1 \neq 0$ for some $a_1 \in X_i'$. Since X_i is non-degenerate and X_i' is neutral, there exists a $b_1 \in X_i''$ such that $[(A-\lambda_i I)^{m-1}a_1,b_1] = 1$. Define sequences $a_1,\cdots,a_m \in X_i'$ and $b_1,\cdots,b_m \in X_i''$ by

$$a_j = (A-\lambda_i I)^{j-1}a_1 \ , \quad b_j = (A-\bar{\lambda}_i I)^{j-1}b_1 \ , \quad j = 1,\cdots,m \ .$$

Observe that $[a_1,b_m] = [a_1,(A-\bar{\lambda}_i I)^{m-1}b_1] = [(A-\lambda_i I)^{m-1}a_1,b_1] = 1$, in particular, $b_m \neq 0$. Further, for every $x \in X_i'$ we have

$$[x,(A-\bar{\lambda}_i I)b_m] = [x,(A-\bar{\lambda}_i I)^m b_1] = [(A-\lambda_i I)^m x,b_1] = 0 \ ;$$

so the vector $(A-\bar{\lambda}_i I)b_m$ is H-orthogonal to X_i'. In view of (3.5) we deduce that $(A-\bar{\lambda}_i I)b_m$ is H-orthogonal to $\mathbb{C}^n$, and hence

$$(A-\bar{\lambda}_i I)b_m = 0 \ .$$

Then clearly $a_m,\cdots,a_1$ (resp. $b_m,\cdots,b_1$) is a Jordan chain of A corresponding to λ_i (resp. $\bar{\lambda}_i$), i.e. for $j = 1,2,\cdots,m-1$,

$$Aa_j - \lambda_i a_j = a_{j+1} \quad \text{and} \quad Aa_m = \lambda_i a_m \ ,$$

with analogous relations for the b_j (replacing λ_i by $\bar{\lambda}_i$). For $j + k = m + 1$ we have

$$[a_j,b_k] = [(A-\lambda_i I)^{j-1}a_1,(A-\bar{\lambda}_i I)^{k-1}b_1] = [(A-\lambda_i I)^{j+k-2}a_1,b_1] = 1 \ ; \tag{3.6}$$

and analogously

$$[a_j, b_k] = 0 \quad \text{if} \quad j + k > m + 1 . \tag{3.7}$$

Now put

$$c_1 = a_1 + \sum_{j=2}^{m} \alpha_j a_j , \quad c_{j+1} = (A-\lambda_i I)c_j , \quad j = 1,\cdots,m-1 ,$$

where $\alpha_2,\cdots,\alpha_m$ are chosen so that

$$[c_1, b_{m-1}] = [c_1, b_{m-2}] = \cdots = [c_1, b_1] = 0 . \tag{3.8}$$

Such a choice is possible, as can be checked easily using (3.6) and (3.7). Now for $j + k \leqslant m$

$$[c_j, b_k] = [(A-\lambda_i I)^{j-1} c_1, b_k] = [c_1, (A-\bar{\lambda}_i I)^{j-1} b_k] = [c_1, b_{k+j-1}] = 0 ,$$

and for $j + k \geqslant m + 1$ we obtain, using $(A-\lambda_i I)^m a_1 = 0$ together with (3.6) and (3.7):

$$\begin{aligned} [c_j, b_k] &= [(A-\lambda_i I)^{j-1} c_1, (A-\bar{\lambda}_i I)^{k-1} b_1] \\ &= [(A-\lambda_i I)^{j+k-2} c_1, b_1] = [(A-\lambda_i I)^{j+k-2} a_1, b_1] \\ &= \begin{cases} 1 , & \text{for} \quad j + k = m + 1 \\ 0 , & \text{for} \quad j + k > m + 1 . \end{cases} \end{aligned}$$

Let $N_1 = \mathrm{Span}\{c_1,\cdots,c_m,b_1,\cdots,b_m\}$. The relations above show that $A|_{N_1} = J_1 \oplus \bar{J}_1$ in the basis $c_m,\cdots,c_1,b_m,\cdots,b_1$ where J_1 is the Jordan block of size m with eigenvalue λ_i;

$$[x,y] = y^* \begin{bmatrix} 0 & P_1 \\ P_1 & 0 \end{bmatrix} x , \quad x,y \in N_1$$

in the same basis, and P_1 is the sip matrix of size m. We see from this representation that N_1 is non-degenerate. By Proposition 1.1, $\mathbb{C}^n = N_1 \dot{+} N_1^{[\perp]}$, and by Proposition 2.2, $N_1^{[\perp]}$ is an invariant subspace for A. If $A|_{N_1^{[\perp]}}$ has non-real eigenvalues, apply the same procedure to construct a subspace $N_2 \subset N_1^{[\perp]}$ with basis $c'_{m'},\cdots,c'_1,b'_{m'},\cdots,b'_1$ such that in this basis $A|_{N_2} = J_2 \oplus \bar{J}_2$, where J_2 is the Jordan block of size m' with non-real eigenvalue, and

$$[x,y] = y^* \begin{bmatrix} 0 & P_2 \\ P_2 & 0 \end{bmatrix} x , \quad x,y \in N_2$$

with the sip matrix P_2 of size m'. Continue this procedure until the nonreal eigenvalues of A are exhausted.

Consider now a fixed X_i, where $i \in \{1,\cdots,\alpha\}$ so that λ_i is real. Again, let m be such that $(A-\lambda_i I)^m|_{X_i} = 0$ but $(A-\lambda_i I)^{m-1}|_{X_i} \neq 0$. Let Q_i be the orthogonal projector on X_i and define $F : X_i \to X_i$ by

$$F = Q_i H(A-\lambda_i I)^{m-1}|_{X_i} .$$

Since λ_i is real, it is easily seen that F is a hermitian linear transformation. Moreover, $F \neq 0$; so there is a nonzero eigenvalue of F (necessarily real) with an eigenvector a_1. Normalize a_1 so that

$$(Fa_1,a_1) = \varepsilon , \quad \varepsilon = \pm 1 .$$

In other words,

$$[(A-\lambda_i I)^{m-1}a_1,a_1] = \varepsilon . \tag{3.9}$$

Let $a_j = (A-\lambda_i I)^{j-1}a_1$, $j = 1,\cdots,m$. It follows from (3.9) that for $j + k = m + 1$

$$[a_j,a_k] = [(A-\lambda_i I)^{j-1}a_1,(A-\lambda_i I)^{k-1}a_1] = [(A-\lambda_i I)^{m-1}a_1,a_1] = \varepsilon . \tag{3.10}$$

Moreover, for $j + k > m + 1$ we have:

$$[a_j,a_k] = [(A-\lambda_i I)^{j+k-2}a_1,a_1] = 0 \tag{3.11}$$

in view of the choice of m. Now put

$$b_1 = a_1 + \alpha_2 a_2 + \cdots + \alpha_m a_m , \quad b_j = (A-\lambda_i I)^{j-1}b_1 , \quad j = 1,\cdots,m ,$$

and choose α_i so that

$$[b_1,b_1] = [b_1,b_2] = \cdots = [b_1,b_{m-1}] = 0 .$$

Such a choice of α_i is possible. Indeed, equality $[b_1,b_j] = 0$ $(j = 1,\cdots,m-1)$ gives, in view of (3.10) and (3.11),

$$0 = [a_1 + \alpha_2 a_2 + \cdots + \alpha_m a_m \ , \ a_j + \alpha_2 a_{j+1} + \cdots + \alpha_{m-j+1} a_m]$$
$$= [a_1, a_j] + 2\varepsilon\alpha_{m-j+1} + (\text{terms in } \alpha_2, \cdots, \alpha_{m-j}) \ .$$

Evidently, these equalities determine unique numbers $\alpha_2, \cdots, \alpha_m$ in succession.

Let $N = \mathrm{Span}\{b_1, \cdots, b_m\}$. In the basis $b_1, \cdots, b_m$ the linear transformation $A|_N$ is represented by the single Jordan block with eigenvalue λ_i, and

$$[x,y] = y^* \varepsilon P_0 x \ , \quad x,y \in N \ ,$$

where P_0 is the sip matrix of size m.

Continue the procedure on the orthogonal companion to N, and so on.

Applying this construction, we find a basis $f_1, \cdots, f_n$ in $\mathbb{C}^n$ such that A is represented by the Jordan matrix J of (3.1) in this basis and, with $P_{\varepsilon,J}$ as defined in (3.2),

$$[x,y] = y^* P_{\varepsilon,J} x \ , \quad x,y \in \mathbb{C}^n \ ,$$

where x and y are represented by their coordinates in the basis $f_1, \cdots, f_n$. Let T be the $n \times n$ invertible matrix whose i-th column is formed by the coordinates of f_i (in the standard orthonormal basis), $i = 1, \cdots, n$. For such a T, the relation $T^{-1}AT = J$ holds because $f_1, \cdots, f_n$ is a Jordan basis for A, and equality $T^* HT = P_{\varepsilon,J}$ follows from the construction of $f_1, \cdots, f_n$. So (A,H) and $(J, P_{\varepsilon,J})$ are unitarily similar.

It remains to prove the uniqueness of the normalized sign characteristic of (A,H). So suppose that (A,H) is unitarily similar to two canonical pairs, $(J, P_{\varepsilon,J})$ and $(J, P_{\delta,J})$ where ε, δ are sets of signs. It is to be proved that ε and δ are the same up to permutation of signs corresponding to Jordan blocks with the same eigenvalue and the same size.

The hypothesis implies that

$$P_{\delta,J} = S^* P_{\varepsilon,J} S \ , \quad J = S^{-1} J S \tag{3.12}$$

for some invertible S. The second of these equations shows that attention can be restricted to the case when $\sigma(J)$ is a singleton. (Indeed, if $J = \mathrm{diag}[K_1, K_2]$ where $\sigma(K_1) \cap \sigma(K_2) = \phi$, then every S commuting with J

has the form $\mathrm{diag}[S_1, S_2]$ with partitioning consistent with that of J; see, for instance, [14], Section VIII. 2.) Thus, it is now assumed that $\sigma(J) = \{\lambda\}$ and $\lambda \in \mathbb{R}$. Let J have k_i blocks $J_{i1}, \cdots, J_{ik_i}$ of size m_i for $i = 1,2,\cdots,t$ and $m_1 > m_2 > \cdots > m_t$. Thus, we may write

$$J = \mathrm{diag}[\mathrm{diag}[J_{ij}]_{j=1}^{k_i}]_{i=1}^{t}$$

and

$$P_{\varepsilon,J} = \mathrm{diag}[\mathrm{diag}[\varepsilon_{i,j} P_i]_{j=1}^{k_i}]_{i=1}^{t}$$

with a similar expression for $P_{\delta,J}$ replacing the $\varepsilon_{i,j}$ by signs $\delta_{i,j}$.

It follows from (3.12) that, for any nonnegative integer k,

$$P_{\delta,J}(I\lambda - J)^k = S^* P_{\varepsilon,J}(I\lambda - J)^k S \,, \tag{3.13}$$

and is a relation between hermitian matrices. Consequently,

$$\mathrm{sig}(P_{\delta,J}(I\lambda - J)^{m_1 - 1}) = \mathrm{sig}(P_{\varepsilon,J}(I\lambda - J)^{m_1 - 1}) \,. \tag{3.14}$$

Observe that $(I\lambda - J_1)^{m_1 - 1}$ is nilpotent of rank one and $(I\lambda - J_i)^{m_1 - 1} = 0$ for $i = 2,3,\cdots,t$. It follows that

$$P_{\varepsilon,J}(I\lambda - J)^{m_1 - 1} = \mathrm{diag}[\varepsilon_{1,1} P_1 (I\lambda - J)^{m_1 - 1}, \cdots, \varepsilon_{1,k_1} P_1 (I\lambda - J_1)^{m_1 - 1}, 0, \cdots, 0] \,,$$

and

$$\mathrm{sig}(P_{\varepsilon,J}(I\lambda - J)^{m_1 - 1}) = \sum_{i=1}^{k_1} \varepsilon_{1,i} \,.$$

Similarly,

$$\mathrm{sig}(P_{\varepsilon,J}(I\lambda - J)^{m_1 - 1}) = \sum_{i=1}^{k_1} \delta_{1,i} \,.$$

Noting (3.14) it is found that

$$\sum_{i=1}^{k_i} \varepsilon_{1,i} = \sum_{i=1}^{k_i} \delta_{1,i} \,. \tag{3.15}$$

Consequently, the subsets $\{\varepsilon_{1,1}, \cdots, \varepsilon_{1,k_1}\}$ and $\{\delta_{1,1}, \cdots, \delta_{1,k}\}$ of ε and δ agree to within normalization.

Now examine the hermitian matrix $P_{\varepsilon,J}(I\lambda-J)^{m_2-1}$. This is found to be block diagonal with non-zero blocks of the form

$$\varepsilon_{1,j}P_1(I\lambda-J_1)^{m_2-1}, \quad j = 1,2,\cdots,k_1$$

and

$$\varepsilon_{2,j}P_2(I\ -J_2)^{m_2-1}, \quad j = 1,2,\cdots,k\ .$$

It follows that the signature of $P_{\varepsilon,J}(I\lambda-J)^{m_2-1}$ is given by

$$\left[\sum_{j=1}^{k_1}\varepsilon_{1,j}\right](\text{sig } P_1(I\lambda-J_1)^{m_2-1}) + \sum_{j=1}^{k_2}\varepsilon_{2,j}\ .$$

But again, in view of (3.13), this must be equal to the corresponding expression formulated using δ instead of ε. Hence, using (3.15), it is found that

$$\sum_{j=1}^{k_2}\varepsilon_{2,j} = \sum_{j=1}^{k_2}\delta_{2,j}$$

and the subsets $\{\varepsilon_{2,1},\cdots,\varepsilon_{2,k_2}\}$ and $\{\delta_{2,1},\cdots,\delta_{2,k_2}\}$ of ε and δ agree within the normalization convention.

Now it is clear that the argument can be continued for t steps after which the uniqueness of the normalized sign characteristic is established. This completes the proof of Theorem 3.3. □

3.5 Classification of Matrices by Unitary Similarity

We now exploit the notion of unitary similarity introduced in Section 3.1 taking advantage of the canonical pairs introduced in Theorem 3.3. Recall the remark in Section 3.1 that if A_1 and A_2 are similar and H_1 and H_2 are congruent it is not necessarily the case that (A_1,H_1) and (A_2,H_2) are unitarily similar. We show first that, if A_j is H_j-selfadjoint for $j = 1$ and 2, then unitary similarity of the two pairs can be characterized quite easily. Let the set $\mathcal{U}$ of pairs of matrices (A,H) be as defined in Section 3.1, i.e. A is arbitrary $n \times n$, and H is $n \times n$, $H^* = H$ and H is invertible.

THEOREM 3.6 *Let* (A_1,H_1) *and* $(A_2,H_2) \in \mathcal{U}$ *and let* A_j *be* H_j-*selfadjoint for* $j = 1$ *and* 2. *Then* (A_1,H_1) *and* (A_2,H_2) *are unitarily similar if and only if* A_1 *and* A_2 *are similar and the pairs* (A_1,H_1), (A_2,H_2) *have the same sign characteristic.*

PROOF. If A_1 and A_2 are similar and (A_1,H_1), (A_2,H_2) have the same sign characteristic then, by Theorem 3.3,

$$A_j = T_j^{-1}JT_j \ , \quad H_j = T^* P_{\varepsilon,J}T_j \ , \quad j = 1 \text{ and } 2 \ .$$

Hence

$$A_1 = (T_2^{-1}T_1)^{-1}A(T_2^{-1}T_1) \ , \quad H_1 = (T_2^{-1}T_1)^* H_2(T_2^{-1}T_1) \ ,$$

i.e. (A_1,H_1) and (A_2,H_2) are unitarily similar and the matrix $T_2^{-1}T_1$ determines the unitary similarity.

Conversely, suppose that (A_1,H_1) and (A_2,H_2) are unitarily similar. Then A_1 and A_2 have a common Jordan form J, which can be arranged as in (3.1). Then, if (A_2,H_2) and $(J,P_{\varepsilon,J})$ are unitarily similar for some set of signs ε, it follows that (A_1,H_1) and $(J,P_{\varepsilon,J})$ are also unitarily similar. In particular, the sign characteristics of (A_1,H_1) and (A_2,H_2) are the same. □

Consider now the important case of unitary similarity in which $H_1 = H_2 = H$. Thus, we say that A_1 and A_2 are H-*unitarily similar* if

$$A_1 = U^{-1}A_2U \quad \text{and} \quad U^*HU = H \ ;$$

in other words, if $A_1 = U^{-1}A_2U$ for some H-unitary matrix U. It is easily seen that H-unitary similarity defines an equivalence relation on the set of all square matrices (with the size of H). Furthermore, if an equivalence class contains an H-selfadjoint matrix it is easily verified that every matrix in the equivalence class is also H-selfadjoint.

Theorem 3.6 shows that H-selfadjoint matrices A_1 and A_2 are H-unitarily similar if and only if they are similar and the pairs (A_1,H) and (A_2,H) have the same sign characteristic. Using this observation, a canonical representative can now be constructed in each equivalence class of H-selfadjoint matrices.

First, a set of normalized Jordan matrices is constructed, all of which have spectrum symmetric with respect to the real axis. Thus, Ξ is defined to be the set of all $n \times n$ Jordan matrices J of the form

$$J_{m_1}(\gamma_1) + \cdots + J_{m_\alpha}(\gamma_\alpha) + J_{m_{\alpha+1}}(\gamma_{\alpha+1}+i\delta_{\alpha+1}) + \cdots + J_{m_\beta}(\gamma_\beta+i\delta_\beta) ,$$

where γ_i $(1 \leq i \leq \beta)$ are real, $\gamma_1 \leq \gamma_2 \leq \cdots \leq \gamma_\alpha$ and blocks with the same eigenvalue are in non-decreasing order of size. Also, $\delta_k > 0$ for $k = \alpha+1,\cdots,\beta$ and $\gamma_{\alpha+1} \leq \gamma_{\alpha+2} \leq \cdots \leq \gamma_\beta$ with $\delta_k \leq \delta_{k+1}$ if $\gamma_k = \gamma_{k+1}$ and, finally, all such blocks with the same eigenvalue are in non-decreasing order of size. (Recall that $J_k(\lambda)$ stands for the $k \times k$ Jordan block with eigenvalue λ if λ is real, and for the direct sum of two $\frac{k}{2} \times \frac{k}{2}$ Jordan blocks with eigenvalues λ and $\bar{\lambda}$ respectively if λ is non-real).

With these conventions it is clear that two matrices $J',J'' \in \Xi$ are similar if and only if $J' = J''$.

Now suppose that H is given and consider the subset

$$\Xi_H = \{J \in \Xi \mid TJT^{-1} \text{ is H-selfadjoint for some } T\} .$$

For each $J \in \Xi_H$ define $\pi(J)$ to be the set of matrices $P_{\varepsilon,J}$ (constructed as in (3.2)) with block structure consistent with that of J, and a set of signs ε such that $P_{\varepsilon,J}$ and H are congruent.

The construction is completed by associating a unique S_P with $P \in \pi(J)$, $(J \in \Xi_H)$, such that $H = S_P^* P S_P$ and defining

$$R_H = \{S_P^{-1} J S_P \mid J \in \Xi_H \text{ and } P \in \pi(J)\} .$$

Then we have:

THEOREM 3.7 *The set* R_H *forms a complete set of representatives of the equivalence classes (under H-unitary similarity) of H-selfadjoint matrices. In other words, for every H-selfadjoint matrix* A *there is an* $A' \in R_H$ *such that* A *and* A' *are H-unitarily similar, and if* $A',A'' \in R_H$ *and* $A' \neq A''$ *then* A' *and* A'' *are not unitarily similar.*

PROOF. Let A be H-selfadjoint, and let $A = S^{-1}JS$, $H = S^* P_{\varepsilon,J} S$ be the canonical form of (A,H), with normalized sign characteristic, and J chosen from Ξ. Clearly, $P_{\varepsilon,J} \in \pi(J)$. Put $A' = S_{P_{\varepsilon,J}}^{-1} J S_{P_{\varepsilon,J}} \in R_H$. Then A and A' are H-unitarily similar by Theorem 3.6.

Conversely, suppose

$$A' = S_{P'}^{-1} J' S_{P'} \in R_H \; ; \; A'' = S_{P''}^{-1} J'' S_{P''} \in R_H ,$$

and A' and A'' are H-unitarily similar. In particular, A' and A'' are similar, and so are J' and J''. But since $J', J'' \in \Xi$ we have $J' = J''$. Now use Theorem 3.6 to deduce that also $P' = P''$ $(\in \pi(J') = \pi(J''))$. Clearly, $A' = A''$ (because the choice of S_P is fixed; so $P' = P''$ implies $S_{P'} = S_{P''}$). □

This section is to be concluded by showing that the equivalence classes of Theorem 3.7 are arcwise connected. But first a lemma is needed.

LEMMA 3.8 *The set of all H-unitary matrices is arcwise connected.*

PROOF. First observe that the set of all H-selfadjoint matrices is arcwise connected because it is a real linear space. The lemma is proved by combining this observation with an appropriate use of the Cayley transformation.

Let U_1, U_2 be H-unitary matrices and let a_1, a_2 be unimodular complex numbers for which $U_1 - a_1 I$ and $U_2 - a_2 I$ are invertible. For $j = 1,2$, define the H-selfadjoint matrices

$$A_j = (U_j - a_j I)^{-1}(wU_j - wa_j I)$$

where $w \in \mathbb{C} \setminus \mathbb{R}$. It follows that $w \notin \sigma(A_1) \cup \sigma(A_2)$.

Let $A(t)$, $t \in [0,1]$, be a continuous path of H-selfadjoint matrices for which $A(0) = A_1$, $A(1) = A_2$, and let $w(t)$, $t \in [0,1]$, be a continuous path in $\mathbb{C} \setminus \mathbb{R}$ for which $w(0) = w = w(1)$ and $w(t) \notin \sigma(A(t))$ for $t \in [0,1]$. For example, $w(t)$ can be chosen to be equal to the constant w, except for neighbourhoods of those points t for which $w \in \sigma(A(t))$.

Then define

$$U(t) = (A(t) - \overline{w(t)}I)(A(t) - w(t)I)^{-1}$$

to obtain a path of H-unitary matrices connecting $a_1^{-1}U_1$ and $a_2^{-1}U_2$. Now it is clear that U_1 and U_2 are also arcwise connected. □

THEOREM 3.9 *Each equivalence class (under H-unitary similarity) of H-selfadjoint matrices is arcwise connected.*

PROOF. Let A_1 and A_2 be H-selfadjoint and $A_1 = U^{-1}A_2U$ where U is H-unitary. From Lemma 3.8, there exists a path $U(t)$, $t \in [0,1]$ of H-unitary matrices for which $U(0) = U$ and $U(1) = I$. Then the path

$$A(t) = U(t)^{-1}A\,U(t)\ , \quad t \in [0,1]$$

connects A_1 and A_2 in the equivalence class. □

3.6 Signature Matrices

Let A be H-selfadjoint and let $(J,P_{\varepsilon,J})$ be the canonical form of the pair (A,H). We have observed that J and $P_{\varepsilon,J}$ satisfy the relations

$$P_{\varepsilon,J}J = J^*P_{\varepsilon,J}\ , \quad P^*_{\varepsilon,J} = P_{\varepsilon,J}\ , \quad P^2_{\varepsilon,J} = I\ . \tag{3.16}$$

The question arises as to what extent these relations determine $P_{\varepsilon,J}$ (for a fixed J of the form (3.1)). In other words, we are interested in solutions of the simultaneous equations

$$PJ = J^*P\ , \quad P^* = P\ , \quad P^2 = I\ . \tag{3.17}$$

The next theorem shows that the set of all solutions is a set of matrices which are unitarily similar to $P_{\varepsilon,J}$; in fact, of the form $SP_{\varepsilon,J}S^*$ where S is unitary and commutes with J. Thus, every solution can be considered as a representation of $P_{\varepsilon,J}$ in some special orthonormal basis for $\mathbb{C}^n$, and this fact casts a new light on the canonical matrix $P_{\varepsilon,J}$.

THEOREM 3.10 *If matrix P satisfies the relations (3.17), then there is a set of signs ε (which is unique up to permutation of signs corresponding to equal Jordan blocks in J) and a unitary matrix S commuting with J such that $P = SP_{\varepsilon,J}S^{-1}$.*

Conversely, if ε is any set of signs and $P = SP_{\varepsilon,J}S^{-1}$ where S is any unitary matrix which commutes with J, then P satisfies the equation (3.17).

PROOF. An easy calculation confirms the converse statement. Indeed, if $P = SP_{\varepsilon,J}S^{-1}$ with $S^*S = I$ and $SJ = JS$, then clearly $P^* = P$ and since $P^2_{\varepsilon,J} = I$, also $P^2 = I$. Furthermore, we have $J^* = (S^{-1}JS)^* = S^{-1}J^*S$ so that

$$PJ = SP_{\varepsilon,J}S^{-1}J = SP_{\varepsilon,J}JS^{-1} = SJ^*P_{\varepsilon,J}S^{-1} = J^*SP_{\varepsilon,J}S^{-1} = J^*P\ .$$

Now consider the direct statement of the theorem and the special case in which $\sigma(J) = \{\alpha\}$ where α is real. Replacing J by $J - \alpha I$ we may assume $\alpha = 0$. Let $J = \mathrm{diag}[J_1,\cdots,J_k]$ be the decomposition of J into Jordan blocks, and let

$$P = (P_{ij})^k_{i,j=1}$$

be the corresponding decomposition of P. Note that $P^* = P$ implies $P^*_{ij} = P_{ji}$, and write

$$P_{ij} = \left[\alpha^{(ij)}_{pq}\right]^{m_i,m_j}_{p,q=1}$$

where m_i and m_j are the sizes of J_i and J_j respectively. It is easily seen that the equation $PJ = J^*P$ implies $P_{ij}J_j = J^*_iP_{ij}$ for $i,j = 1,2,\cdots,k$, and that the last relation implies that each block of PJ has lower triangular Toeplitz form. More precisely,

$$\alpha^{(ij)}_{pq} = 0 \qquad \text{if} \quad p + q \leq \max(m_i,m_j)$$

$$\alpha^{(ij)}_{pq} = \alpha^{(ij)}_{rs} \qquad \text{if} \quad p + q = r + s \ .$$

Now assume, without loss of generality, that the sizes of the Jordan blocks are ordered so that $m_1 = \cdots = m_s > m_{s+1} \geq \cdots \geq m_k$ for some s. Let $m = \sum_{i=1}^{k} m_i$ and consider the $s \times m$ submatrix A of P formed by rows $1,(m_1+1),\cdots,(m_1+\cdots+m_{s-1}+1)$. Let B be the submatrix of P formed by the last $m_{s+1} + \cdots + m_k$ columns. Since $P^2 = I$ the product AB is zero.

Because of the ordering of the m_j's the last $m - s$ columns of A are zero and so if A_0 is the $m \times m$ leading submatrix of A, we have $AB = A_0B$. Then the invertibility of P is easily seen to imply that of A_0 so that $AB = 0$ implies $B = 0$. In other words, $\alpha^{(ij)}_{pq} = 0$ for $i = 1,2,\cdots,s$ and $j = s+1,\cdots,k$.

Apply the same argument to the next group of Jordan blocks of equal size, and repeat as often as possible. The result of this process shows that P must be block diagonal with each diagonal block corresponding to a group of Jordan blocks of the same size. Therefore the problem is reduced to the case in which $m_1 = m_2 = \cdots = m_k = \ell$, say.

Let C be the $k \times k(\ell-1)$ submatrix of P formed by rows $\ell,2\ell,\cdots,k\ell$ and all columns except columns $1,\ell+1,\cdots,(k+1)\ell+1$. If $A = [\alpha^{(ij)}_{1\ell}]^k_{i,j=1}$ then $P^2 = I$ implies $AC = 0$. Again A is invertible and we deduce $C = 0$. Thus, P is reduced to the form

$$P = [\gamma_{ij}P_0]^k_{i,j=1}$$

where P_0 is the sip matrix of size ℓ, and the matrix $\Gamma = [\gamma_{ij}]_{i,j=1}^k$ is hermitian. Then $P^2 = I$ implies also that $\Gamma^2 = I$. Hence there exists a $k \times k$ unitary matrix T such that

$$T^{-1}\Gamma T = \operatorname{diag}[\varepsilon_1,\cdots,\varepsilon_k]$$

where $\varepsilon_j = \pm 1$ for each j.

Now define $S = [t_{ij}I_\ell]_{i,j=1}^k$ where $T = [t_{ij}]_{i,j=1}^k$ and I_ℓ is the unit matrix of size ℓ. Then S is unitary and

$$S^{-1}PS = \operatorname{diag}[\varepsilon_1 P_0,\cdots,\varepsilon_k P_0] .$$

Moreover, $SJ = JS$ so the theorem is proved in the case $\sigma(J) = \{\alpha\}$, $\alpha \in \mathbb{R}$.

Consider now the case

$$J = J_0 \oplus \bar{J}_0 , \tag{3.18}$$

where J_0 is a Jordan matrix with single non-real eigenvalue, and let

$$P = \begin{bmatrix} P_{11} & P_{12} \\ P_{21} & P_{22} \end{bmatrix}$$

be the corresponding decomposition of P. The condition $PJ = J^*P$ can be written in the form

$$\begin{bmatrix} P_{11}J_0 & P_{12}\bar{J}_0 \\ P_{21}J_0 & P_{22}\bar{J}_0 \end{bmatrix} = \begin{bmatrix} J_0^*P_{11} & J_0^*P_{12} \\ J_0^TP_{21} & J_0^TP_{22} \end{bmatrix} .$$

Since $\sigma(J_0) \cap \sigma(\bar{J}_0) = \phi$, the equations $P_{11}J_0 = J_0^*P_{11}$ and $P_{22}\bar{J}_0 = J_0^TP_{22}$ imply $P_{11} = P_{22} = 0$. Also

$$P_{12}\bar{J}_0 = J_0^*P_{12} \ ; \quad P_{21}J_0 = J_0^TP_{21} . \tag{3.19}$$

But $P = P^*$ implies $P_{21} = P_{12}^*$, so the two relations in (3.19) are equivalent.

Let $S_0 = P_{21}^*P_0$, where $P_0 = \operatorname{diag}[P_1,\cdots,P_k]$, and P_i are sip matrices with sizes equal to the sizes of Jordan blocks in J_0. Clearly, S_0 is unitary. We prove that

$$S_0J_0 = J_0S_0 .$$

Indeed, (3.19) implies $J_0P_{21}^{-1} = P_{21}^{-1}J_0^T$ and since $P_{21}^{-1} = P_{21}^*$, we have

$P_{21}^* J_0^T = J_0 P_{21}^*$. Now, using the fact that $P_0 J_0 = J_0^T P_0$, we have

$$S_0 J_0 = P_{21}^* P_0 J_0 = P_{21}^* J_0^T P_0 = J_0 P_{21}^* P_0 = J_0 S_0 .$$

Put $S = \mathrm{diag}[S_0, I]$, and clearly, $SJ = JS$, S is unitary, and

$$S^{-1}PS = \begin{bmatrix} S_0^{-1} & 0 \\ 0 & I \end{bmatrix} \begin{bmatrix} 0 & P_{21}^* \\ P_{21} & 0 \end{bmatrix} \begin{bmatrix} S_0 & 0 \\ 0 & I \end{bmatrix} = \begin{bmatrix} 0 & S_0^{-1} P_{21}^* \\ P_{21} S_0 & 0 \end{bmatrix}$$

$$= \begin{bmatrix} 0 & P_0 \\ P_0 & 0 \end{bmatrix} .$$

So the theorem is proved in the case when J has the form (3.18) with $\sigma(J_0) = \{\alpha\}$, α non-real.

We turn now to the general case of the theorem. Write as in (3.1):

$$J = J(\lambda_1) \oplus \cdots \oplus J(\lambda_\alpha) \oplus J(\lambda_{\alpha+1}) \oplus \cdots \oplus J(\lambda_\beta) ,$$

and partition P accordingly: $P = [P_{ij}]_{i,j=1}^{\beta}$. The equality $PJ = J^* P$ implies

$$P_{ik} J(\lambda_k) = J(\lambda_i)^* P_{ik} . \tag{3.20}$$

It is well-known, and easily verified, that (3.20) implies $P_{ik} = 0$ whenever $\sigma(J(\lambda_k)) \cap \sigma(J(\lambda_i)^*) = \phi$. So the proof of the general case is reduced to the two cases considered above. □

Theorem 3.10 and its proof allow us to establish some particular cases in which every P satisfying (3.17) will be in the form $P_{\varepsilon,J}$ *in the same basis*, i.e. when one can always choose $S = I$ in Theorem 3.10. This will be the case, for instance, when there is a unique (apart from multiplication of vectors of each Jordan chain of J by a unimodular complex number, which may be different for different chains) orthonormal basis in $\mathbb{C}^n$ in which J has the fixed form (3.1). This property is easily seen to be true for a non-derogatory matrix J, i.e. when there is just one Jordan block associated with each distinct eigenvalue. So we obtain the first part of the following Corollary. The second part can be traced from the proof of Theorem 3.10.

COROLLARY 3.11 *Assume that* J *is given by* (3.1) *and at least one of the following conditions holds:*

1) J *is non-derogatory;*
2) $\sigma(J)$ *is real and each distinct eigenvalue has no two associated Jordan blocks of the same size.*
Then P *satisfies equations* (3.17) *if and only if* $P = P_{\varepsilon,J}$ *for some choice of the signs* ε.

In the language of Theorem 3.3, Theorem 3.10 has the consequence:

COROLLARY 3.12 *Let* H *be an invertible hermitian matrix and* A *be a matrix with Jordan form* J. *Then* A *is* H*-selfadjoint if and only if* (A,H) *and* (J,P) *are unitarily similar for any matrix* P *satisfying the equations* (3.17).

Suppose now that J and ε are fixed, and consider corresponding solutions P_1 and P_2 of (3.17). Theorem 3.6 implies that (J,P_1) and (J,P_2) are unitarily similar; which can be expressed by saying that P_1 and P_2 are congruent, say $P_1 = T^*P_2T$, for some invertible T commuting with J. In fact, Theorem 3.10 implies that the transforming matrix T is unitary so that the congruence becomes a similarity.

THEOREM 3.13 *Solutions* P_1 *and* P_2 *of equations* (3.17) *are unitarily similar with a unitary transforming matrix which commutes with* J *if and only if* P_1 *and* P_2 *have the same sign characteristic.*

PROOF. Since J is fixed in this argument we abbreviate $P_{\varepsilon,J}$ to P_ε. If P_1 and P_2 are solutions of (3.17) with the same sign characteristic ε then there are unitary matrices S_1 and S_2 commuting with J such that

$$P_1 = S_1P_\varepsilon S_1^* , \quad P_2 = S_2P_\varepsilon S_2^* .$$

Defining $U = S_1S_2^*$ it is easily verified that $P_2 = U^*P_1U$ and that U is unitary and commutes with J.

Conversely, let P_1, P_2 be solutions of (3.17) with sign characteristics ε_1 and ε_2, respectively, and $P_1 = U^*P_2U$ where $U^*U = I$ and $UJ = JU$. Then Theorem 3.10 implies $P_2 = S^*P_{\varepsilon_2}S$ where $S^*S = I$ and $SJ = JS$. Consequently,

$$P_1 = (SU)^*P_{\varepsilon_2}(SU)$$

where $(SU)^*(SU) = I$ and $(SU)J = J(SU)$. Then (J,P_1) and (J,P_{ε_2}) are unitarily similar so, by Theorem 3.6, $\varepsilon_1 = \varepsilon_2$. □

3.7 The Structure of H-Selfadjoint Matrices When H Has a Small Number of Negative Eigenvalues

In this section we shall find the structure of H-selfadjoint matrices A in the cases when H has 0, 1 or 2 negative eigenvalues (counting multiplicities). The results of this section are easily obtained by inspecting the canonical form $(J,P_{\varepsilon,J})$ of (A,H), and by sorting out the cases when $P_{\varepsilon,J}$ has the required number of negative eigenvalues. Note also that the number of negative eigenvalues of the $n \times n$ sip matrix is m if $n = 2m$ or $2m + 1$.

The first case is obvious: if A is H-selfadjoint with positive definite H, then the spectrum of A is real, all elementary divisors of $I\lambda - A$ are linear (i.e. A is similar to a diagonal matrix), and all the signs in the sign characteristic are +1's.

When H has exactly one negative eigenvalue, for an H-selfadjoint matrix A one of the following holds:

(i) $\sigma(A)$ is real, all elementary divisors of $I\lambda - A$ are linear, and all but one sign in the sign characteristic are +1's;

(ii) $\sigma(A)$ is real, one elementary divisor of $I\lambda - A$ is quadratic (with arbitrary sign in the sign characteristic), and the rest of the elementary divisors are linear with signs +1;

(iii) $\det(I\lambda-A)$ has exactly 2 non-real zeros (counting multiplicities); all elementary divisors are linear, and all signs are +1;

(iv) $\sigma(A)$ is real, one elementary divisor is cubic, the rest are linear, and all the signs are +1.

It is easily seen that for every hermitian invertible H with $n \geq 3$ and exactly 1 negative eigenvalue all four possibilities can be realized, i.e. for each case (i)-(iv) there will be an H-selfadjoint A for which this case occurs.

The case of 2 negative eigenvalues of H is somewhat more complicated. In this case, for an H-selfadjoint matrix A of size at least six, one of 12 possibilities holds. For convenience, they are tabulated on the following page.

Again, all 12 possibilities can be realized for every $n \times n$ hermitian invertible H with exactly 2 negative eigenvalues (counting multiplicities) and $n \geq 6$.

Case Number	No. of non-real eigenvalues of A (counting multiplicities)	Number and signs (in the sign characteristic) of elementary divisors of $I\lambda - A$ with real eigenvalues			
		of degree 1	of degree 2	of degree 3	of degree 4
1	0	n all but 2 signs are +1	0	0	0
2	0	n - 4 all signs +1	2 signs arbitrary	0	0
3	0	n - 6 all signs +1	0	2 both signs +1	0
4	0	n - 5 all signs +1	1 sign arbitrary	1 sign +1	0
5	0	n - 4 all signs +1	0	0	1 sign arbitrary
6	0	n - 3 all signs +1	0	1 sign -1	0
7	0	n - 3 all signs but one are +1	0	1 sign +1	0
8	0	n - 1 all signs but one are +1	1 sign arbitrary	0	0
9	2	n - 2 all signs but one are +1	0	0	0
10	2	n - 4 all signs +1	1 sign arbitrary	0	0
11	2	n - 3 all signs +1	0	1 sign +1	0
12	4	n - 4 all signs +1	0	0	0

3.8 H-Definite Matrices

Another interesting class of H-selfadjoint matrices are "H-definite" matrices which are defined as follows. Let $H = H^*$ be an invertible matrix, and let $[.,.] = (Hx,y)$ be the indefinite scalar product defined by H. An $n \times n$ matrix A is called H-*nonnegative* (resp. H-*positive*) if $[Ax,x] \geq 0$ for all $x \in \mathbb{C}^n$ (resp. $[Ax,x] > 0$ for all $x \in \mathbb{C}^n \setminus \{0\}$). Clearly, these classes of matrices contain only (not all) H-selfadjoint matrices. So Theorem 3.3 is applicable, and it is easily found that A is H-nonnegative (resp. H-positive) if and only if the matrix $P_{\varepsilon,J}J$ is nonnegative (resp. positive) with respect to the usual scalar product. Examining the product $P_{\varepsilon,J}J$ leads to the conclusions:

THEOREM 3.14 *Matrix* A *is* H-*positive if and only if the following conditions hold:*

(i) A *is* H-*selfadjoint and invertible;*

(ii) *the spectrum* $\sigma(A)$ *is real;*

(iii) *all elementary divisors of* $\lambda I - A$ *are linear;*

(iv) *the sign (in the sign characteristic of* (A,H) *attached to all elementary divisors of* $\lambda I - A$ *corresponding to an eigenvalue* λ_0, *is* $\operatorname{sgn} \lambda_0$.

THEOREM 3.15 *Matrix* A *is* H-*nonnegative if and only if the following conditions hold:*

(i) A *is* H-*selfadjoint;*

(ii) $\sigma(A)$ *is real;*

(iii) *for non-zero eigenvalues of* A *all elementary divisors are linear; for the zero eigenvalue of* A *(if zero is an eigenvalue) all elementary divisors are either linear or quadratic;*

(iv) *the sign attached to the elementary divisors of a non-zero eigenvalue* λ_0 *is* $\operatorname{sgn} \lambda_0$; *the sign attached to the quadratic elementary divisors (if any) of the zero eigenvalue, is* +1.

An $n \times n$ matrix A is called H-*nonpositive* (resp. H-*negative*) is $[Ax,x] \leq 0$ for all $x \in \mathbb{C}^n$ (resp. $[Ax,x] < 0$ for all $x \in \mathbb{C}^n \setminus \{0\}$). We leave to the reader the description of H-nonpositive and H-negative matrices. It is, of course, closely analogous to Theorem 3.14 and 3.15.

3.9 Second Description of the Sign Characteristic

The sign characteristic of a pair (A,H), where A is H-selfadjoint, was described in Section 3.2 in terms of the canonical form. In this section the sign characteristic will be defined directly in terms of the Jordan chains of the H-selfadjoint matrix A.

Let λ_0 be a fixed real eigenvalue of A, and let $\Psi_1 \in \mathbb{C}^n$ be the subspace spanned by the eigenvectors of A corresponding to λ_0. For $x \in \Psi_1 \setminus 0$ denote by $\nu(x)$ the maximal length of a Jordan chain of A beginning with the eigenvector x. Let Ψ_i, $i = 1,2,\cdots,\gamma$ $(\gamma = \max\{\nu(x) \mid x \in \Psi_1 \setminus \{0\}\})$ be the subspace of Ψ_1 spanned by all $x \in \Psi_1$ with $\nu(x) \geq i$. Then

$$\operatorname{Ker}(I\lambda_0 - A) = \Psi_1 \supset \Psi_2 \supset \cdots \supset \Psi_\gamma .$$

The following result describes the sign characteristic of the pair (A,H) in terms of certain bilinear forms defined on the subspaces Ψ_i.

THEOREM 3.16 *For* $i = 1,\cdots,\gamma$, *let*

$$f_i(x,y) = (x,Hy^{(i)}) , \quad x,y \in \Psi_i ,$$

where $y = y^{(1)}, y^{(2)}, \cdots, y^{(i)}$ *is a Jordan chain of* A *corresponding to real* λ_0 *with the eigenvector* y. *Then:*

(i) $f_i(x,y)$ *does not depend on the choice of* $y^{(2)},\cdots,y^{(i)}$;

(ii) *for some selfadjoint linear transformation* $G_i : \Psi_i \to \Psi_i$,

$$f_i(x,y) = (x,G_i y) , \quad x,y \in \Psi_i ;$$

(iii) *for the transformation* G_i *of* (ii), $\Psi_{i+1} = \operatorname{Ker} G_i$ (*by definition* $\Psi_{\gamma+1} = \{0\}$);

(iv) *the number of positive (negative) eigenvalues of* G_i, *counting multiplicities, coincides with the number of positive (negative) signs in the sign characteristic of* (A,H) *corresponding to the elementary divisors* $(\lambda-\lambda_0)^i$ *of* $I\lambda - A$.

PROOF. Let $(J,P_{\varepsilon,J})$ be the canonical form for (A,H) as described in (3.1) and (3.2). By Theorem 3.3, (A,H) and $(J,P_{\varepsilon,J})$ are unitarily similar for some set of signs ε, i.e.

$$H = T^* P_{\varepsilon,J} T , \quad A = T^{-1} J T$$

for some nonsingular T. Hence, for $x,y \in \Psi_i$,

$$f_i(x,y) = (Tx, P_{\varepsilon,J}Ty^{(i)}) .$$

Clearly, Tx, and $Ty^{(1)},\cdots,Ty^{(i)}$ are eigenvector and Jordan chain, respectively, of J corresponding to λ_0. In this way the proof is reduced to the case $A = J$ and $H = P_{\varepsilon,J}$. But in this case the assertions (i)-(iv) are easily verified. □

Consider once more the Jordan matrix J of the proof above. The argument shows that the basis in which the quadratic form $f_i(x,x)$ on Ψ_i is reduced to the sum of squares, consists of vectors $T^{-1}x_i,\cdots,T^{-1}x_{i,q_i}$, where x_{ij}, $j = 1,\cdots,q_i$ are all the coordinate unit vectors in $\mathbb{C}^n$ such that $x_{ij} = (I\lambda_0-J)^{i-1}y_{ij}$ for some y_{ij}, and $(I\lambda_0-J)x_{ij} = 0$.

For the case that $H = P_{\varepsilon,J}$ and $A = J$ it is not hard to obtain a formula for the transformations G_i of (ii). Namely, in the standard orthonormal basis (consisting of unit coordinate vectors) in Ψ_i,

$$G_i = P_{\Psi_i}(-I\lambda_0+J)^{i-1}P_{\varepsilon,J}\big|_{\Psi_i} , \quad i = 1,\cdots,\gamma , \tag{3.21}$$

where P_{Ψ_i} is the orthogonal projector on Ψ_i. Indeed, it is sufficient to prove that

$$f_i(x,y) = (x, P_{\Psi_i}(-I\lambda_0+J)^{i-1}P_{\varepsilon,J}y)$$

for any coordinate unit vectors $x,y \in \Psi_i$. This is an easy exercise bearing in mind the special structure of $P_{\varepsilon,J}$ and J.

As an application of Theorem 3.16 we have the following description of the connected components in the set of all invertible hermitian matrices H such that a given matrix A is H-selfadjoint.

THEOREM 3.17 *Let* A *be an* $n \times n$ *matrix similar to* A^*. *Suppose that there are* b *different Jordan blocks* $J_1,\cdots,J_b$ *with real eigenvalues in the Jordan form* J *of* A, *and let* k_i $(i = 1,\cdots,b)$ *be the number of times* J_i *appears in* J. *Then the set* Ω *of all invertible hermitian* $n \times n$ *matrices* H *such that* A *is H-selfadjoint, is a disconnected union of* $k = \prod_{i=1}^{b}(k_i+1)$ *(arcwise) connected components* $\Omega = \bigcup_{i=1}^{k}\Omega_i$, *where each* Ω_i *consists of all matrices* H *with the same sign characteristic.*

PROOF. In view of Corollary 3.4 we have

$$\Omega = \{H | H = T^* P_{\varepsilon,J} T \text{ for some } \varepsilon \text{ and invertible } T \text{ such that } TA = JT\} . \quad (3.22)$$

By Theorem 3.3 we can suppose that the set of signs ε in (3.22) is normalized. It is easily seen that the number of all normalized sets of signs is just k; designate them $\varepsilon^{(1)},\cdots,\varepsilon^{(k)}$. Let

$$\Omega_i = \{H \mid H = T^* P_{\varepsilon^{(i)},J} T \quad \text{for invertible } T \text{ such that } TA = JT\} .$$

We prove now that Ω_i is connected. It is sufficient to prove that the set of all invertible matrices T such that $TJ = JT$ is connected. Indeed, this set can be represented as $\{UT_0 \mid U \text{ is invertible and } UJ = JU\}$, where T_0 is a fixed invertible matrix such that $T_0A = JT_0$. Now from the structure of the set of matrices commuting with J (see, for instance, Chapter VIII in [14]) it is easy to deduce that the invertible matrices U commuting with J form a connected set. Hence Ω_i is connected for every $i = 1,\cdots,k$.

It remains to prove that, for $H \in \Omega_j$, a sufficiently small neighbourhood of H in Ω will contain elements from Ω_j only. This can be done using Theorem 3.16. In the notation of Theorem 3.16 the bilinear forms $f_1(x,y),\cdots,f_\gamma(x,y)$ depend continuously on H; therefore the same is true for $G_1,\cdots,G_\gamma$. So $G_i = G_i(H) : \Psi_i \to \Psi_i$ is a selfadjoint linear transformation which depends continuously on $H \in \Omega$ and such that $\operatorname{Ker} G_i = \Psi_{i+1}$ is fixed (i.e. independent of H). It follows that the number of positive eigenvalues and the number of negative eigenvalues of each G_i remain constant in a neighbourhood of $H \in \Omega_j$. By Theorem 3.16(iv), this neighbourhood belongs to Ω_j. □

3.10 Canonical Forms for Pairs of Hermitian Matrices

Let G_1 and G_2 be hermitian $n \times n$ matrices (invertible or not). Consider the following problem: reduce G_1 and G_2 simultaneously by a congruence transformation to as simple a form as possible. In other words, by transforming the pair G_1, G_2 to the pair X^*G_1X, X^*G_2X with some invertible $n \times n$ matrix X we would like to bring G_1, G_2 to the simplest possible form.

If one of the matrices G_1 or G_2 is positive (or negative) definite, it is well known that G_1 and G_2 can be reduced simultaneously to a diagonal form. More generally, such a simple reduction cannot be achieved. However, for the case when one of the matrices G_1 and G_2 is invertible, Theorem 3.3 leads to the following result.

THEOREM 3.18 *Let* G_1 *and* G_2 *be hermitian* $n \times n$ *matrices with* G_2 *invertible. Then there is an invertible* $n \times n$ *matrix* X *such that* X^*G_1X *and* X^*G_2X *have the form:*

$$X^*G_1X = \varepsilon_1 K_1 \oplus \cdots \oplus \varepsilon_a K_a \oplus \begin{bmatrix} 0 & K_{a+1} \\ \bar{K}_{a+1} & 0 \end{bmatrix} \oplus \cdots \oplus \begin{bmatrix} 0 & K_\beta \\ \bar{K}_\beta & 0 \end{bmatrix}, \tag{3.23}$$

where

$$K_q = \begin{bmatrix} 0 & \cdot & \cdot & \cdot & 0 & \lambda_q \\ \cdot & & & \cdot & \lambda_q & 1 \\ \cdot & & \cdot & \cdot & 1 & 0 \\ \cdot & \cdot & \cdot & \cdot & & \cdot \\ 0 & \lambda_q & 1 & & & \cdot \\ \lambda_q & 1 & 0 & \cdot & \cdot & 0 \end{bmatrix},$$

λ_q *is real for* $q = 1,\cdots,a$; λ_q *is non-real for* $q = a+1,\cdots,\beta$ *and* $\varepsilon_q = \pm 1$ $(q = 1,\cdots,a)$;

$$X^*G_2X = \varepsilon_1 P_1 \oplus \cdots \oplus \varepsilon_a P_a \oplus \begin{bmatrix} 0 & P_{a+1} \\ P_{a+1} & 0 \end{bmatrix} \oplus \cdots \oplus \begin{bmatrix} 0 & P_\beta \\ P_\beta & 0 \end{bmatrix}, \tag{3.24}$$

where P_q *is the sip matrix whose size is equal to that of* K_q *for* $q = 1,2,\cdots,\beta$. *The representation* (3.23) *and* (3.24) *are uniquely determined by* G_1 *and* G_2, *up to simultaneous permutation of the same blocks in each formula* (3.23) *and* (3.24).

PROOF. Observe that $G_2^{-1}G_1$ is G_2-selfadjoint and let $(J, P_{\varepsilon,J})$ be the canonical form for $(G_2^{-1}G_1, G_2)$. Thus

$$G_2^{-1}G_1 = XJX^{-1}, \quad G_2 = X^{*-1}P_{\varepsilon,J}X^{-1} \tag{3.25}$$

for some invertible matrix X. Now (3.25) implies that $G_1 = G_2XJX^{-1} = X^{*-1}P_{\varepsilon,J}JX^{-1}$. Taking into account the form of $P_{\varepsilon,J}J$ (see formulas (3.1) and (3.2)) we obtain the representations (3.23) and (3.24). The uniqueness of representations (3.23) and (3.24) follows from the uniqueness of the canonical form (Theorem 3.3). □

Note that the sizes of blocks K_q $(q = 1,\cdots,a)$, in (3.23) and the corresponding numbers λ_q are just the sizes and eigenvalues of those Jordan blocks in the Jordan form of $G_2^{-1}G_1$ which correspond to the real eigenvalues. The sizes of blocks K_q $(q = a+1,\cdots,\beta)$ and the corresponding numbers λ_q may be chosen as the sizes and eigenvalues of those Jordan blocks of $G_2^{-1}G_1$ which correspond to a maximal set of eigenvalues $C \subset \sigma(G_2^{-1}G_1)$ which does not contain a conjugate pair of complex numbers.

The problem of the simultaneous reduction of a pair of hermitian matrices can also be stated as reduction (under congruence transformations) of a linear pencil of matrices. Namely, given a linear pencil $\lambda G_1 + G_2$ of $n \times n$ hermitian matrices G_1 and G_2, find the simplest form $\lambda G_1' + G_2'$ of this pencil which can be obtained from $\lambda G_1 + G_2$ by congruency transformation: $\lambda G_1' + G_2 = X^*(\lambda G_1+G_2)X$, where X is an invertible $n \times n$ matrix. We leave it to the reader to restate Theorem 3.18 in terms of linear pencils of hermitian matrices.

3.11 Third Description of the Sign Characteristic

The two ways of describing the sign characteristic already introduced result in the statements of Theorems 3.3 and 3.16. The third approach is apparently quite different in character and relies on fundamental ideas of analytic perturbation theory. This approach will be developed more fully for more general problems in Chapter 3 of Part II with the full proofs, and so we present only a brief account of the conclusions for H-selfadjoint matrices in this section.

If A is H-selfadjoint we have described the associated sign characteristic as that of the pair (A,H). The discussion of the preceding section suggests that it would not be unnatural to associate the sign characteristic with the pair of hermitian matrices (H,HA), or with the pencil $\lambda H - HA$. In fact, our new description is to be formulated in terms of the (λ-dependent) eigenvalues of $\lambda H - HA$.

The fundamental perturbation theorem concerning these eigenvalues (ref. Theorem II.3.3) asserts that there is a function $U(\lambda)$ of the real variable λ with values in the unitary matrices such that, for all real λ

$$\lambda H - HA = U(\lambda)\, \mathrm{diag}[\mu_1(\lambda),\cdots,\mu_n(\lambda)]U(\lambda)^* ,$$

and, moreover, the functions $\mu_j(\lambda)$ and $U(\lambda)$ can be chosen to be analytic functions of the real parameter λ (although they are not generally polynomials in λ). Thus, the eigenvalue functions $\mu_j(\lambda)$ (which are, of course, the zeros of $\det(I\mu - (\lambda H\text{-}HA)) = 0$) are analytic in λ. This is the essential prerequistite for understanding the next characterization of the sign characteristic. Note also that λ is a real eigenvalue of A if and only if $\mu_j(\lambda) = 0$ for at least one j.

THEOREM 3.19 *Let* A *be an* $n \times n$ *H-selfadjoint matrix, let* $\mu_1(\lambda),\cdots,\mu_n(\lambda)$ *be the zeros of the scalar polynomial* $\det(\mu I\text{-}(\lambda H\text{-}HA))$ *of degree* n *arranged so that* $\mu_j(\lambda)$, $j = 1,\cdots,n$ *are real analytic functions of real* λ, *and let* $\lambda_1,\cdots,\lambda_r$ *be the different real eigenvalues of* A.

For every $i = 1,2,\cdots,r$ *write*

$$\mu_j(\lambda) = (\lambda-\lambda_i)^{m_{ij}}\nu_{ij}(\lambda)$$

where $\nu_{ij}(\lambda_i) \neq 0$ *and is real. Then the non-zero numbers among* $m_{i1},\cdots,m_{in}$ *are the sizes of Jordan blocks with eigenvalue* λ_0 *in the Jordan form of* A *and* $\mathrm{sgn}\, \nu_{ij}(\lambda_i)$ *(for* $m_{ij} \neq 0$*) is the sign attached to* m_{ij} *associated with* λ_i *in the (possibly unnormalized) sign characteristic of* (A,H).

EXAMPLE 3.2 Let

$$H = \begin{bmatrix} 0 & \varepsilon_1 & 0 & 0 \\ \varepsilon_1 & 0 & 0 & 0 \\ 0 & 0 & 0 & \varepsilon_1 \\ 0 & 0 & \varepsilon_1 & 0 \end{bmatrix}, \quad A = \begin{bmatrix} 0 & 1 & 0 & 0 \\ 0 & 0 & 0 & 0 \\ 0 & 0 & 1 & 1 \\ 0 & 0 & 0 & 1 \end{bmatrix}$$

(cf. Example 3.1) where $\varepsilon_1 = \pm 1$, $\varepsilon_2 = \pm 1$. Then we have $\lambda_1 = 0$, $\lambda_2 = 1$. It is easily seen that the matrix $\lambda H - HA$ has just one eigenvalue $\mu_1(\lambda)$ vanishing at $\lambda = 0$ and just one eigenvalue, say $\mu_2(\lambda)$, vanishing at $\lambda = 1$. Furthermore, in sufficiently small neighbourhoods of these two points

$$\mu_1(\lambda) = (\lambda^2-\lambda^4+\cdots)\varepsilon_1 , \quad \mu_2(\lambda) = ((1-\lambda)^2-(1-\lambda)^4+\cdots)\varepsilon_2 .$$

3.12 Maximal Nonnegative Invariant Subspaces

Invariant subspaces of an H-selfadjoint matrix A will play a very important role in subsequent parts of this book; particularly in connection with factorization problems for polynomial and rational matrix-valued functions. More especially, we shall be concerned with A-invariant subspaces which are nonnegative (or nonpositive) with respect to H (see Section 1.3) and of maximal dimension (ref. Theorem 1.3). This section is devoted to the analysis of such subspaces.

Some notations and definitions will first be set up. Let A be an $n \times n$ H-selfadjoint matrix, and let $m_1, \cdots, m_r$ be the sizes of Jordan blocks in a Jordan form for A corresponding to the real eigenvalues. Let the corresponding signs in the sign characteristic for (A,H) be $\varepsilon_1, \cdots, \varepsilon_r$, and let k be the number of positive eigenvalues of H, counting multiplicities.

A set C of non-real eigenvalues of A is called a c-*set* if $C \cap \bar{C} = \phi$ and $C \cup \bar{C}$ is the set of all non-real eigenvalues of A. If M is an H-nonnegative subspace, define the *index of positivity* of M, written $p(M)$, to be the maximal dimension of an H-positive subspace $M_0 \subset M$. It was noted after the proof of Theorem 1.4 that $p(M)$ does not depend on the choice of M_0.

THEOREM 3.20 *For every* c-*set* C *there exists a* k-*dimensional,* A-*invariant,* H-*nonnegative subspace* N *such that the non-real part of* $\sigma(A|_N)$ *coincides with* C. *The subspace* N *is maximal* H-*nonnegative and*

$$p(N) = \frac{1}{2} \sum_{i=0}^{r} [1-(-1)^{m_i}]\delta_i \ , \tag{3.26}$$

where $\delta_i = 1$ *if* $\varepsilon_i = 1$, *and* $\delta_i = 0$ *if* $\varepsilon_i = -1$.

PROOF. We make use of Theorem 3.3 and write $A = T^{-1}JT$, $H = T^* P_{\varepsilon,J} T$ for some invertible T. Note that the columns of T^{-1} corresponding to each Jordan block in J form a Jordan chain for A. So J has exactly $q = \frac{1}{2}\left[n - \sum_{i=1}^{r} m_i\right]$ eigenvalues (counting multiplicities) in C, and exactly r Jordan blocks $J_1, \cdots, J_r$ of sizes $m_1, \cdots, m_r$ with signs $\varepsilon_1, \cdots, \varepsilon_r$ respectively corresponding to real eigenvalues. Let N be the A-invariant subspace spanned by the following columns of T^{-1} : a) the columns corresponding to the Jordan blocks with eigenvalues in C (the number of such columns is q); b) for even m_i, the first $m_i/2$ columns corresponding to the block J_i; c) for odd m_i, the first $(m_i+1)/2$ or $(m_i-1)/2$ columns corresponding to J_i, according as $\varepsilon_i = +1$ or $\varepsilon_i = -1$. Then $\dim N$ is found to be

$$q + \sum_{m_i \text{ even}} \frac{m_i}{2} + \sum_{\substack{m_i \text{ odd} \\ \varepsilon_i = 1}} \frac{m_i+1}{2} + \sum_{\substack{m_i \text{ odd} \\ \varepsilon_i = -1}} \frac{m_i-1}{2} = \frac{1}{2} + \frac{1}{4} \sum_{i=1}^{r} [1-(-1)^{m_i}]\varepsilon_i .$$

Comparing this with equation (3.4) it is seen that $\dim N = k$. Furthermore, it is easily seen from the structure of $P_{\varepsilon,J}$ that N is H-nonnegative.

To check that $p(N)$ is given by (3.26), let P_N be the orthogonal projection on N. Then in the basis for N formed by the chosen columns of T^{-1} we have

$$P_N H|_N = \operatorname{diag}[\zeta_1, \zeta_2, \cdots, \zeta_k] : N \to N ,$$

where $\zeta_i = 0$ or 1 and $\zeta_i = 1$ if and only if the i-th chosen column of T^{-1} is the $\frac{1}{2}(m_j+1)$-th generalized eigenvector in a chain of length m_j for which m_j is odd and $\zeta_m = +1$. It is clear that, with this construction, $p(N) = \sum_{i=1}^{k} \zeta_i$ and the formula (3.26) follows.

Finally, the maximality of N follows from Theorem 1.3. □

For the case of H-nonpositive invariant subspaces of A there is, of course, another statement dual to that of Theorem 3.20. This can be obtained by considering $-H$ in place of H in the last theorem.

In general, the subspace N is not unique (for a given c-set C).

EXAMPLE 3.3 Let

$$A = \begin{bmatrix} 0 & 1 & 0 & 0 \\ 0 & 0 & 0 & 0 \\ 0 & 0 & 0 & 1 \\ 0 & 0 & 0 & 0 \end{bmatrix} ; \quad H = \begin{bmatrix} 0 & 1 & 0 & 0 \\ 1 & 0 & 0 & 0 \\ 0 & 0 & 0 & -1 \\ 0 & 0 & -1 & 0 \end{bmatrix} .$$

The following 2-dimensional subspaces of $\mathbb{C}^4$ are A-invariant and H-nonnegative (here the c-set C is empty):

$$\operatorname{Span}\{\langle 1\ 0\ 0\ 0\rangle , \langle 0\ 0\ 1\ 0\rangle\} , \quad \operatorname{Span}\{\langle 1\ 0\ 1\ 0\rangle , \langle 0\ 1\ 0\ 1\rangle\} .$$

Note that both subspaces are not only H nonnegative, but also H-neutral. □

The following result includes extra conditions needed to ensure the uniqueness of the subspace N from Theorem 3.20.

THEOREM 3.21 *Let* A *be an H-selfadjoint* $n \times n$ *matrix such that the sizes* $m_1, \cdots, m_r$ *of the Jordan blocks of* A *corresponding to real eigenvalues are all even. Then for every c-set* C *there exists a unique H-neutral A-invariant subspace* N *with* $\dim N = \frac{1}{2} n$, $\sigma(A|_N) \setminus \mathbb{R} = C$, *and the sizes of the Jordan blocks of* $A|_N$ *corresponding to the real eigenvalues are* $\frac{1}{2} m_1, \cdots, \frac{1}{2} m_r$.

PROOF. Observe first, using (3.3) for example, that when $m_1, \cdots, m_r$ are all even the number of positive eigenvalues of H is just $k = \frac{1}{2} n$. So choosing a c-set C, Theorem 3.20 ensures the existence of a subspace N with all the properties required by the theorem.

Let N_C be the sum of root subspaces of $A|_N$ corresponding to the non-real eigenvalues of $A|_N$ (i.e. the eigenvalues in C), and let N_r be the sum of root subspaces of $A|_N$ corresponding to the real eigenvalues of $A|_N$. Clearly

$$N = N_C \dotplus N_r . \tag{3.27}$$

Now form the decomposition

$$\mathbb{C}^n = X_C \dotplus X_{\bar{C}} \dotplus X_r , \tag{3.28}$$

where X_C (resp. $X_{\bar{C}}$) is the sum of the root subspaces of A corresponding to the eigenvalues in C (resp. $\bar{C} = \{\lambda \in \mathbb{C} \mid \bar{\lambda} \in C\}$).

It is clear from the construction that $N_C = X_C$ and is uniquely determined, and also $N \cap X_{\bar{C}} = \{0\}$. Consequently, the only possible non-uniqueness in the determination of N arises in forming N_r of (3.27).

Let $\lambda_1, \cdots, \lambda_t$ be the distinct real eigenvalues of A $(t \leq r)$ and $E_A(\lambda_j)$ be the (uniquely determined) root subspace of A corresponding to λ_j, $j = 1,2,\cdots,t$. Then

$$X_r = E_A(\lambda_1) \dotplus \cdots \dotplus E_A(\lambda_t)$$

and there is a corresponding decomposition of N_r :

$$N_r = N_{\lambda_1} \dotplus \cdots \dotplus N_{\lambda_r} ,$$

with $N_{\lambda_j} \subset E_A(\lambda_j)$ for each j and, moreover the dimension of N_{λ_j} is just half that of $E_A(\lambda_j)$. These inclusions mean that uniqueness of N now depends on that of each N_{λ_j}. Thus the proof is reduced to the case when A has just one eigenvalue, say λ_0, and λ_0 is real.

Without loss of generality assume that $\lambda_0 = 0$ and that (A,H) is in the canonical form (J,P), where $P = P_{\varepsilon,J}$. Let $m = \sum_{j=1}^{r} m_j = \frac{1}{2} n$, and let N_1, N_2 be subspaces with all the properties listed in Theorem 3.21, and we have to prove that $N_1 = N_2$.

The proof is by induction on m. Consider first the case that $m_1 = \cdots = m_r = 2$. Then there exists a unique J-invariant k-dimensional subspace N such that the partial multiplicities of $J|_N$ are $1,\cdots,1$ (namely, N is spanned by all the eigenvectors of J). So in that case Theorem 3.21 is evident (even without the condition of P-neutrality). Consider now the general case.

Consider the subspace Ker J. From the properties of N_i it is clear that $\text{Ker } J \subset N_1 \subset N_2$. Let $g_1,\cdots,g_r$ be the coordinate unit vectors with a one in the positions $m_1, m_1+m_2, \cdots, m_1+\cdots+m_r$, respectively. We show that $N_i \perp \text{Span}\{g_1,\cdots,g_r\}$, $i = 1,2$, where the orthogonality is understood in the sense of the usual scalar product $(.,.)$. Suppose the contrary and there exists $x \in N_i$ such that $(x,g_j) \neq 0$ for some j. Let f_j be the eigenvector of J corresponding to the j-th Jordan block. Clearly, $f_j \in N_i$, but $(Pf_j,x) = (g_j,x) \neq 0$; a contradiction with the P-neutrality of N_i. So

$$N_i \perp \text{Span}\{g_1,\cdots,g_r\}\ , \quad i = 1,2\ .$$

Consider now the subspaces $N_i' \subset \mathbb{C}^n$ defined as follows: for every $x \in \mathbb{C}^n$, let $\varphi x \in \mathbb{C}^n$ be the vector obtained from x by putting zero instead of the coordinates of x in the positions 1, m and $m_1 + \cdots + m_p$ and $m_1 + \cdots + m_p+1$ for $p = 1,\cdots,r-1$: the rest of the coordinates of x remain unchanged under φ. Now put

$$N_i' = \{\varphi x \mid x \in N_i\}\ , \quad i = 1,2\ .$$

It is clear that $N_i' \perp (\text{Ker } J + \text{Span}\{g_1,\cdots,g_r\})$. Moreover, since $N_i \perp$ $\perp \text{Span}\{g_1,\cdots,g_r\}$ and $N_i \supset \text{Ker } J$, the equalities

$$N_i' \dotplus \text{Ker } J = N_i\ , \quad i = 1,2$$

hold. Evidently, it suffices to prove the equality $N_1' = N_2'$. To this end, consider the Jordan matrix $\tilde{J}$ which is obtained from J by crossing out the first and last column and row in each Jordan block. So $\tilde{J}$ has blocks of sizes $(m_1-2),\cdots,(m_r-2)$. Further, define the subspaces $\tilde{N}_i \subset C^{n-2r}$, as follows

$$\tilde{N}_i = \{\tilde{x} \in ¢^{n-2r} \mid x \in N_i'\}\ , \quad i = 1,2$$

where $\tilde{x}$ is obtained from x by crossing out its $2r$ zero coordinates in the above mentioned positions. Clearly, the subspaces $\tilde{N}_i$ are $\tilde{J}$-invariant and $\tilde{P}$-neutral, where $\tilde{P}$ is obtained from P by crossing out the first and last column and row in each of $P_1,\cdots,P_r$. So we can apply the induction hypothesis to deduce that $\tilde{N}_1 = \tilde{N}_2$ and therefore also $N_1' = N_2'$. □

Note that in Theorem 3.21 the requirement that the multiplicities of $A|_N$ are $\frac{1}{2}m_1,\cdots,\frac{1}{2}m_r$ cannot be omitted in general, as Example 3.3 shows. However, it turns out that if we require an additional condition that the signs in the sign characteristic of (A,H) which correspond to Jordan blocks with the same eigenvalue are equal, then for every c-set C there exists an unique H-neutral A-invariant $\frac{n}{2}$-dimensional subspace N with $(A|_N) \setminus \mathbb{R} = C$. (Obviously, this additional condition is violated in Example 3.3.) Moreover, in this case we are able to describe all H-neutral A-invariant $\frac{n}{2}$-dimensional subspaces, as follows.

THEOREM 3.22 *Let* A *be an H-selfadjoint matrix such that the sizes* $m_1,\cdots,m_r$ *of the Jordan blocks of* A *corresponding to the real eigenvalues are all even, and the signs in the sign characteristic of* (A,H) *corresponding to the same eigenvalue are all equal. Let* M_+ *be the sum of the root subspaces of* A *corresponding to the eigenvalues of* A *in the open upper half-plane. Then for every A-invariant subspace* $N_+ \subset M_+$ *there exists a unique A-invariant H-neutral* $\frac{n}{2}$*-dimensional subspace* N *such that*

$$N \cap M_+ = N_+ \ . \tag{3.29}$$

In other words, Theorem 3.22 gives a one-to-one correspondence between the set of A-invariant subspaces N_+ of M_+ and the set of A-invariant H-neutral $\frac{n}{2}$-dimensional subspaces N, which is given by equality (3.29).

PROOF. Passing to the canonical form of (A,H), we see that it is sufficient to consider only the two following cases (cf. the proof of Theo-

rem 3.21):
(i) $\sigma(A) = \{0\}$;
(ii) $\sigma(A) = \{\lambda_0, \bar{\lambda}_0\}$, $\lambda_0 \in \mathbb{R}$.
Consider case (i). In this case $M_+ = 0$, so Theorem 3.22 asserts that there is a unique A-invariant H-neutral $\frac{n}{2}$-dimensional subspace N. To prove this, assume that $(A,H) = (J,P_{\varepsilon,J})$ is in the canonical form; so

$$J = J_1 \oplus \cdots \oplus J_r ,$$

where J_i is the nilpotent Jordan block of size m_i, $i = 1,\cdots,r$. Suppose, for instance, that the signs in the sign characteristic of $(J,P_{\varepsilon,J})$ are all +1's. Denote by x_{ij} the $(m_1+\cdots+m_{i-1})$-th unit coordinate vector in $\mathbb{C}^n$, $j = 1,\cdots,m_i$ (recall that $n = m_1 + \cdots + m_r$); so the vectors $x_{i1},\cdots,x_{im_i}$ form a Jordan chain of $\lambda I - J$. Let N be a $P_{\varepsilon,J}$-neutral J-invariant subspace with $\dim N = \frac{1}{2} n$, and let

$$x = \sum_{i,j} a_{ij}x_{ij} \in N , \quad a_{ij} \in \mathbb{C} .$$

We claim that the coefficients a_{ij} with $j > \frac{1}{2} m_i$ are zeros. Suppose not; let K be the set of all indices i $(1 \leqslant i \leqslant r)$ for which the set $\{j \mid \frac{1}{2} m_i < j \leqslant m_i , a_{ij} \neq 0\}$ is non-void and the difference

$$\max\{j \mid \frac{1}{2} m_i < j \leqslant m_i , a_{ij} \neq 0\} - \frac{1}{2} m_i$$

is maximal. Denote this difference by γ. Since N is J-invariant, the vectors

$$y_1 = J^{\gamma}x , \quad y_2 = J^{\gamma-1}x$$

are again in N. A computation shows that

$$(P_{\varepsilon,J}y_1,y_2) = \sum_{i\in K} |a_{i,\frac{1}{2}m_i+\gamma}|^2 = 0$$

because N is $P_{\varepsilon,J}$-neutral. So $a_{i,\frac{1}{2}m_i+\gamma} = 0$ for all $i \in K$, which contradicts the choice of K. Thus, $a_{ij} = 0$ for $j > \frac{1}{2} m_i$. Since $\dim N = \frac{1}{2} n$, this leaves only one possibility for N; namely, $N = \mathrm{Span}\{x_{ij} \mid 1 \leqslant j \leqslant \frac{1}{2} m_i ; i = 1,\cdots,r\}$.

Now consider case (ii). Again, assume $(A,H) = (J,P_{\varepsilon,J})$ is the canonical form. Rearranging blocks in J and $P_{\varepsilon,J}$ (which amounts to a unitary similarity), we can write J and $P_{\varepsilon,J}$ in the following form:

$$J = \begin{bmatrix} J_+ & 0 \\ 0 & \bar{J}_+ \end{bmatrix}$$

where $J_+ = J_1 \oplus \cdots \oplus J_r$ and J_i is the Jordan block of size p_i with eigenvalue λ_0;

$$P_{\varepsilon,J} = \begin{bmatrix} 0 & P \\ P & 0 \end{bmatrix}$$

where $P = P_1 \oplus \cdots \oplus P_r$, and P_i is the sip matrix of size p_i. Observe that $\bar{J}_+ = P^{-1}J_+^*P$; so the pair $(J,P_{\varepsilon,J})$ is unitarily similar to the pair (K,Q), where $K = \begin{bmatrix} J_+ & 0 \\ 0 & J_+^* \end{bmatrix}$, $Q = \begin{bmatrix} 0 & I \\ I & 0 \end{bmatrix}$. It is sufficient to verify Theorem 3.22 for $A = K$, $H = Q$.

Given J_+-invariant subspace N_+, put $N = \left\{ \begin{bmatrix} x \\ y \end{bmatrix} \in \mathbb{C}^n \mid x \in N_+ , y \in N_+^\perp \right\}$. Clearly, N is K-invariant, $\dim N = \frac{n}{2}$ and $N \cap E_K(\lambda_0) = \begin{bmatrix} N_+ \\ 0 \end{bmatrix}$, where $E_K(\lambda_0)$ is the root subspace of K corresponding to λ_0. Further, N is Q-neutral; indeed, for $x_1,x_2 \in N_+$, $y_1,y_2 \in N_+^\perp$ we have

$$\left(\begin{bmatrix} 0 & I \\ I & 0 \end{bmatrix} \begin{bmatrix} x_1 \\ y_1 \end{bmatrix} , \begin{bmatrix} x_2 \\ y_2 \end{bmatrix} \right) = (y_1,x_1) + (x_2,y_2) = 0 .$$

Now let N' be a K-invariant, $\frac{n}{2}$-dimensional, Q-neutral subspace such that

$$N' \cap E_K(\lambda_0) = \begin{bmatrix} N_+ \\ 0 \end{bmatrix} . \tag{3.30}$$

As N' is K-invariant, $N' = (N' \cap E_K(\lambda_0)) \dotplus (N' \cap E_K(\bar{\lambda}_0))$. In fact

$$N' \cap E_K(\bar{\lambda}_0) \subset \left\{ \begin{bmatrix} 0 \\ x \end{bmatrix} \middle| \, x \in N_+ \right\} . \tag{3.31}$$

Indeed, if $\begin{bmatrix} 0 \\ x \end{bmatrix} \in N'$ with some $x \in \mathbb{C}^{n/2} \setminus N_+^\perp$, then there exists $y \in N_+$ such that $(x,y) \neq 0$, and

$$\left(\begin{bmatrix} 0 & I \\ I & 0 \end{bmatrix} \begin{bmatrix} 0 \\ x \end{bmatrix} , \begin{bmatrix} y \\ 0 \end{bmatrix} \right) = (x,y) \neq 0 ,$$

a contradiction with Q-neutrality of N'. Now (3.30) and (3.31) together with $\dim N' = \frac{n}{2}$, imply that $N' = N$.

Theorem 3.22 is proved. □

Let A, H be as in Theorem 3.22. Observe that for every A-invariant H-neutral $\frac{n}{2}$-dimensional subspace N the sizes of Jordan blocks of the restriction $A|_N$ which correspond to a real eigenvalue are just half the sizes of Jordan blocks of A corresponding to the same eigenvalue (see Theorem 3.21). In particular, the restriction of $A|_N$ to the sum of the root subspaces of $A|_N$ corresponding to the real eigenvalues is independent of N (in the sense that any two such restrictions, for different subspaces N, are similar).

A special case of Theorem 3.22 is sufficiently important to justify a separate statement:

COROLLARY 3.23 *Let* A *be an* H-*selfadjoint matrix as in Theorem* 3.22, *and suppose, in addition, that* $\sigma(A) \subset \mathbb{R}$. *Then there exists a unique* A-*invariant*, H-*neutral*, $\frac{n}{2}$-*dimensional subspace.*

3.13 Inverse Problems

It has been noted in Corollary 3.5 that if an $n \times n$ matrix A is similar to A^* then there is an H with $\det H \neq 0$ and $H^* = H$ such that A is H-selfadjoint. Now we ask more specifically: if A is similar to A^*, is there an H with a prescribed number of negative eigenvalues such that A is H-selfadjoint? In this connection, recall equation (3.4)

$$N = \frac{1}{2} n - \frac{1}{4} \sum_{j=1}^{a} [1-(-1)^{m_j}]\varepsilon_j \tag{3.32}$$

which expresses N, the number of negative eigenvalues of H, in terms of the sizes of Jordan blocks with the real eigenvalues, $m_1,\cdots,m_a$ and the associated signs $\varepsilon_1,\cdots,\varepsilon_a$ of the sign characteristic. Observe that if the sign characteristic is prescribed along with A then N is fixed by this relation. When the sign characteristic is not prescribed the next result holds.

THEOREM 3.24 *Let* A *be an* $n \times n$ *matrix which is similar to* A^* *and let* $m_1,\cdots,m_a$ *be the degrees of all the elementary divisors of* $I\lambda - A$ *associated with real eigenvalues. Then there exists a nonsingular hermitian matrix* H *with exactly* N *negative eigenvalues (counting multiplicities) for which* A *is* H-*selfadjoint if and only if*

$$\frac{1}{2} n - \frac{1}{4} \sum_{j=1}^{a} [1-(-1)^{m_j}] \leq N \leq \frac{1}{2} n + \frac{1}{4} \sum_{j=1}^{a} [1-(-1)^{m_j}] . \tag{3.33}$$

PROOF. If, given A, an H exists as required by the theorem, then (3.33) follows immediately from (3.32).

Conversely, if (3.33) holds then signs $\varepsilon_{01},\cdots,\varepsilon_{0a}$ can be chosen in such a way that

$$N = \frac{1}{2} n - \frac{1}{4} \sum_{j=1}^{a} [1-(-1)^{m_j}]\varepsilon_{0j}$$

and a matrix H is found accordingly. □

Note that Theorem 3.24 can be re-formulated on replacing N by the number of *positive* eigenvalues of H. This is because A is H-selfadjoint if and only if A is $(-H)$-selfadjoint and the number of positive eigenvalues of H is just the number of negative eigenvalues of $-H$.

Now let 2ν be the number of non-real eigenvalues of A. The inequalities (3.33) can also be expressed in terms of ν. Thus, provided $N \leq \frac{1}{2} n$, (3.33) is equivalent to the two conditions:

a) $\nu + \sum_{j=1}^{a} [\frac{1}{2} m_j] \leq N$,

b) the number of Jordan blocks of A corresponding to real eigenvalues of odd sizes is not less than

$$N - \nu - \sum_{j=1}^{a} [\frac{1}{2} m_j] .$$

Here, $[x]$ denotes the integer part of x. This reformulation is established by examination of (3.33) making use of the relation $\nu = \frac{1}{2} (n- \sum_{j=1}^{a} m_j)$.

In particular, this reformulation leads to the corollary:

COROLLARY 3.25 *Assume that the numbers* $m_1,\cdots,m_a$ *in Theorem 3.24 are all even. Then there exists an* $n \times n$ *hermitian* H *with exactly* N *negative eigenvalues (counting multiplicities) and for which* A *is* H-*selfadjoint, if and only if* $N = \frac{1}{2} n$, *i.e.* $\operatorname{sig} H = 0$.

Note that, under the conditions of this result, n is necessarily even.

CHAPTER 4

CANONICAL FORMS FOR H-UNITARY MATRICES

For a pair of matrices (U,H) in which U is H-unitary a canonical form is obtained in this chapter under transformations like those considered in Chapter 3, i.e. $U \to T^{-1}UT$ and $H \to T^*HT$. The problem for H-unitary matrices is solved by transforming to the H-selfadjoint case and applying the results of Chapter 3. Properties of H-normal matrices are also briefly discussed.

4.1 First Examples of Canonical Forms

The main result of Chapter 3 was the reduction of a pair of matrices (A,H), for which A is H-selfadjoint, to a canonical form $(J,P_{\varepsilon,J})$. This is described in Theorem 3.3 and the rest of that chapter is devoted to direct deductions from this result. The program for this chapter is similar, but for pairs (A,H) in which A is H-unitary. In addition, the concept of H-normal matrices is to be discussed.

First recall the basic ideas. Let H be an $n \times n$ invertible hermitian matrix and $[x,y] = (Hx,y)$ be the associated indefinite scalar product on $\mathbb{C}^n$. An $n \times n$ matrix A is said to be H-unitary if A is invertible and $[A^{-1}x,y] = [x,Ay]$ for $x,y \in \mathbb{C}^n$, or, what is equivalent, $A^{-1} = A^{[*]}$, or $A^{-1} = H^{-1}A^*H$. Some simple properties of H-unitary matrices have already been noted in Section 2.3. In particular, the spectrum of such a matrix is symmetric with respect to the unit circle, and the root subspaces of A are orthogonal in the indefinite scalar product $[.,.]$.

It has been seen in Chapter 3 that, if J is an $n \times n$ Jordan block with *real* eigenvalue λ and if P is the $n \times n$ sip matrix, then (J,P) form a primitive matrix pair for which J is P-selfadjoint. It is natural to search for a primitive P-*unitary* matrix by examining matrices of the form $f(J)$ where f is a fractional transformation of the form (2.12). It is certainly the case that with λ real $\mu = f(\lambda)$ will be the only eigenvalue of $f(J)$, and it will be unimodular.

It will be convenient to write the Jordan block J in the form $J = \lambda I + D$ (so that D has the "super-diagonal" made up of "ones", all other elements zero, and $D^n = 0$). Let $|\eta| = 1$, suppose $w \neq \bar{w}$, $w \neq \lambda$, $w \neq \bar{\lambda}$ (for the time being λ may be real or non-real), and define the Moebius transformation

$$f(z) = \eta(z-\bar{w}) / (z-w) , \tag{4.1}$$

and $K = f(J)$. Then

$$f(J) = \eta(\lambda I+D-\bar{w}I)(\lambda I+D-wI)^{-1} = \eta\left[\frac{\lambda-\bar{w}}{\lambda-w}\right](I - \frac{1}{\bar{w}-\lambda} D)(I - \frac{1}{w-\lambda} D)^{-1} .$$

Let $\mu = f(\lambda)$ and $q = (w-\lambda)^{-1}$. Then $(\bar{w}-\lambda)^{-1} = \mu^{-1}\eta q$ and

$$K = f(J) = (\mu I-\eta qD)(I-qD)^{-1} . \tag{4.2}$$

Note that, since

$$K = (\mu I-\eta qD)(I+qD+\cdots+q^{n-1}D^{n-1}) ,$$
$$= \mu I + (\mu-\eta)qD + (\mu-\eta)q^2D^2 + \cdots + (\mu-\eta)q^{n-1}D^{n-1} ,$$

it follows immediately that K is an upper triangular Toeplitz matrix. Furthermore, it is easily seen that K is similar to $\mu I + D$ (preserving the Jordan chain structure of J) if and only if the coefficient of D in the above expansion is non-zero. But this coefficient is just $(\mu-\eta)q$ and since $q \neq 0$, and $\mu = f(\lambda)$, it is found that $\mu = \eta$ only if $w = \bar{w}$, and this possibility has been excluded in the definition of f. Consequently, *the matrix* K *of* (4.2) *is similar to* $\mu I + D$ *and* (*from Proposition* 2.8) *is* P*-unitary*.

Consider in more detail the matrices obtained by this construction from a real λ, and from a conjugate pair $\lambda,\bar{\lambda}$ (with $\bar{\lambda} \neq \lambda$).

<u>Case (i)</u> ($\lambda = \bar{\lambda}$, $|\mu| = 1$).

From the definitions of q and μ it is easily seen that $\eta q = \mu\bar{q}$ and the representation (4.2) becomes

$$K = \mu(I-\bar{q}D)(I-qD)^{-1} . \tag{4.3}$$

We are left with the possibility of choosing the parameters w (and hence q) and η so as to simplify (4.3) as far as possible. The choice $w = \lambda - i$ (implying $q = i$) is legitimate and gives

$$K = \mu(I+iD)(I-iD)^{-1} = \mu \begin{bmatrix} 1 & 2i & 2i^2 & \cdot & \cdot & \cdot & 2i^{n-1} \\ 0 & 1 & 2i & \cdot & & & \cdot \\ \cdot & & 1 & \cdot & \cdot & & \cdot \\ \cdot & & & & \cdot & \cdot & \cdot \\ \cdot & & & & & 1 & 2i \\ 0 & \cdot & \cdot & \cdot & \cdot & 0 & 1 \end{bmatrix} \tag{4.4}$$

which is just the matrix of Example 2.3. Note that η is still undetermined.

Case (ii) ($\lambda \neq \bar{\lambda}$, $\mu_1 = f(\lambda)$, $\mu_2 = f(\bar{\lambda})$, $\mu_1\bar{\mu}_2 = 1$, $|\mu| \neq 1$).

Let $J_\lambda = \lambda I + D$, $J_{\bar{\lambda}} = \bar{\lambda}I + D$ and, as above, form $K_1 = f(J_\lambda)$, $K_2 = f(J_{\bar{\lambda}})$. If $q_1 = (w-\lambda)^{-1}$, $q_2 = (w-\bar{\lambda})^{-1}$, then (4.2) gives

$$\left.\begin{aligned} K_1 &= (\mu_1 I-\eta q_1 D)(I-q_1D)^{-1} \\ K_2 &= (\mu_2 I-\eta q_2 D)(I-q_2D)^{-1} \end{aligned}\right\} \tag{4.5}$$

If Q is the matrix $\begin{bmatrix} 0 & P \\ P & 0 \end{bmatrix}$, then it follows from Proposition 2.8 that $\operatorname{diag}[K_1,K_2]$ is Q-unitary. In this case, one verifies that

$$\eta q_1 = \mu_1\bar{q}_2 , \quad \eta q_2 = \mu_2\bar{q}_1$$

so that equations (4.5) can be written

$$\left.\begin{aligned} K_1 &= \mu_1(I-\bar{q}_2D)(I-q_1D)^{-1} \\ K_2 &= \mu_2(I-\bar{q}_1D)(I-q_2D)^{-1} \end{aligned}\right\} \tag{4.6}$$

and it is apparent that $K_2 = \bar{K}_1^{-1}$.

Now consider "simple" choices for w, and hence q_1 and q_2. If $w = a - i$ where a is real, then

$$q_1 = \frac{1}{w-\lambda} = \frac{i}{2}\left[\frac{1}{w-\lambda}\right](w-\bar{w}) = \frac{i}{2}\left[\frac{1}{w-\lambda}\right](w-\lambda-(\bar{w}-\lambda)) = \frac{i}{2}(1-\bar{\eta}\mu_1) ,$$

and similarly, $q_2 = \frac{i}{2}(1-\bar{\eta}\mu_2)$.

One can simplify further by putting $\eta = -1$ to obtain

$$q_1 = \frac{1}{2}\, i(1+\mu_1) , \quad q_2 = \frac{1}{2}\, i(1+\mu_2) .$$

Matrices K_1 and K_2 of (4.6) with this choice of parameters are just those of Example 2.4, namely

$$K_1 = \begin{bmatrix} \mu_1 & k_1 & k_2 & \cdot & \cdot & \cdot & k_{n-1} \\ 0 & \mu_1 & k_1 & & & & \cdot \\ \cdot & & \cdot & & & & \cdot \\ \cdot & & & \cdot & & & \cdot \\ \cdot & & & & \cdot & & \cdot \\ \cdot & & & & & \mu_1 & k_1 \\ 0 & \cdot & \cdot & \cdot & \cdot & 0 & \mu_1 \end{bmatrix}, \quad K_2 = \bar{K}_1^{-1} = \begin{bmatrix} \mu_2 & \kappa_1 & \kappa_2 & \cdot & \cdot & \cdot & \kappa_{n-1} \\ \cdot & \mu_2 & \kappa_1 & & & & \cdot \\ \cdot & & \cdot & & & & \cdot \\ \cdot & & & \cdot & & & \cdot \\ \cdot & & & & \cdot & & \cdot \\ \cdot & & & & & \mu_2 & \kappa_1 \\ 0 & \cdot & \cdot & \cdot & \cdot & 0 & \mu_2 \end{bmatrix} \quad (4.7)$$

where $k_r = \mu_1 q_1^{r-1}(q_1-\bar{q}_2)$, $\kappa_r = \mu_2 q_2^{r-1}(q_2-\bar{q}_1)$, $r = 1,2,\cdots,n-1$, and, of course $\mu_1\bar{\mu}_2 = 1$.

4.2 Canonical Forms in the General Case

It has been observed that, for any H-unitary matrix A, the spectrum, $\sigma(A)$, is symmetric relative to the unit circle. The blocks of a Jordan form for A can therefore be arranged as follows:

$$J = \text{diag}[J_1, J_2, \cdots, J_\alpha, J_{\alpha+1}, \cdots, J_{\alpha+2\beta}] , \quad (4.8)$$

where each J_i is a Jordan block, $J_1,\cdots,J_\alpha$ each have their associated eigenvalue on the unit circle, the eigenvalues of $J_{\alpha+1}, J_{\alpha+3}, \cdots, J_{\alpha+2\beta-1}$ are outside the unit circle and the eigenvalue of $J_{\alpha+2j}$ $(j = 1,2,\cdots,\beta)$ is obtained from that of $J_{\alpha+2j-1}$ by inversion in the unit circle.

Construct a block diagonal matrix

$$K_J = \text{diag}[K_1, K_2, \cdots, K_\alpha, K_{\alpha+1}, \cdots, K_{\alpha+2\beta}] \quad (4.9)$$

in the following way. If J_j of (4.8) has a unimodular eigenvalue μ_j, let

$$K_j = \mu_j(I+iD)(I-iD)^{-1}$$

as in equation (4.4). For the pair $J_{\alpha+2j-1}$, $J_{\alpha+2j}$ having non-unimodular eigenvalues $\mu_1, \mu_2 = \bar{\mu}_1^{-1}$, define $K_{\alpha+2j-1}$, $K_{\alpha+2j}$ by (4.7), with $q_1 = \frac{i}{2}(1+\mu_1)$, $q_2 = \frac{i}{2}(1+\mu_2)$.

It is clear from our study of primitive canonical forms in the preceding section that, with these definitions, K_J is $P_{\varepsilon,J}$-unitary. With these preparations the first important result of this section can be stated as follows.

THEOREM 4.1 *Let* A *be H-unitary, and let* J *be the Jordan normal form of* A *arranged as in* (4.8). *Then* (A,H) *is unitarily similar to a pair* $(K_J, P_{\varepsilon,J})$, *where*

$$P_{\varepsilon,J} = \mathrm{diag}\left[\varepsilon_1 P_1, \cdots, \varepsilon_\alpha P_\alpha, \begin{bmatrix} 0 & P_{\alpha+1} \\ P_{\alpha+1} & 0 \end{bmatrix}, \cdots, \begin{bmatrix} 0 & P_{\alpha+\beta} \\ P_{\alpha+\beta} & 0 \end{bmatrix}\right],$$

P_j *is the sip matrix with size equal to that of* J_j *(and* K_j*) for* $j = 1,\cdots,\alpha$, *and equal to that of* $J_{\alpha+2(j-\alpha)}$ *(and* $K_{\alpha+2(j-\alpha)}$*) for* $j > \alpha$, *and* $\varepsilon = (\varepsilon_1,\cdots,\varepsilon_\alpha)$ *is an ordered set of signs* ± 1. *The set of signs* ε *is uniquely determined by* (A,H) *up to permutation of signs corresponding to equal blocks* K_j.

PROOF. Let $\mu_1,\cdots,\mu_\alpha$ be the distinct unimodular eigenvalues of A and $\mu_{\alpha+1},\cdots,\mu_{\alpha+\beta}$ be distinct non-unimodular eigenvalues chosen one from each conjugate pair μ_j, $\bar{\mu}_j^{-1}$. Note that the inverse transformation of (4.1) is $(w\zeta-\bar{w}\eta)/(\zeta-\eta)$ (as a function of ζ).

For $j = 1,2,\cdots,\alpha$ define functions $g_j(\zeta)$ on $\sigma(A)$ by writing $w_j = \lambda_j - i$, $\lambda_j \in \mathbb{R}$, and $\eta_j = -\mu_j$ (cf. case (i) of the preceding section) and

$$g_j(\zeta) = [(\lambda_j - i)\zeta + (\lambda_j+i)\mu_j] \,/\, (\zeta+\mu_j) \tag{4.10}$$

in a (sufficiently small) neighbourhood of μ_j with $g_j(\zeta) \equiv 0$ in neighbourhoods of every other point of $\sigma(A)$. Assume also that $\lambda_1,\cdots,\lambda_\alpha$ are chosen so that $\mu_j \neq \mu_k$ $(j,k = 1,2,\cdots,\alpha)$ implies $\lambda_j \neq \lambda_k$.

For a pair of non-unimodular eigenvalues μ_j and $\bar{\mu}_j^{-1}$ (cf. case (ii) of the preceding section) let $w_j = -i$, $\eta_j = -1$ and, in a neighbourhood of μ_j and $\bar{\mu}_j^{-1}$

$$g_j(\zeta) = -i(\zeta-1) \,/\, (\zeta+1) \tag{4.11}$$

with $g_j(\zeta) \equiv 0$ on the remainder of $\sigma(A)$.

Now define a function $g(\zeta)$ on $\sigma(A)$ by $g(\zeta) = \sum_{j=1}^{p} g_j(\zeta)$. Let $T_j = g_j(A)$ and $T = g(A) = \sum_{j=1}^{p} T_j$. We show that T is H-selfadjoint.

Observe first that, since $A^{*-1} = HAH^{-1}$,

$$Hg_j(A) = (Hg_j(A)H^{-1})H = g_j(HAH^{-1})H = g_j(A^{*-1})H .$$

So that T_j is H-selfadjoint if $g_j(A) = g_j(A^{*-1})$. The latter fact is readily established using the functional calculus and the fact that $\bar{g}_j(z) = g_j(z^{-1})$ when μ_j is unimodular (as in equation (4.10) and, otherwise with g_j given by (4.11), $\bar{g}_j(z) = g_j(\bar{z}^{-1})$. Consequently, T_j is H-selfadjoint for each j and so T is H-selfadjoint.

Furthermore, the Jordan form J_T of T is obtained from the Jordan form J of A by replacing the eigenvalue μ_j in each block of J by λ_j if $|\mu_j| = 1$ and by $i(1-\mu_j)/(1+\mu_j)$ if $|\mu_j| \neq 1$. Thus, for a fixed set of signs ε we may write $P_{\varepsilon,J_T} = P_{\varepsilon,J}$. Then, applying Theorem 3.3 it is found that (T,H) and $(J_T,P_{\varepsilon,J})$ are unitarily similar for some set of signs ε. Thus,

$$T = S^{-1}J_T S \;, \quad H = S^* P_{\varepsilon,J} S$$

for some nonsingular S.

By construction, $A = f(T)$ where the function $f(z)$ is the inverse of $g(\zeta)$ and is defined in a neighbourhood of a real eigenvalue λ_j of T by

$$f(z) = -\mu_j(z-\bar{w}_j) \;/\; (z-w_j)$$

where $w_j = \lambda_j - i$, and μ_j is the corresponding unimodular eigenvalue of A. In a neighbourhood of a pair of non-real eigenvalues for T, f is defined by

$$f(z) = -(z-\bar{w}) \;/\; (z-w)$$

with $w = -i$. Furthermore, the construction of K_J shows that $K_J = f(J_T)$. Now we have

$$A = f(T) = S^{-1}f(J_T)S = S^{-1}K_J S$$

and the relations $A = S^{-1}K_J S$, $H = S^* P_{\varepsilon,J} S$ show that (A,H) and $(K_J,P_{\varepsilon,J})$ are unitarily similar, as required. The uniqueness of the set of signs ε will be discussed later, making use of Theorem 4.3. □

The set of signs ε associated with the pair (A,H), where A is H-unitary is, naturally, called the *sign characteristic* of (A,H). It is clear that the sign characteristic associated with a unimodular eigenvalue

μ of A is just the sign characteristic of the real eigenvalue $\lambda = g(\mu)$ of the H-selfadjoint matrix $T = g(A)$. It also follows, from the discussion of Section 4.1, that the set of J-invariant subspaces coincides with the set of K_J-invariant subspaces.

The following example confirms that the notion of the sign characteristic of (A,H) when A is H-unitary is consistent with that for the sign characteristic of (A,H) when A is H-selfadjoint, as introduced in Section 3.2.

EXAMPLE 4.1 Let A be a matrix which is both H-unitary and H-selfadjoint. Thus,

$$A = H^{-1}A^{*}H = A^{-1} .$$

In particular $A^2 = I$ and $\sigma(A) \subset \{1,-1\}$. The sign characteristic of A as an H-selfadjoint matrix is $\varepsilon^{(1)}$ say, and (A,H) is unitarily similar to the canonical pair (J_1,P_1).

Now let $\varepsilon^{(2)}$ be the sign characteristic of A as an H-unitary matrix. Thus, by definition, $\varepsilon^{(2)}$ is the sign characteristic of an H-selfadjoint matrix $T = g(A)$ where g is given by

$$g(\zeta) = \begin{cases} [(\lambda_1-i)\zeta + (\lambda_1+i)] \,/\, (\zeta+1) & \text{near} \quad \zeta = 1 \\ [(\lambda_2-i)\zeta - (\lambda_2+i)] \,/\, (\zeta-1) & \text{near} \quad \zeta = -1 \end{cases}$$

where $\lambda_1,\lambda_2 \in \mathbb{R}$ and $\lambda_1 \neq \lambda_2$. Note that the possible eigenvalues of T are $\lambda_1 = g(1)$ and $\lambda_2 = g(-1)$. Let (T,H) be unitarily similar to a canonical pair (J_2,P_2).

Now we have

$$A = S_1^{-1}J_1S_1 , \quad H = S_1^{*}P_1S_1 ,$$
$$T = S_2^{-1}J_2S_2 , \quad H = S_2^{*}P_2S_2 .$$

and since $T = g(A)$, $S_2^{-1}J_2S_2 = S_1^{-1}g(J_1)S_1$, i.e.

$$g(J_1) = S_3^{-1}J_2S_3$$

where $S_3 = S_1S_2^{-1}$. But it is easily seen that $P_1 = S_3^{*}P_2S_3$ and so $(g(J_1),P_1)$ and (J_2,P_2) are unitarily similar. It follows from Theorem 3.6 that P_1 and P_2 have the same sign-characteristic, i.e. $\varepsilon^{(1)} = \varepsilon^{(2)}$. □

The result of Theorem 4.1 can be seen as providing a canonical form $(K_J, P_{\varepsilon,J})$ for the pair (A,H) but we prefer to reserve the phrase "canonical form" for a unitarily similar pair $(J, Q_{\varepsilon,J})$ which is the subject of the next theorem.

Consider first a lemma which specifies a reduction of a typical block of K_J (as given by (4.2)) to a Jordan block.

LEMMA 4.2 *Let* K *be an* $n \times n$ *complex matrix of the form*

$$K = \mu \begin{bmatrix} 1 & \alpha & \alpha q & \alpha q^2 & \cdots & \alpha q^{n-2} \\ 0 & 1 & \alpha & \alpha q & \cdots & \alpha q^{n-3} \\ 0 & 0 & 1 & & & \cdot \\ \cdot & & & \ddots & & \cdot \\ \cdot & & & & 1 & \alpha \\ 0 & & \cdots & & 0 & 1 \end{bmatrix} = \mu(I - (q-\alpha)D)(I-qD)^{-1}$$

where $\mu\alpha q \neq 0$ *and let* J *be the Jordan block with eigenvalue* μ. *Then* $KX = XJ$ *where*

$$X = \mathrm{diag}[1, q^{-1}, \cdots, q^{-n+1}]\hat{X}_n \, \mathrm{diag}[1, \tfrac{q}{\mu\alpha}, \cdots, (\tfrac{q}{\mu\alpha})^{n-1}]$$

and

$$\hat{X}_n = \begin{bmatrix} 1 & 0 & 0 & 0 & 0 & \cdots & 0 \\ 0 & 1 & -1 & 1 & -1 & \cdots & (-1)^n \\ 0 & 0 & 1 & -2 & 3 & \cdots & \cdot \\ 0 & 0 & 0 & 1 & -3 & \cdots & \cdot \\ 0 & 0 & 0 & 0 & 1 & \cdots & \cdot \\ \cdot & & & & & \ddots & \\ \cdot & & & & & 1 & -n+2 \\ 0 & & \cdots & & & 0 & 1 \end{bmatrix}. \qquad (4.12)$$

Note that the columns of $\hat{X}_n$ are made up of the signed binomial coefficients and $\hat{X}_n^{-1}$, can also be written down in simple explicit form. Thus, X is invertible, and the columns of X form a Jordan chain for K. The proof of the lemma is an exercise in familiar properties of the binomial coefficients and is omitted.

The implications of the lemma for the reduction of blocks of matrix K_J of (4.9) will now be examined for the cases of eigenvalues which are unimodular or not unimodular.

<u>Case (a)</u> Suppose $|\mu| = 1$ and (as in the construction of K_J) set $q = i$

and $\alpha = 2i$. A pair (J,Q) is unitarily similar to $(K,\varepsilon P)$ (where $\varepsilon = +1$ or -1 and P is the $n \times n$ sip matrix) if $J = X^{-1}KX$ and $Q = \varepsilon X^* PX$. The lemma indicates the choice

$$X = \mathrm{diag}[1, i^{-1}, \cdots, i^{-n+1}]\hat{X}_n \,\mathrm{diag}[1, (2\mu)^{-1}, \cdots, (2\mu)^{-n+1}] .$$

It is easily seen that $Q = [q_{jk}]_{j,k=1}^n$ is of triangular form with elements $q_{jk} = 0$ when $j + k \leqslant n$.

Case (b) With K_1 and K_2 $(= \bar{K}_1^{-1})$ as defined in (4.7) we are to find a matrix X which will determine the unitary similarity between pairs

$$\left(\begin{bmatrix} K_1 & 0 \\ 0 & K_2 \end{bmatrix}, \begin{bmatrix} 0 & P \\ P & 0 \end{bmatrix}\right) \text{ and } \left(\begin{bmatrix} J_1 & 0 \\ 0 & J_2 \end{bmatrix}, Q\right)$$

and hence the structure of Q. Here, J_j is the $n \times n$ Jordan block with eigenvalue μ_j for $j = 1$ and 2 where $\mu_1\bar{\mu}_2 = 1$.

Using the lemma construct a matrix X_1 for which $K_1X_1 = X_1J_1$ by setting $q = q_1 = \frac{1}{2} i(1+\mu_1)$ and $\alpha = q_1 - \bar{q}_2 = \frac{1}{2} i(2+\mu_1+\bar{\mu}_2)$ $= \frac{1}{2} i(1+\mu_1)(1+\bar{\mu}_2)$. Similarly, on setting $q = q_2 = \frac{1}{2} i(1+\mu_2)$, $\alpha = q_2 - \bar{q}_1$ $= \frac{1}{2} i(1+\bar{\mu}_1)(1+\mu_2)$ a matrix X_2 is obtained for which $K_2X_2 = X_2J_2$. Then

$$\begin{bmatrix} K_1 & 0 \\ 0 & K_2 \end{bmatrix}\begin{bmatrix} X_1 & 0 \\ 0 & X_2 \end{bmatrix} = \begin{bmatrix} X_1 & 0 \\ 0 & X_2 \end{bmatrix}\begin{bmatrix} J_1 & 0 \\ 0 & J_2 \end{bmatrix}$$

and Q is defined by the congruence

$$Q = \begin{bmatrix} X_1^* & 0 \\ 0 & X_2^* \end{bmatrix}\begin{bmatrix} 0 & P \\ P & 0 \end{bmatrix}\begin{bmatrix} X_1 & 0 \\ 0 & X_2 \end{bmatrix} = \begin{bmatrix} 0 & X_1^*PX_2 \\ X_2^*PX_1 & 0 \end{bmatrix} .$$

Here, the matrix $X_1^*PX_2$ (and hence Q) is found to be nonsingular and of triangular form with zero elements above the secondary diagonal. Note also that $\dfrac{q_1}{\mu_1\alpha_1} = \dfrac{1}{\mu_1(1+\bar{\mu}_2)} = \dfrac{1}{1+\mu_1}$.

The next step is to put the conclusions of steps (a) and (b) together to formulate a canonical form for (A,H) where A is H-unitary. However, note first that by using the lemma a preferred choice of Jordan chain for K is implied. To investigate reduction by *any* Jordan chain the matrix X can be replaced by XT where T is an arbitrary nonsingular upper-triangular Toeplitz matrix. It may be imagined that there is a propitious choice for T giving rise to a matrix Q in cases (a) and (b) of particularly simple structure, but the choice $T = I$ of the lemma seems, in fact, to be the least intractable.

In formulating the canonical form suppose that block J_j of J in (4.8) has size n_j and P_j is the sip matrix of size n_j for $j = 1,2,\cdots,\alpha+\beta$.

THEOREM 4.3 *Let* A *be H-unitary and let* J *be a Jordan form for* A *arranged as in* (4.8). *Then* (A,H) *is unitarily similar to a pair* $(J,Q_{\varepsilon,J})$ *where*

$$Q_{\varepsilon,J} = \text{diag}\left[\varepsilon_1 Q_1,\cdots,\varepsilon_\alpha Q_\alpha, \begin{bmatrix} 0 & Q_{\alpha+1} \\ Q_{\alpha+1}^* & 0 \end{bmatrix}, \cdots, \begin{bmatrix} 0 & Q_{\alpha+\beta} \\ Q_{\alpha+\beta}^* & 0 \end{bmatrix}\right] \tag{4.13}$$

and for $j = 1,2,\cdots,\alpha$, $Q_j = X_{n_j}^* P_j X_{n_j}$,

$$X_{n_j} = \text{diag}[1,i^{-1},i^{-2},\cdots]\hat{X}_{n_j}\ \text{diag}[1,(2\mu)^{-1},(2\mu)^{-2},\cdots] ,$$

$\hat{X}_{n_j}$ *is given by* (4.12) *and* $\varepsilon = (\varepsilon_1,\cdots,\varepsilon_\alpha)$ *is an ordered set of signs* ± 1.

For $j = \alpha+1,\cdots,\alpha+\beta$, $Q_j = X_{1,j}^* P_j X_{2,j}$ *where*

$$X_{1,j} = \text{diag}[1,\{\tfrac{1}{2}\, i(1+\mu_j)\}^{-1},\{\tfrac{1}{2}\, i(1+\mu_j)\}^{-2},\cdots]\hat{X}_{n_j}\text{diag}[1,(1+\mu_j)^{-1},(1+\mu_j)^{-2},\cdots],$$

X_{n_j} *is given by* (4.12) *and* $X_{2,j}$ *is obtained from* $X_{1,j}$ *on replacing* μ_j *by* $\bar{\mu}_j^{-1}$.

To illustrate, observe that the canonical blocks Q_j of sizes 2, 3 and 4 for a unimodular eigenvalue μ are as follows:

$$\frac{i}{2}\begin{bmatrix} 0 & -\bar{\mu} \\ \mu & 0 \end{bmatrix}, \quad -\frac{1}{4}\begin{bmatrix} 0 & 0 & \bar{\mu}^2 \\ 0 & -1 & \frac{1}{2}\bar{\mu} \\ \mu^2 & \frac{1}{2}\mu & -\frac{1}{4} \end{bmatrix}, \quad \frac{i}{8}\begin{bmatrix} 0 & 0 & 0 & \bar{\mu}^3 \\ 0 & 0 & -\bar{\mu} & \bar{\mu}^2 \\ 0 & \mu & 0 & -\frac{1}{4}\bar{\mu} \\ -\mu^3 & -\mu^2 & \frac{1}{4}\mu & 0 \end{bmatrix}$$

In each case, we have $J_j^* Q_j J_j = Q_j$ where J_j is the Jordan block with the size of Q_j and eigenvalue μ.

4.3 Correctness of the Sign Characteristic

In this section we shall show that the sign characteristic of (A,H), where A is H-unitary, was properly defined in the proof of Theorem 4.1, i.e. that the choice of signs associated with the eigenvalues of A does not depend on the choice of the parameter w in the Moebius transformation (4.1). It was, of course, essential for the proof of Theorem 4.1 that different choices be made for w at different points of spectrum. Also, the choice of signs was independent of the parameter η in (4.1).

For the purpose of this theorem consider two Moebius transformations

$$f_j(\zeta) = \eta_j(z-\bar{w}_j) \,/\, (z-w_j) \;, \quad j = 1,2$$

and their inverses

$$g_j(\zeta) = (w_j\zeta-\bar{w}_j\eta_j) \,/\, (\zeta-\eta_j) \;,$$

where $|\eta_1| = |\eta_2| = 1$, $\eta_1 \neq \eta_2$, and $w_j \neq \bar{w}_j$. Thus, each f_j maps the reals onto the unit circle, etc.

THEOREM 4.4 *Let* A *be* H-*unitary and suppose that, with the definitions above,* $T_1 = g_1(A)$, $T_2 = g_2(A)$. *Then* T_1 *and* T_2 *are* H-*selfadjoint and if* $(Im\ w_1)(Im\ w_2) > 0$ *then the sign characteristic of* (T_1,H) *corresponding to a real eigenvalue* λ_1 *of* T_1 *coincides with the sign characteristic of* (T_2,H) *corresponding to the real eigenvalue* $\lambda_2 = g_2(f(\lambda_1))$.

In the proof of Theorem 4.3 the condition $Im\ w_j = -1$ was used for each Moebius transformation employed so the hypothesis of the present theorem applies there.

PROOF. It has been seen in Section 4.1 that the conditions $w_j \neq \bar{w}_j$ $(j = 1,2)$ ensure that the sizes of the Jordan blocks of A are preserved under the transformations $A \to g_1(A) = T_1$ and $A \to g_2(A) = T_2$. Consequently, it suffices to consider a matrix A which is similar to a single $n \times n$ Jordan block with unimodular eigenvalue μ.

Choose a basis $x_1,x_2,\cdots,x_n$ of (generalized) eigenvectors of $T_1 = g_1(A)$ as a basis for $\mathbb{C}^n$ and, in this basis the representation $\hat{T}_1$ of T_1 is an $n \times n$ Jordan block, say J, with real eigenvalue $\lambda_1 = g_1(\mu)$. Similarly, $T_2 = g_2(A)$ has just one real eigenvalue $\lambda_2 = g_2(\mu) = g_2(f_1(\lambda_1))$. Writing $h(\lambda) = g_2(f_1(\lambda))$ the representation of T_2 in the Jordan basis constructed for T_1 has the form

$$\hat{T}_2 = h(\hat{T}_1) = \begin{bmatrix} \gamma_0 & \gamma_1 & \gamma_2 & \cdot & \cdot & \cdot & \gamma_{n-1} \\ 0 & \gamma_0 & \gamma_1 & & & & \cdot \\ \cdot & & \gamma_0 & \cdot & & & \cdot \\ \cdot & & & \cdot & \cdot & \cdot & \cdot \\ \cdot & & & & \cdot & \cdot & \gamma_1 \\ 0 & & \cdot & \cdot & \cdot & 0 & \gamma_0 \end{bmatrix}$$

where $j!\gamma_j = h^{(j)}(\lambda_1)$ for $j = 0,1,\cdots,n-1$. Since $h : \mathbb{R} \to \mathbb{R}$ it is clear that $\hat{T}_2$ is a real matrix.

It is necessary to examine the element $\gamma_1 = h^{(1)}(\lambda_1)$ in some detail. First, as observed in Section 4.1, $\gamma_1 \neq 0$. Then note that the function h has just one singularity and this is at

$$\lambda_c = (\eta_1\bar{w}_1-\eta_2 w_1) \,/\, (\eta_1-\eta_2) \quad (= \bar{\lambda}_c) .$$

Now some computation shows that

$$\gamma_1 = h^{(1)}(\lambda_1) = \frac{(\bar{w}_1-w_1)(\bar{w}_2-w_2)\eta_1\eta_2}{(\lambda_1-\lambda_c)^2(\eta_1-\eta_2)^2} .$$

It is easily verified that $\eta_1\eta_2(\eta_1-\eta_2)^{-2} < 0$ and, consequently,

$$\gamma_1 = \kappa(\mathit{Im}\ w_1)(\mathit{Im}\ w_2) \tag{4.14}$$

where $\kappa > 0$.

To compare the sign characteristics of (T_1,H) and (T_2,H) (which both consist of a single sign) we shall take advantage of Theorem 3.16 and, to this end, note that given the Jordan basis $x_1,\cdots,x_n$ for T_1 there is a Jordan basis for T_2 of the form

$$\begin{aligned} y_1 &= x_1 , \quad y_2 = \gamma_1^{-1}x_2 , \\ y_3 &= \gamma^{-2}x_3 + \beta_{32}x_2,\cdots,y_n = \gamma^{-n+1}x_n + \beta_{n,n-1}x_{n-1} + \cdots + \beta_{n,2}x_2 \end{aligned} \tag{4.15}$$

This statement is readily verified (cf. Lemma 4.2). Consider linear transformations $G_1^{(j)},\cdots,G_n^{(j)}$ of Theorem 3.16 associated with the matrices T_j, for $j = 1$ and 2. In this case all of the $G_i^{(j)}$ act on a one-dimensional space and it is easily seen that, in fact, $G_1^{(j)} = \cdots = G_{n-1}^{(j)} = 0$ for $j = 1,2$ and, because of (4.15), $G_n^{(1)}$ and $G_n^{(2)}$ correspond to multiplication by constants k_1 and k_2, respectively, where $k_2 = \gamma_1^{-n+1}k_1$.

Thus, the sign characteristics agree whenever $\gamma_1 > 0$; but this is ensured by equation (4.14). □

The line of argument used in the preceding proof shows that there are other sufficient conditions guaranteeing preservation of the sign characteristics of $g_1(A)$, $g_2(A)$. For example, if all the unimodular eigenvalues of A have only linear elementary divisors, then the difficulties of the theorem "go away" and the sign characteristics of (T_1,H) and (T_2,H) will agree. More generally, it can be seen that the sign characteristic is preserved if and only if

$$[(Im\ w_1)(Im\ w_2)]^{-n_j+1} > 0$$

for each partial multiplicity n_j associated with a unimodular eigenvalue of A.

Theorem 4.4 demonstrates that the technique used in Theorem 4.1 and 4.3 to arrive at a sign-characteristic for an H-unitary A leads to a unique definition. More simply, the sign characteristic of (A,H), where A is H-unitary can be defined as follows: Let $g(\zeta) = (w\zeta-\bar{w}\eta)/(\zeta-\eta)$ where $|\eta| = 1$, $\eta \notin \sigma(A)$, and $Im\ w > 0$. Define the H-selfadjoint matrix $T = g(A)$, and then the sign-characteristic of (A,H) at a unimodular $\mu_0 \in \sigma(A)$ is just the sign characteristic of (T,H) at the real eigenvalue $\lambda_0 = g(\mu_0)$ of T.

4.4 First Applications of the Canonical Form

The points to be made in this section are an exact parallel for those made in Section 3.3 for H-selfadjoint matrices. Comparing the conclusions of Theorem 3.3 for H-selfadjoint matrices, and Theorem 4.3 for H-unitary matrices, one sees that matrix $Q_{\varepsilon,J}$ of (4.13) inherits sufficient properties from $P_{\varepsilon,J}$ of (3.2) to admit conclusions parallel to those of Section 3.3. The vital characteristics of $Q_{\varepsilon,J}$ are that each block Q_j is congruent to P_j and is of triangular form with all elements zero above the secondary diagonal.

The conclusions to be drawn will simply be summarized here; the arguments justifying them are essentially the same as those used in Section 3.3. First, if A is H-unitary and $\mu \in \sigma(A)$ with $|\mu| = 1$ then $\mathrm{Ker}(\mu I-A)$ can be written as a direct sum

$$\mathrm{Ker}(\mu I-A) = L_1 \dotplus L_2$$

where L_1 is H-non-degenerate and L_2 is H-neutral. The spaces L_1 and L_2 are spanned by eigenvectors of μ associated with linear and nonlinear divisors of λ_0, respectively.

If, on the other hand, $\mu \in \sigma(A)$ and $|\mu| \neq 1$ then the whole root-subspace X_μ is H-neutral, and the sum $X_\mu \dot{+} X_{\bar\mu}$ is H-non-degenerate.

If H and A are $n \times n$ and H has N negative eigenvalues it follows (as in (3.4)) that

$$N = \frac{1}{2}n - \frac{1}{4}\sum_{j=1}^{\alpha}[1-(-1)^{m_j}]\varepsilon_j$$

and, from this, it is readily concluded that the number of unimodular eigenvalues of A is at least $|\text{sig } H| = |\text{sig } Q_{\varepsilon,J}|$. Also, H has $\frac{1}{2}n$ positive eigenvalues if and only if the signs associated with elementary divisors of odd degree for unimodular eigenvalues of A (if any) are equally divided between +1's and -1's.

4.5 Further Deductions from the Canonical Form

It has been seen that H-selfadjoint and H-unitary matrices have in common the notion of a sign-characteristic, as well as common geometrical properties of root subspaces. As a result, and in addition to the parallels drawn in the preceding section, there are other deeper analogues for theorems obtained in Chapter 3. Three of these will be considered here.

Denote by k the number of positive eigenvalues of H (counting multiplicities) and recall that for an H-non-negative subspace L, the index of positivity $p(L)$ is the maximal dimension (≥ 0) of an H-positive subspace $L_0 \subset L$.

THEOREM 4.5 *Let* A *be an H-unitary matrix, and let* C *be a set of non-unimodular eigenvalues of* A *which is maximal with respect to the property that if* $\lambda_0 \in C$ *then* $\bar\lambda_0^{-1} \notin C$. *Then there exists a* k*-dimensional* A*-invariant H-nonnegative subspace* N *with the following properties:*

(i) N *is maximal H-nonnegative;*

(ii) *the non-unimodular part of* $\sigma(A|_N)$ *coincides with* C;

(iii) $p(N) = \frac{1}{2}\sum_{j=0}^{r}[1-(-1)^{m_j}]\delta_j$

where $m_1,\cdots,m_r$ *are the sizes of Jordan blocks in the Jordan form for* A *having unimodular eigenvalues and associated signs* $\varepsilon_1,\cdots,\varepsilon_r$ *in the sign characteristic of* (A,H), *and* $\delta_j = 1$ *if* $\varepsilon_j = 1$, $\delta_j = 0$ *if* $\varepsilon_j = -1$.

This theorem is the analogue of Theorem 3.20 and can be obtained from it by an application of the Cayley transform. Such a procedure will be demonstrated in proving the analogue of Theorem 3.6.

THEOREM 4.6 *Let* U_1 *be* H_1*-unitary and* U_2 *be* H_2*-unitary. Then*

(U_1,H_1) and (U_2,H_2) are unitarily similar if and only if U_1 and U_2 are similar and the pairs (U_1,H_1) and (U_2,H_2) have the same sign characteristic.

PROOF. Let $\alpha, w \in \mathbb{C}$ with $|\alpha| = 1$, $Im\ w > 0$; let $g(\zeta) = (w\zeta-\bar{w}\alpha)/(\zeta-\alpha)$, and let $A_1 = g(U_1)$, $A_2 = g(U_2)$. Then, by Proposition 2.8, A_1 is H_1-selfadjoint and A_2 is H_2-selfadjoint.

The theorem is proved using Theorem 3.6 if it can be shown that (A_1,H_1) and (A_2,H_2) are unitarily similar if and only if the same is true of (U_1,H_1) and (U_2,H_2). So suppose that (A_1,H_1) and (A_2,H_2) are unitarily similar; i.e. there is a nonsingular T such that

$$A_1 = T^{-1}A_2T \ , \quad H_1 = T^*H_2T \ .$$

Then

$$g(U_1) = T^{-1}g(U_2)T = g(T^{-1}U_2T)$$

and applying the inverse transformation to g (which is well-defined on $\sigma(U_1)$ and $\sigma(U_2)$) it is found that $U_1 = T^{-1}U_2T$ and since $H_1 = T^*H_2T$ it follows that (U_1,H_1) and (U_2,H_2) are unitarily similar. But this argument is reversible and so the theorem is proved. □

Finally, we quote the analogue of Theorem 3.9 for future reference. The proof follows just the same lines as that of Theorem 3.9.

THEOREM 4.7 *Each equivalence class (under H-unitary similarity) of H-unitary matrices is arcwise connected.*

Recall that H-unitary similarity defines an equivalence relation on the complex square matrices with the size of H (ref. Section 3.5). It is easily verified that if an equivalence class defined by this relation contains an H-unitary matrix, then every matrix of the class is H-unitary.

4.6 H-normal Matrices

In this section some preliminary remarks and results concerning H-normal matrices are presented. The major question of finding canonical forms for pairs (A,H) where A is H-normal remains open, however.

First recall the definitions. If H is hermitian and nonsingular and generates an indefinite scalar product on $\mathbb{C}^n$, then $A^{[*]}$ denotes the H-adjoint of a matrix A, and A is said to be H-normal if $AA^{[*]} = A^{[*]}A$

or, what is equivalent, $AH^{-1}A^*H = H^{-1}A^*HA$.

Our first example turns out to be fundamental.

EXAMPLE 4.2 Let J be an $m \times m$ Jordan block with eigenvalue $\alpha \in \mathbb{C}$ and let $H = P$, the $m \times m$ sip matrix. It is easily verified that J is P-normal. It is important to note that, in contrast to Example 2.1 (of an H-selfadjoint matrix), there is no restriction on the complex number α. □

Example 4.2 leads directly to the first result.

THEOREM 4.8 *Every complex square matrix is* H-*normal for some non-singular* $H = H^*$.

PROOF. Let $J = \mathrm{diag}[J_1,\cdots,J_k]$ be a Jordan form for matrix A with $J_1,\cdots,J_k$ all Jordan blocks, and let $A = S^{-1}JS$. If P_j is the sip matrix with the size J_j for $j = 1,2,\cdots,k$, we write $P = \mathrm{diag}[P_1,\cdots,P_k]$. As in Example 4.2, it follows immediately that J is P-normal and, using Proposition 2.3, it follows that A is S^*PS-normal. □

The proof of Theorem 4.8 allows us to answer partially the following question: given an $n \times n$ matrix A what is the minimal possible number $N(A)$ of negative eigenvalues of a hermitian invertible $n \times n$ matrix H such that A is H-normal? For instance, $N(A) = 0$ if and only if A is similar to a diagonal matrix. The proof of Theorem 4.8 shows that in general

$$N(A) \leq \sum_{i=1}^{r} [\tfrac{1}{2} m_i] \tag{4.16}$$

where $m_1,\cdots,m_r$ are the sizes of all Jordan blocks in the Jordan form of A, and $[x]$ is the integer part of x. Whether equality holds in (4.16) remains an open question.

It is easily verified that if A is H-normal, then every matrix H-unitarily similar to A is also H-normal. Consequently, it is natural to seek canonical forms for the equivalence classes of H-normal matrices generated in this way. Since H-selfadjoint and H-unitary matrices are also H-normal, such a canonical form should include those described in Theorems 3.3 and 4.3. However, an essentially new feature arises. The next example shows that a pair (A,H) with A H-normal cannot be reduced to block diagonal forms as simple as those which are possible if A is either H-selfadjoint or H-unitary.

EXAMPLE 4.3 Let $J = \begin{bmatrix} 0 & 0 \\ 0 & 1 \end{bmatrix}$, $H = \begin{bmatrix} 0 & 1 \\ 1 & 0 \end{bmatrix}$ and verify that J is H-normal. Now J is already in Jordan form and if (J,H) is unitarily similar to (J,R) then

$$S^{-1}JS = J\ , \quad S^*HS = R\ ,$$

and there is *no* such S which will reduce H to a diagonal matrix R, as the examples of H-selfadjoint and H-unitary matrices might lead one to expect. In other words, the root subspace for a real (or unimodular) eigenvalue may now be neutral, and root subspaces for distinct real eigenvalues need not be H-orthogonal. □

We conclude the section with some easily verified properties of H-normal matrices.

(1) If A is H-normal then so is $A + \alpha I$ for any $\alpha \in \mathbb{C}$.

(2) Suppose $H^2 = I$ (and $H^* = H$, $\det H \neq 0$). Then A is H-normal if and only if A^* is H-normal.

(3) Suppose $H^2 = I$ and $\det A \neq 0$. Then A is H-normal if and only if A^{-1} is H-normal.

CHAPTER 5

REAL MATRICES

We now turn attention to the real space $\mathbb{R}^n$ and to an indefinite scalar product $[.,.]$ on $\mathbb{R}^n$ defined by a real symmetric invertible matrix H, as described in Chapter 1. Attention is focussed on real $n \times n$ matrices acting on $\mathbb{R}^n$ together with such an indefinite scalar product, and several results obtained in Chapters 1 to 4 are to be re-examined in this context. In particular, the reader will quickly verify that all the results and observations of Chapter one on the geometry of indefinite scalar product spaces (when properly understood) are also valid for the real case.

5.1 Real H-Selfadjoint Matrices and Canonical Forms

Let $[.,.]$ denote an indefinite scalar product defined on $\mathbb{R}^n$ by a real symmetric invertible matrix H of size n i.e. $[x,y] = (Hx,y)$. The adjoint $A^{[*]}$ of a real $n \times n$ matrix A is defined just as in (2.1) and, as in (2.3), it is easily seen that $A^{[*]} = H^{-1}A^*H$. Since A^* is now just the transpose of A, $A^{[*]}$ is obviously real. The following facts and definitions are all formally identical with predecessors in earlier chapters:
A real matrix A is H-selfadjoint if $A^{[*]} = A$, i.e. if $HA = A^*H$.
A real matrix A is H-unitary if $A^{[*]}A = I$, i.e. if $A^*HA = H$.
A real matrix A is H-normal if $A^{[*]}A = AA^{[*]}$, i.e. if $(H^{-1}A^*H)A = A(H^{-1}A^*H)$.

This section is devoted to properties of real H-selfadjoint matrices. Some simple examples follow and, following a trend set in earlier chapters, they will form the basic blocks in the real canonical form (see Theorem 5.3, below).

EXAMPLE 5.1 Let A be a Jordan block with real eigenvalue and let P be sip matrix of the same size. Then A is real $\pm P$-selfadjoint. □

EXAMPLE 5.2 If $\sigma,\tau \in \mathbb{R}$ then $\begin{bmatrix} \sigma & \tau \\ -\tau & \sigma \end{bmatrix}$ is real P_2-selfadjoint where P_2 is the 2×2 sip matrix. More generally, the matrix A of even size given by

$$A = \begin{bmatrix} \sigma & \tau & 1 & 0 & \cdot & \cdot & \cdot & 0 \\ -\tau & \sigma & 0 & 1 & \cdot & & & \cdot \\ \cdot & & \sigma & \tau & & \cdot & & \cdot \\ \cdot & & -\tau & \sigma & \cdot & & \cdot & \cdot \\ \cdot & & & & & \cdot & 1 & 0 \\ \cdot & & & & & \cdot & 0 & 1 \\ \cdot & & & & & & \sigma & \tau \\ 0 & \cdot & \cdot & \cdot & \cdot & & -\tau & \sigma \end{bmatrix}$$

is real P-selfadjoint where P is the sip matrix with the size of A. □

For any real matrix A the polynomial $\det(\lambda I-A)$ has real coefficients and can therefore be written in the form

$$\det(\lambda I-A) = \prod_{j=1}^{t} (\lambda-\lambda_j)^{\alpha_j} \prod_{j=1}^{s} \{(\lambda-\sigma_j)^2 + \tau_j^2\}^{\beta_j} ,$$

where $\lambda_1,\cdots,\lambda_t$ are the distinct real zeros of $\det(\lambda I-A)$, and for $j = 1,2,\cdots,s$, $\tau_j \neq 0$ and $\sigma_j \pm i\tau_j$ are the distinct non-real zeros of $\det(\lambda I-A)$. Corresponding to this factorization there is a decomposition of $\mathbb{R}^n$ into a direct sum of real A-invariant subspaces X_{ri}:

$$\mathbb{R}^n = X_{r1} \dotplus \cdots \dotplus X_{r,t} \dotplus X_{r,t+1} \dotplus \cdots \dotplus X_{r,t+s} ,$$

where the minimal polynomial of $A|_{X_{rj}}$ is a positive integer power of $\lambda - \lambda_j$, for $j = 1,\cdots,t$, and the minimal polynomial of $A|_{X_{rj}}$ is a positive integer power of $((\lambda-\sigma_j)^2 + \tau_j^2)$, for $j = t+1,\cdots,t+s$ (see, e.g., Section 6.34 in [42]). Note that $\dim X_{rj}$ is even for $j = t+1,\cdots,t+s$. Considering A as a linear transformation acting in $\mathbb{C}^n$, we also obtain the following decomposition:

$$\mathbb{C}^n = \hat{X}_{r1} \dotplus \cdots \dotplus \hat{X}_{r,t} \dotplus \hat{X}_{r,t+1} \dotplus \cdots \dotplus \hat{X}_{r,t+s} ,$$

where the complex subspace $\hat{X}_{rj}$ is equal to $X_{rj} + iX_{rj}$. Then

$$\sigma(A|_{\hat{X}_{rj}}) = \{\lambda_j\} \quad \text{if} \quad j = 1,\cdots,t$$

and

$$\sigma(A|_{\hat{X}_{rj}}) = \{\sigma_j + i\tau_j , \sigma_j - i\tau_j\} \quad \text{if} \quad j = t+1,\cdots,t+s .$$

(See [42], section 6.62 for more detail of this construction). Using Theorem 2.5, we immediately obtain the following.

THEOREM 5.1 *Let* A *be a real* H*-selfadjoint matrix, where* H *is a real symmetric invertible* $n \times n$ *matrix. Let* M_1 *(resp.* M_2*) be a real* A*-invariant subspace of* $\mathbb{R}^n$ *such that the minimal polynomial of the restriction* $A|_{M_1}$ *(resp.* $A|_{M_2}$*) is a power of an irreducible (over the real field) real polynomial* $p_1(\lambda)$ *(resp.* $p_2(\lambda)$*). Then* M_1 *and* M_2 *are* H*-orthogonal provided* $p_1(\lambda) \neq p_2(\lambda)$*. Moreover, the maximal real* A*-invariant subspace* M_1 *with the property that the minimal polynomial of* $A|_{\hat{M}_1}$ *is a power of* $p_1(\lambda)$*, is* H*-nondegenerate.*

We shall need the following real Jordan form of a real matrix A (not necessarily H-selfadjoint). There exists a real $n \times n$ matrix S such that

$$SAS^{-1} = J ,$$

where J (a real Jordan form of A) is a block diagonal matrix, each block being of one of the two following forms (see [42], Theorem 6.65):

$$\begin{bmatrix} \lambda_0 & 1 & \cdot & \cdot & \cdot & 0 \\ 0 & \lambda_0 & \cdot & & & \cdot \\ \cdot & & \cdot & \cdot & & \cdot \\ \cdot & & & \cdot & \cdot & \cdot \\ \cdot & & & & \lambda_0 & 1 \\ 0 & \cdot & \cdot & \cdot & 0 & \lambda_0 \end{bmatrix}, \quad \lambda_0 \in \mathbb{R} \tag{5.1}$$

$$\begin{bmatrix} \sigma & \tau & 1 & 0 & \cdot & \cdot & \cdot & 0 \\ -\tau & \sigma & 0 & 1 & & & & \cdot \\ \cdot & & \sigma & \tau & & & & \cdot \\ \cdot & & -\tau & \sigma & & & & \cdot \\ \cdot & & & & & & 1 & 0 \\ \cdot & & & & & & 0 & 1 \\ \cdot & & & & & & \sigma & \tau \\ 0 & \cdot & \cdot & \cdot & \cdot & \cdot & -\tau & \sigma \end{bmatrix}, \quad \sigma,\tau \in \mathbb{R} , \quad \tau \neq 0 . \tag{5.2}$$

Note that the minimal polynomial of the Jordan block in (5.1) is $(\lambda-\lambda_0)^p$ where p is the size of the block, and for the matrix in (5.2) the minimal polynomial is $\{(\lambda-\sigma)^2+\tau^2\}^{p/2}$ where p is again the size of the matrix.

As in Chapter 3 we construct a matrix $P_{\varepsilon,J}$ with the same block-diagonal structure as J. Assume that

$$J = J_1 \oplus \cdots \oplus J_t \oplus J_{t+1} \oplus \cdots \oplus J_{t+s} \tag{5.3}$$

where $J_1,\cdots,J_t$ are of type (5.1) and $J_{t+1},\cdots,J_{t+s}$ are of type (5.2). Then define

$$P_{\varepsilon,J} = \varepsilon_1 P_1 \oplus \cdots \oplus \varepsilon_t P_t \oplus P_{t+1} \oplus \cdots \oplus P_{t+s} \tag{5.4}$$

where P_j is the sip matrix with size equal to that of J_j for $j = 1,2,\cdots,t+s$ and $\varepsilon = \{\varepsilon_1,\cdots,\varepsilon_t\}$ and $\varepsilon_j = \pm 1$ for each j. It is easily verified that J is $P_{\varepsilon,J}$ selfadjoint.

This observation together with the real Jordan form has an immediate corollary. Let $A = S^{-1}JS$ where S is real and let ε be *any* set of signs. On forming the matrix $H = S^* P_{\varepsilon,J} S$, which is also real, it is found that A is real H-selfadjoint.

COROLLARY 5.2 *Every real matrix* A *is* H-*selfadjoint for some invertible, real, symmetric matrix* H, *i.e. there is such an* H *for which* $A = H^{-1}A^T H$.

Note that Corollary 3.5 asserts that A is similar to its transpose A^T with a hermitian transforming matrix H. This result says that there is a real symmetric transforming matrix H.

Let (A_1,H_1) and (A_2,H_2) be pairs of real $n \times n$ matrices, where H_1, H_2 are hermitian and invertible. We say that (A_1,H_1) and (A_2,H_2) are *real unitarily similar* (or *r-unitarily similar* for short) if $A_1 = S^{-1}A_2 S$, $H_1 = S^* H_2 S$ for some real invertible matrix S.

The following theorem is a real version of Theorem 3.3.

THEOREM 5.3 *A pair* (A,H) *of real matrices, where* $H = H^*$ *is real and invertible and* A *is* H-*selfadjoint, is r-unitarily similar to a pair* $(J,P_{\varepsilon,J})$, *where* J *is the real Jordan form of* A *given by* (5.3), *and* $P_{\varepsilon,J}$ *is given by* (5.4). *The signs* ε_i *are determined uniquely by* (A,H) *up to permutation of signs in the blocks of* $P_{\varepsilon,J}$ *corresponding to the Jordan blocks of* J *with the same real eigenvalue and the same size.*

As in the case of complex matrices, we call the ordered set of signs $\varepsilon = (\varepsilon_1,\cdots,\varepsilon_t)$ the *sign characteristic* of (A,H).

5.2 Proof of Theorem 5.3

Let $\det(\lambda I-A) = p_1(\lambda)^{\alpha_1} \cdots p_m(\lambda)^{\alpha_m}$, where $p_i(\lambda)$ are real irreducible polynomials over $\mathbb{R}$. Write

$$\mathbb{R}^n = X_{r1} \dot{+} \cdots \dot{+} X_{rm} ,$$

where X_{ri} is a real A-invariant subspace of $\mathbb{R}^n$, and the minimal polynomial of $A|_{X_{rj}}$ is a power of $p_j(\lambda)$, $j = 1,\cdots,m$. Using Theorem 5.1 we can assume (as in the proof of Theorem 3.3) that $m = 1$, i.e. $\det(\lambda I-A) = p(\lambda)^\alpha$ for some irreducible real polynomial $p(\lambda)$. Two cases can occur:

(1) $p(\lambda) = \lambda - \lambda_0$, $\lambda_0 \in \mathbb{R}$. In this case we can repeat the argument from the proof of Theorem 3.3 word for word.

(2) $p(\lambda) = (\lambda-\sigma)^2 + \tau^2$, $\tau \neq 0$. In this case n is even. Consider A as a linear transformation acting in $\mathbb{C}^n$; then $\sigma(A) = \{\sigma+i\tau,\sigma-i\tau\}$. Let $\lambda_0 = \sigma + i\tau$ and

$$X' = \{x \in \mathbb{C}^n \mid (A-\lambda_0 I)^n x = 0\} ;$$

$$X'' = \{x \in \mathbb{C}^n \mid (A-\bar{\lambda}_0 I)^n x = 0\} ;$$

be complex subspaces of $\mathbb{C}^n$. Let m be the least positive integer such that $(A-\lambda_0 I)^{m-1}|_{X'} \neq 0$, and note that m is also the least positive integer such that $(A-\bar{\lambda}_0 I)^{m-1}|_{X''} \neq 0$.

It will be convenient to introduce the (nonlinear) map $K : \mathbb{C}^n \to \mathbb{C}^n$ as follows: $K \langle x_1,x_2,\cdots,x_n\rangle = \langle \bar{x}_1,\bar{x}_2,\cdots,\bar{x}_n\rangle$. Since A and H are real, we have $AK = KA$, $HK = KH$.

Note that $KX' \subset X''$. Indeed, if $y \in KX'$ then $y = Kx$, $x \in X'$ and, since $AK = KA$,

$$(A-\bar{\lambda}_0 I)^m y = (A-\bar{\lambda}_0 I)^m Kx = K(A-\lambda_0 I)^m x = 0 ,$$

so that $y \in X''$. In fact, since we also have $KX'' \subset X'$, it follows immediately that $KX' = X''$.

It is now to be shown that

$$[(A-\lambda_0 I)^{m-1}x,Kx] \neq 0$$

for some $x \in X'$ and, of course $[x,y] = (Hx,y)$. Assuming the contrary, we have for every $x,y \in X'$,

$$0 = [(A-\lambda_0 I)^{m-1}(x+y)\ ,\ K(x+y)]$$
$$= [(A-\lambda_0 I)^{m-1}x, Ky] + [(A-\lambda_0 I)^{m-1}y, Kx]\ .$$

But using the fact that K commutes with A and H a direct calculation shows that

$$[(A-\lambda_0 I)^{m-1}x, Ky] = [(A-\lambda_0 I)^{m-1}y, Kx]\ .$$

Consequently, $[(A-\lambda_0 I)^{m-1}y, Kx] = 0$. In other words, $[(A-\lambda_0 I)^{m-1}y, z] = 0$ for every $y \in X'$ and $z \in X''$. Taking $y \in X'$ such that $(A-\lambda_0 I)^{m-1}y \neq 0$, we observe that $(A-\lambda_0 I)^{m-1}y$ is perpendicular (with respect to H) to $X' + X''$. But this contradicts Proposition 1.1 since (see the proof of Theorem 3.3) $X' + X''$ is non-degenerate.

Thus, there exists an $a_1 \in X'$ such that

$$\theta \overset{\text{def}}{=} [(A-\lambda_0 I)^{m-1}a_1, Ka_1] \neq 0\ .$$

Replacing a_1 by αa_1 ($\alpha \in \mathbb{C}$) will replace θ by $\alpha^2\theta$. So we can (and will) assume that

$$[(A-\lambda_0 I)^{m-1}a_1, Ka_1] = 2i\ .$$

Let $b_1 = Ka_1$ and for $j = 1,2,\cdots,m$ define $a_j = (A-\lambda_0 I)^{j-1}a_1$, $b_j = (A-\bar{\lambda}_0 I)^{j-1}b_1$. As in the proof of Theorem 3.3 it will be shown that $a_m,\cdots,a_1$ and $b_m,\cdots,b_1$ are Jordan chains of A corresponding to λ_0 and $\bar{\lambda}_0$, respectively, and

$$[a_j, b_k] = \begin{cases} 2i & \text{if} \quad j + k = m + 1 \\ 0 & \text{if} \quad j + k > m + 1\ . \end{cases} \tag{5.5}$$

For $j = 1,2,\cdots,m$ let $g_j = \frac{1}{2}(a_j+b_j)$ and $b_j = \frac{1}{2i}(a_j-b_j)$. Then all the vectors g_j and h_j are real (i.e. $g_j = Kg_j$ and $h_j = Kh_j$) and linearly independent. The real subspace $\mathcal{X}_r$ spanned by $g_1,h_1,\cdots,g_m,h_m$ is A-invariant and A has as matrix representation in the basis $g_m,h_m,\cdots,g_1,h_1$:

$$J = \begin{bmatrix} \sigma & \tau & 1 & 0 & \cdot & \cdot & \cdot & 0 \\ -\tau & \sigma & 0 & 1 & & & & \cdot \\ \cdot & & & & \cdot & & & \cdot \\ \cdot & & & & & \cdot & & \cdot \\ \cdot & & & & & & \cdot\ 1 & 0 \\ \cdot & & & & & & 0 & 1 \\ & & & & & & \sigma & \tau \\ 0 & & & \cdot & \cdot & \cdot & -\tau & \sigma \end{bmatrix} \tag{5.6}$$

On the other hand, the relations (5.5) together with $[a_i,a_k] = [b_i,b_k] = 0$ for $i,k = 1,2,\cdots,m$ (as in Theorem 2.5) imply

$$[g_j,b_k] = \begin{cases} 0 & \text{if} \quad j + k > m + 1 \\ 1 & \text{if} \quad j + k = m + 1 \, . \end{cases} \tag{5.7}$$

Note also that $[g_j,g_k] = \frac{1}{2}[a_j,b_k] + \frac{1}{2}[b_j,a_k] = 0$ for $j + k \geq m + 1$, and similarly $[h_j,h_k] = 0$ for $j + k \geq m + 1$. Further,

$$[g_j,h_k] = [h_j,g_k] \quad \text{for} \quad j,k = 1,\cdots,m \, ;$$

$$[g_j,g_k] = -[h_j,h_k] \quad \text{for} \quad j,k = 1,\cdots,m \, ;$$

and for $j_1 + k_1 = j_2 + k_2$, $1 \leq j_1,k_1,j_2,k_2 \leq m$ we have:

$$[g_{j_1},g_{k_1}] = [g_{j_2},g_{k_2}] \, ; \qquad [g_{j_1},h_{k_1}] = [g_{j_2},h_{k_2}] \, ;$$

$$[h_{j_1},g_{k_1}] = [h_{j_2},g_{k_2}] \, ; \qquad [h_{j_1},h_{k_1}] = [h_{j_2},h_{k_2}] \, .$$

Then the hermitian matrix $H_0 \overset{\text{def}}{=} [f_i,f_j]_{i,j=1}^{2m}$, where $f_1 = g_m$, $f_2 = h_m,\cdots,f_{2m-1} = g_1$, $f_{2m} = h_1$, has the following structure:

$$H_0 = \begin{bmatrix} 0 & & & & P_1 \\ & & & \cdot & P_2 \\ & & \cdot & & \vdots \\ & P_1 & & & \\ P_1 & P_2 & \cdot & \cdot & \cdot & P_m \end{bmatrix},$$

where $P_1 = \begin{bmatrix} 0 & 1 \\ 1 & 0 \end{bmatrix}$, $P_j = \begin{bmatrix} x_j & y_j \\ y_j & -x_j \end{bmatrix}$, $j = 2,\cdots,m$, for some real numbers x_j and y_j. For brevity, denote by U (resp. V) the class of all 2×2 real matrices of the form $\begin{bmatrix} x & y \\ y & -x \end{bmatrix}$ (resp. $\begin{bmatrix} x & y \\ -y & x \end{bmatrix}$); thus $P_j \in U$, $j = 1,\cdots,m$. We claim that there exist matrices $Z_2,\cdots,Z_m \in V$ such that

$$\begin{bmatrix} I & & & \\ Z_2^* & I & & 0 \\ \cdot & \cdot & & \\ \cdot & & \cdot & \\ \cdot & & & \cdot \\ Z_m^* & \cdots & Z_2^* & I \end{bmatrix} H_0 \begin{bmatrix} I & Z_2 & \cdots & Z_m \\ & I & \cdots & Z_{m-1} \\ & & \cdot & \\ & & \cdot & \\ & & \cdot & \\ & 0 & & Z_2 \\ & & & I \end{bmatrix} = \begin{bmatrix} & & & P_1 \\ & 0 & \cdot & \\ & & \cdot & \\ & \cdot & & \\ & P_1 & & 0 \\ P_1 & & & \end{bmatrix} \tag{5.8}$$

Indeed, (5.8) is equivalent to

$$\sum_{\substack{j+k+\ell=q \\ 1\leq j,k,\ell\leq m}} Z_j^* P_k Z_\ell = 0 \ , \quad q = 4,5,\cdots,m+2 \ , \tag{5.9}$$

where we write $Z_1 = I$. Rewrite (5.9) in the form

$$P_1 Z_{q-2} + Z_{q-2}^* P_1 = Q_q \ , \quad q = 4,5,\cdots,m+2 \tag{5.10}$$

where $Q_q = -\Sigma\, Z_j^* P_k Z_\ell$, and the sum is taken over all triples (j,k,ℓ) such that $j + k + \ell = q$; $1 \leq j,k,\ell \leq m$; $j < q - 2$; $\ell < q - 2$. Equations (5.10) can be solved for $Z_2,\cdots,Z_m$ successively; indeed, (5.10) with $q = 4$ is

$$P_1 Z_2 + Z_2^* P_1 = -P_2 \ ,$$

and one can take $Z_2 = \frac{1}{2}\begin{bmatrix} -y_2 & x_2 \\ -x_2 & -y_2 \end{bmatrix} \in V$ to be a solution. Suppose (5.10) to be already solved for $q \leq q_0$ to obtain solutions $Z_2,\cdots,Z_{q_0-2} \in V$. Then, as one checks easily, $Q_{q_0+1} \in U$, and take $Z_{q_0-1} = -\frac{1}{2}\begin{bmatrix} 0 & 1 \\ 1 & 0 \end{bmatrix} Q_{q_0+1} \in V$ to solve (5.10) with $q = q_0 + 1$.

Having constructed $Z_2,\cdots,Z_m \in V$ with the property (5.8), observe that

$$S = \begin{bmatrix} I & Z_2 & \cdot & \cdot & \cdot & Z_m \\ & I & \cdot & \cdot & \cdot & Z_{m-1} \\ & & \cdot & & & \vdots \\ & & & \cdot & & \vdots \\ & & & & \cdot & Z_2 \\ & 0 & & & & I \end{bmatrix}$$

commutes with J_0. Let s_{ij} be the (i,j)-th entry of S^{-1}, and denote $g_i = \sum_{j=1}^{2m} s_{ji} f_j$, $i = 1,\cdots,2m$. Then the real subspace L spanned by $g_1,\cdots,g_{2m}$ is A-invariant, and in the basis $g_1,\cdots,g_{2m}$, the matrix representing A is just J_0, while $[g_i,g_j] = 0$ if $i + j \neq 2m + 1$; $[g_i,g_j] = 1$

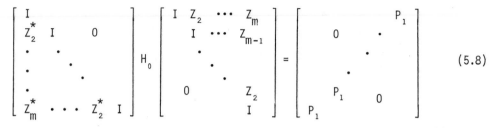

if $i + j = 2m + 1$.

Apply this construction to the pair of real linear transformations $(A|_{L_1}, P_{L_1}H|_{L_1})$, where P_{L_1} is the orthogonal projector on $L_1 = \{x \in \mathbb{R}^n \mid [x,y] = 0 \text{ for all } y \in L\}$, and so on.

Finally, the uniqueness of the sign characteristic is verified as in the proof of Theorem 3.3. □

5.3 Comparison with Results in the Complex Case

As in the case of complex matrices, the basic Theorem 5.3 allows us to consider various problems concerning real H-selfadjoint operators. In many cases the results and their proofs are the same for the real and for the complex cases. For instance, all results and their proofs of Sections 3.5, 3.6, 3.10 are also valid in the real case. Another description of the sign characteristic of an H-selfadjoint operator A, given in Theorem 3.16, is valid also for the real case. Theorem 3.6 on classification of pairs (A,H), where A is an H-selfadjoint operator, up to unitary similarity, holds also in the real case (in this case the classification is up to real unitary similarity, of course). One can obtain the real version of Theorem 3.6 from the complex one, using the fact that for a pair of real matrices (A,H), where A is H-selfadjoint, the sign characteristics of (A,H) as a pair of real matrices and as a pair of complex matrices are the same.

Theorem 3.18 on the simultaneous reduction of pairs of hermitian matrices has an analogue for pairs of real symmetric matrices. Some reformulation is required but the short proof is an exact parallel of that used for Theorem 3.18.

THEOREM 5.4 *Let* G_1 *and* G_2 *be real symmetric* $n \times n$ *matrices with* G_2 *invertible. Then there is an invertible* $n \times n$ *real matrix* X *such that* X^*G_iX, $i = 1,2$ *have the following forms:*

$$X^*G_1X = \varepsilon_1K_1 \oplus \cdots \oplus \varepsilon_\alpha K_\alpha \oplus K_{\alpha+1} \oplus \cdots \oplus K_\beta , \tag{5.11}$$

where, for $q = 1,2,\cdots,\alpha$, λ_α *is real and*

$$K_q = \begin{bmatrix} & & & & \lambda_q \\ & & & \cdot & 1 \\ & & \cdot & \cdot & \\ & \lambda_q & \cdot & & \\ \lambda_q & 1 & & & \end{bmatrix};$$

for $q = \alpha+1,\cdots,\beta$, σ_q *and* τ_q *are real with* $\tau_q \neq 0$ *and*

$$K_q = \begin{bmatrix} 0 & 0 & 0 & 0 & \cdots & -\tau_q & \sigma_q \\ 0 & 0 & 0 & 0 & \cdots & \sigma_q & \tau_q \\ \cdot & & & & & 0 & 1 \\ \cdot & & & & . & 1 & 0 \\ 0 & 0 & -\tau_q & \sigma_q & \vdots & & \cdot \\ 0 & 0 & \sigma_q & \tau_q & & & \cdot \\ -\tau_q & \sigma_q & 0 & 1 & & 0 & 0 \\ \sigma_q & \tau_q & 1 & 0 & \cdots & 0 & 0 \end{bmatrix};$$

and $\varepsilon_1,\cdots,\varepsilon_\alpha$ *are* ± 1;

$$X^*G_2X = \varepsilon_1 P_1 \oplus \cdots \oplus \varepsilon_\alpha P_\alpha \oplus P_{\alpha+1} \oplus \cdots \oplus P_\beta\ , \tag{5.12}$$

where, for $q = 1,2,\cdots,\beta$ P_q *is the sip matrix with size equal to that of* K_q. *The representations* (5.11) *and* (5.12) *are uniquely determined by* G_1 *and* G_2 *up to simultaneous permutation of equal blocks in* (5.11) *and* (5.12).

Concerning the results of Section 3.12 note that a real H-selfadjoint matrix does not always have a non-trivial invariant subspace, much less an invariant subspace which is also maximal H-nonnegative, or H-positive. For example, $\begin{bmatrix} 0 & 1 \\ -1 & 0 \end{bmatrix}$ has no non-trivial subspace and is selfadjoint with respect to $\begin{bmatrix} 0 & 1 \\ 1 & 0 \end{bmatrix}$. The situation is clearer if all the eigenvalues of the real matrix A are real. If such a matrix is real H-selfadjoint, then all the conclusions of Section 3.12 can be applied with the understanding that the only candidate for a c-set is the empty set.

5.4 Connected Components of Real Unitary Similarity Classes

In Section 3.1 the set of pairs of complex $n \times n$ matrices (A,H) for which A is H-selfadjoint was introduced in order to study the equivalence classes under unitary similarity. To handle the corresponding problem in the real case consider S_r, the set of all pairs of *real* $n \times n$ matrices (A,H) for which A is H-selfadjoint. Then pairs (A_1,H_1), $(A_2,H_2) \in S_r$ are said to be unitarily similar (or r-unitarily similar) if there is a *real* invertible T such that $A_1 = T^{-1}A_2T$ and $H_1 = T^*H_2T$. Also, (A_1,H), $(A_2,H) \in S_r$ are H-unitarily similar (H is now fixed) if $A_1 = U^{-1}A_2U$ for some real H-unitary matrix U, i.e. for which $U^*HU = H$.

In contrast to the conclusions for the complex case described in Theorems 3.1 and 3.9, it will be shown in this section that the real unitary similarity classes and the real H-unitary similarity classes are not generally connected. The basic reason for this is the fact that the group of invertible *real* $n \times n$ matrices is not connected (see Lemma 5.6).

THEOREM 5.5 *The* r-*unitary similarity class of* S_r *containing a pair* (A,H) *is connected if any real Jordan form* J *of* A *has a Jordan block of odd size with real eigenvalue and, otherwise, the class consists of exactly two connected components. In the latter case, the two connected components consist of those* $(B,G) \in S_r$ *for which the relations* $B = T^{-1}AT$, $G = T^*HT$ *hold with real matrices* T *having positive determinant in one case, and negative determinant in the other.*

The proof of Theorem 5.5 is based on the following well-known fact which it will be convenient to present with full proof.

LEMMA 5.6 *The set* $GL_r(n)$ *of all real invertible* $n \times n$ *matrices has two connected components; one contains the matrices with determinant* $+1$, *the other contains those with determinant* -1.

PROOF. Let T be a real matrix with $\det T > 0$ and let J be a real Jordan form for T. It will first be shown that J can be connected in $GL_r(n)$ to a diagonal matrix with diagonal entries ± 1. Indeed, J may have blocks J_p of two types: first as in (5.1) with non-zero eigenvalue λ_p in which case we define

$$J_p(t) = \begin{bmatrix} \lambda_p(t) & 1-t & \cdot & \cdot & \cdot & 0 \\ 0 & \lambda_p(t) & \cdot & & & \cdot \\ \cdot & & \cdot & \cdot & & \cdot \\ \cdot & & & \cdot & \cdot & \cdot \\ \cdot & & & & \cdot & 1-t \\ 0 & \cdot & \cdot & \cdot & 0 & \lambda_p(t) \end{bmatrix} \tag{5.13}$$

for any $t \in [0,1]$, where $\lambda_p(t)$ is a continuous path of non-zero real numbers such that $\lambda_p(0) = \lambda_p$, and $\lambda_p(1) = 1$ or -1 according as $\lambda_p > 0$ or $\lambda_p < 0$.

Second, a Jordan block J_p may have the form (5.2) when $J_p(t)$ is defined to have the same zero blocks as J_p, while the diagonal and super-diagonal blocks are replaced by

$$\begin{bmatrix} (1-t)\sigma+t & (1-t)\tau \\ -(1-t)\tau & (1-t)\sigma+t \end{bmatrix}, \quad \begin{bmatrix} 1-t & 0 \\ 0 & 1-t \end{bmatrix}, \tag{5.14}$$

respectively, for $t \in [0,1]$. Then $J_p(t)$ determines a continuous path of real invertible matrices such that $J_p(0) = J_p$ and $J_p(1)$ is an identity matrix.

Applying the above procedures to every diagonal block in J, J is connected to J_1 by a path in $GL_r(n)$. Now observe that the path in $GL_r(2)$ defined for $t \in [0,2]$ by

$$\begin{bmatrix} -(1-t) & t \\ -t & -(1-t) \end{bmatrix} \text{ when } t \in [0,1], \quad \begin{bmatrix} t-1 & 2-t \\ -(2-t) & t-1 \end{bmatrix} \text{ when } t \in [1,2],$$

connects $\begin{bmatrix} -1 & 0 \\ 0 & -1 \end{bmatrix}$ to $\begin{bmatrix} 1 & 0 \\ 0 & 1 \end{bmatrix}$. Consequently J_1, and hence J, is connected in $GL_r(n)$ with either I or $\operatorname{diag}[-1,1,1,\cdots,1]$. But $\det T > 0$ implies $\det J > 0$ and so the latter case is excluded. Since $T = S^{-1}JS$ for some invertible real S, we can hold S fixed and observe that the path in $GL_r(n)$ connecting J and I will also connect T and I.

Now assume $T \in GL_r(n)$ and $\det T < 0$. Then $\det T' > 0$, where $T' = T\operatorname{diag}[-1,1,\cdots,1]$. Using the argument above, T' is connected with I in $GL_r(n)$. Hence T' is connected with $\operatorname{diag}[-1,1,\cdots,1]$ in $GL_r(n)$. □

Proof of Theorem 5.5 Without loss of generality we can assume that $(A,H) = (J,P_{\varepsilon,J})$ is in the (real) canonical form, so that J is a real Jordan form of A. Denote by US_+ (resp. US_-) the set of all pairs (B,G) such that $B = T^{-1}JT$, and $G = T^*P_{\varepsilon,J}T$ for some real matrix T with $\det T > 0$ (resp. $\det T < 0$). Clearly, the r-unitary similarity class containing $(J,P_{\varepsilon,J})$ is a union of US_+ and US_-. By Lemma 5.6 each set US_+ and US_- is connected. Moreover, $US_+ = US_-$ if and only if there is a $(B,G) \in S_r$ which can be transformed to $(J,P_{\varepsilon,J})$ by both T_+ and T_-, say, with $\det T_+ > 0$ and $\det T_- < 0$. Thus,

$$B = T_+^{-1}JT_+ = T_-^{-1}JT_-, \quad G = T_+^*P_{\varepsilon,J}T_+ = T_-^*P_{\varepsilon,J}T_-,$$

and it follows immediately that $US_+ = US_-$ if and only if there is a real T with negative determinant such that

$$J = T^{-1}JT, \quad P_{\varepsilon,J} = T^*P_{\varepsilon,J}T. \tag{5.15}$$

Thus, it remains to prove that there exists a real T with

$\det T < 0$ such that (5.15) hold if and only if J has a Jordan block of odd size with a real eigenvalue.

Assume J has such a block, J_0. We can write

$$J = J_1 \oplus J_0 , \quad P_{\varepsilon,J} = P_1 \oplus P_0 , \tag{5.16}$$

in an obvious notation, where J_1 is the "rest" of J, and $P_{\varepsilon,J}$ is partitioned accordingly. Put $T = I \oplus (-I)$ with partition consistent with that in (5.16). Evidently, (5.15) holds and, since J_0 has odd size, $\det T < 0$.

Conversely, assume that J does not have a Jordan block of odd size with real eigenvalue. It will be proved that $\det T > 0$ for every real invertible T satisfying $JT = TJ$, showing thereby that (5.15) never holds for a T with negative determinant.

It is sufficient to consider two cases separately:

(1) $\det(\lambda I-J) = (\lambda-\lambda_0)^{\alpha}$, $\lambda_0 \in \mathbb{R}$;

(2) $\det(\lambda I-J) = [(\lambda-\sigma)^2 + \tau^2]^{\alpha}$, $\sigma,\tau \in \mathbb{R}$, $\tau \neq 0$.

Consider case (1); then J is a Jordan matrix with eigenvalue λ_0. Let $m_1 = \cdots = m_{k_1} > m_{k_1+1} = \cdots = m_{k_2} > m_{k_2+1} = \cdots = m_{k_3} > \cdots > m_{k_{r-1}+1} = \cdots = m_{k_r}$ be the sizes of the Jordan blocks of J and, by assumption, all m_i are even. Let T be a real invertible matrix commuting with J. Then T has the following form (see [14]):

$$T = (T_{ij})_{i,j=1}^{k_r} ,$$

where each T_{ij} is defined by an upper triangular Toeplitz matrix in the following way. Let

$$T'_{ij} = \begin{bmatrix} t_{ij1} & t_{ij2} & \cdots & t_{ij\gamma} \\ 0 & t_{ij1} & \ddots & \vdots \\ \vdots & & \ddots & \vdots \\ \vdots & & \ddots & \vdots \\ \vdots & & t_{ij1} & t_{ij2} \\ 0 & \cdots & 0 & t_{ij1} \end{bmatrix} \tag{5.17}$$

be a real $\gamma \times \gamma$ Toeplitz matrix, then

$$(1) \quad \text{if } m_i < m_j , \quad T_{ij} = [0 \;\; T'_{ij}] \tag{5.18}$$

$$(2)\quad \text{if}\quad m_i > m_j\ ,\qquad T_{ij} = \begin{bmatrix} T'_{ij} \\ 0 \end{bmatrix} \tag{5.19}$$

$$(3)\quad \text{if}\quad m_i = m_j\ ,\qquad T_{ij} = T'_{ij}\ , \tag{5.20}$$

and in each case, $\gamma = \min(m_i, m_j)$.

An easy determinantal computation shows that

$$\det T = \left(\det[t_{ij1}]_{i,j=1}^{k_1}\right)^{m_{k_1}} \left(\det[t_{ij1}]_{i,j=k_1+1}^{k_2}\right)^{m_{k_2}} \cdots \left(\det[t_{ij1}]_{i,j=k_{r-1}+1}^{k_r}\right)^{m_{k_r}} \tag{5.21}$$

which is positive because the m_i are all even and T is invertible.

For the second case write $J = \text{diag}[J_1 J_2 \cdots J_s]$ where $J_1 = J_2 = \cdots = J_s$ and each matrix is of the form (5.2) with size $2m \times 2m$ for $j = 1,2,\cdots,s$. Together with J consider the (complex) Jordan matrix

$$\hat{J} = \text{diag}[\hat{J}_1, \hat{J}_2, \cdots, \hat{J}_{2s-1}, \hat{J}_{2s}]\ ,$$

where $\hat{J}_{2j-1}$ (resp. $\hat{J}_{2j}$) is the Jordan block of size m_j with eigenvalue $\sigma - i\tau$ (resp. $\sigma + i\tau$). Then, as one checks easily,

$$J = S\hat{J}S^{-1}\ ,$$

where the invertible complex matrix S has block diagonal form: $S = \text{diag}[S_1, \cdots, S_s]$, where

$$S_j = \begin{bmatrix} 1 & 0 & \cdot & \cdot & \cdot & 1 & 0 & \cdot & \cdot & \cdot & 0 \\ -i & 0 & & & & i & 0 & & & & 0 \\ 0 & 1 & \cdot & \cdot & \cdot & 0 & 1 & \cdot & \cdot & \cdot & 0 \\ 0 & -i & & & & 0 & i & & & & 0 \\ \cdot & & & & & & & & & & \cdot \\ \cdot & & & & & & & & & & \cdot \\ \cdot & & & & & & & & & & \cdot \\ 0 & 0 & & & 1 & 0 & 0 & & & & 1 \\ 0 & 0 & \cdot & \cdot & -i & 0 & 0 & \cdot & \cdot & \cdot & i \end{bmatrix}$$

is a $2m_j \times 2m_j$ matrix. (For the reader's convenience note that $S^{-1} = \text{diag}[S'_1, \cdots, S'_s]$ where

$$S_j' = \frac{1}{2i}\begin{bmatrix} i & -1 & 0 & 0 & \cdot & \cdot & \cdot & 0 & 0 \\ 0 & 0 & i & -1 & & & & 0 & 0 \\ \cdot & & & \cdot & \cdot & \cdot & \cdot & \cdot & \cdot \\ 0 & 0 & 0 & 0 & & & & i & -1 \\ i & 1 & 0 & 0 & \cdot & \cdot & \cdot & 0 & 0 \\ 0 & 0 & i & 1 & & & & 0 & 0 \\ \cdot & & & \cdot & \cdot & \cdot & \cdot & \cdot & \cdot \\ 0 & 0 & 0 & 0 & \cdot & \cdot & \cdot & i & 1 \end{bmatrix}) .$$

It is well-known (see [14]) that any complex matrix $\hat{T}$ commuting with $\hat{J}$ has the form

$$\hat{T} = \begin{bmatrix} \hat{T}_{11} & 0 & \hat{T}_{13} & 0 & \cdot\ \cdot\ \cdot & \hat{T}_{1,2S-1} & 0 \\ 0 & \hat{T}_{22} & 0 & \hat{T}_{24} & & 0 & \hat{T}_{2,2S} \\ \cdot & & & & & & \cdot \\ \cdot & & & & & & \cdot \\ \cdot & & & & & & \cdot \\ \hat{T}_{2S-1,1} & 0 & \hat{T}_{2S-1,3} & 0 & & \hat{T}_{2S-1,2S-1} & 0 \\ 0 & \hat{T}_{2S,2} & 0 & \hat{T}_{2S,4} & \cdot\ \cdot & 0 & \hat{T}_{2S,2S} \end{bmatrix} \tag{5.22}$$

where $\hat{T}_{uv}$ is an $m_{[\frac{u+1}{2}]} \times m_{[\frac{v+1}{2}]}$ complex matrix (here u and v are either both even or both odd), which is upper triangular and Toeplitz (a complex analogue of the description in formulas (5.17)-(5.20)). Denote by $t_{uv} \in \mathbb{C}$ the entry on the main diagonal of $\hat{T}_{uv}$.

Now every matrix commuting with J has the form $T = S\hat{T}S^{-1}$, where $\hat{T}$ commutes with $\hat{J}$. Write

$$T = \begin{bmatrix} T_{11} & T_{12} & \cdot\ \cdot\ \cdot & T_{1S} \\ \cdot & & & \cdot \\ \cdot & & & \cdot \\ \cdot & & & \cdot \\ T_{S1} & T_{S2} & \cdot\ \cdot\ \cdot & T_{SS} \end{bmatrix},$$

where T_{jk} is a $2m_j \times 2m_k$ matrix. A computation shows that given $\hat{T}$ as in (5.22), the first column of T_{jk} is

$$\begin{bmatrix} \frac{1}{2}(t_{2j-1,2k-1} + t_{2j,2k}) \\ \frac{1}{2i}(t_{2j-1,2k-1} - t_{2j,2k}) \\ * \\ \vdots \\ * \end{bmatrix}, \tag{5.23}$$

where $t_{uv} \in \mathbb{C}$ is the entry on the main diagonal of $\hat{T}_{uv}$.

Assume now that $TJ = JT$ and T is real. Then (5.23) implies $t_{2j-1,2k-1} = \bar{t}_{2j,2k}$, $1 \leq j,k \leq s$. Since $\det(\hat{T}_{2j-1,2k-1})_{j,k=1}^{s}$ is a polynomial in $t_{2j-1,2k-1}$ $(1 \leq j,k \leq s)$ with real coefficients (cf. (5.21)) and $\det(\hat{T}_{2j,2k})_{j,k=1}^{s}$ is obtained from $\det(\hat{T}_{2j-1,2k-1})_{j,k=1}^{s}$ on replacing $t_{2j-1,2k-1}$ by $t_{2j,2k}$, it follows that

$$\det(\hat{T}_{2j,2k})_{j,k=1}^{s} = \overline{\det(\hat{T}_{2j-1,2k-1})_{j,k=1}^{s}} \; .$$

Now it is clear that

$$\det T = \det \hat{T} = \det[\hat{T}_{2j,2k}]_{j,k=1}^{s} \det[\hat{T}_{2j-1,2k-1}]_{j,k=1}^{s} \geq 0 \; ,$$

so that $\det T > 0$ provided T is invertible. □

5.5 Connected Components of Real Unitary Similarity Classes (H fixed)

Consider now the real unitary similarity classes in the real $n \times n$ matrices obtained when the real hermitian invertible matrix H is kept fixed. The analogue for Theorem 3.10 turns out to be:

THEOREM 5.7 *Let* A *be a real* H*-selfadjoint* $n \times n$ *matrix. Then the real* H*-unitary similarity class* $US_H(A)$ *which contains* A *has either* 1, 2 *or* 4 *connected components.*

The proof of Theorem 5.7 will follow from results on the connected components of the group of real H-unitary matrices which are to be presented in Theorem 5.8 below. Observe that all 3 possibilities ($US_H(A)$ connected, or has 2 or 4 connected components) may occur. In many cases one can find the exact number of connected components of $US_H(A)$ in terms of the structure of the real Jordan form of A and the sign characteristic of A with respect to H.

We noticed (Lemma 3.8) that in the case of complex matrices, the set of H-unitary matrices is connected. In the real case the situation is completely different (as one can see in the scalar case $n = 1$, where the set of H-unitary matrices consists of two points: 1 and -1). The following result describes the connected components of the set $\mathbb{U}_r(H)$ of all real H-unitary matrices.

THEOREM 5.8 *Let* $H = H^*$ *be a real invertible matrix.*
(i) *If* H *is neither positive nor negative definite, then* $\mathbb{U}_r(H)$ *has exactly* 4 *(arcwise) connected components, whose representatives can be described as follows: Let* x_1 *(resp.* x_2*) be a real eigenvector of* H *corresponding to a positive (resp. negative) eigenvalue. For every choice of the signs* $\eta_1 = \pm 1$, $\eta_2 = \pm 1$ *consider the real* H*-unitary matrix* $A(\eta_1, \eta_2)$ *which maps* x_1 *to* $\eta_1 x_1$, x_2 *to* $\eta_2 x_2$ *and* x *to itself, for every* x *belonging to an orthogonal complement (in the standard sense) to* $\text{Span}\{x_1, x_2\}$. *The* 4 *matrices* $A(1,1)$, $A(-1,1)$, $A(1,-1)$, $A(-1,-1)$ *belong to the different connected components in* $\mathbb{U}_r(H)$.
(ii) *If* H *is either positive definite or negative definite, then* $\mathbb{U}_r(H)$ *has exactly* 2 *connected components, one consisting of matrices with determinant* 1, *and the second consisting of matrices with determinant* -1.

PROOF. We start with (ii). Considering the case of positive definite H, we assume without loss of generality that $H = I$ (indeed, write $H = S^*S$ for some real invertible $n \times n$ matrix S; then A is H-unitary if and only if SAS^{-1} is I-unitary). So the group $\mathbb{U}_r(H)$ becomes just the group Φ_n of (real) orthogonal $n \times n$ matrices.

Since the determinant is a continuous function of a matrix, clearly the subsets Φ_n^+ and Φ_n^- consisting of all $n \times n$ orthogonal matrices with determinant 1 and -1, respectively, are disconnected in Φ_n. We shall prove now that Φ_n^+ is connected. Pick $A \in \Phi_n^+$. There exists an orthogonal matrix S such that $K \overset{\text{def}}{=} S^{-1}AS$ is in the real Jordan canonical form (see [42], Section 6.6):

$$K = \text{diag}[K_1, K_2, \cdots, K_r] ,$$

where K_i is either the scalar 1, or a 2×2 matrix of the form

$$\begin{bmatrix} \cos\theta & \sin\theta \\ -\sin\theta & \cos\theta \end{bmatrix}, \quad 0 \leq \theta \leq 2\pi . \tag{5.24}$$

Since $\det K = 1$, the number of -1's is even. The 2×2 matrix $\begin{bmatrix} -1 & 0 \\ 0 & -1 \end{bmatrix}$ is of type (5.24) (with $\theta = \pi$), so we can assume that K_i is either 1 or has the form (5.24). There exists a continuous path in Φ_n^+ connecting K and I; indeed, it is sufficient to connect a block (5.24) with $\begin{bmatrix} 1 & 0 \\ 0 & 1 \end{bmatrix}$ by a continuous path. But this is easy: take

$$\begin{bmatrix} \cos t & \sin t \\ -\sin t & \cos t \end{bmatrix}, \quad \theta \leqslant t \leqslant 2\pi .$$

So there exists a continuous path $K(t)$, $t \in [0,1]$ in ϕ_n^+ such that $K(0) = K$; $K(1) = I$. Now the continuous path $S^{-1}K(t)S$ in ϕ_n^+ connects A and I. Hence ϕ_n^+ is arcwise connected. The arcwise connectedness of ϕ_n^- is proved similarly.

The proof of part (i) is more complicated. It can be obtained by an argument similar to that used in the proof of Theorem IV.3.1 on the structure of the group of complex H-unitary matrices (assuming without loss of generality that

$$H = \begin{bmatrix} I_k & 0 \\ 0 & -I_\ell \end{bmatrix}, \quad 0 < k < k+\ell = n) . \quad \square$$

Let $H = H^*$ be real, invertible and neither positive definite nor negative definite. Denote the 4 connected components of $\mathbb{U}_r(H)$ as follows, where $A(\eta_1,\eta_2)$ are taken from Theorem 5.8: $\mathbb{U}_r^{++}(H) \ni A(1,1)$; $\mathbb{U}_r^{-+}(H) \ni$ $\ni A(-1,1)$; $\mathbb{U}_r^{+-}(H) \ni A(1,-1)$; $\mathbb{U}_r^{--}(H) \ni A(-1,-1)$. Observe that $I \in \mathbb{U}_r^{++}(H)$. Also

$$\det S = 1 \quad \text{for} \quad S \in \mathbb{U}_r^{++}(H) \cup \mathbb{U}_r^{--}(H) ;$$

$$\det S = -1 \quad \text{for} \quad S \in \mathbb{U}_r^{+-}(H) \cup \mathbb{U}_r^{-+}(H) .$$

The multiplication between the different components of $\mathbb{U}_r(H)$ is given by the following table:

	+ +	+ -	- +	- -
+ +	+ +	+ -	- +	- -
+ -	+ -	+ +	- -	- +
- +	- +	- -	+ +	+ -
- -	- -	- +	+ -	+ +

. (5.25)

The table means, for instance, that $S_1S_2 \in \mathbb{U}_r^{-+}(H)$ for every $S_1 \in \mathbb{U}_r^{+-}(H)$ and $S_2 \in \mathbb{U}_r^{--}(H)$.

<u>Proof of Theorem 5.7</u>. Assume that H is neither positive definite

nor negative definite. We shall use the notation $\mathbb{U}_r^{\pm\pm}(H)$ introduced above. Four cases can occur depending on the relation of the set $\mathbb{U}_r(H;A)$ of all real H-unitary matrices which commute with A to the sets $\mathbb{U}_r^{\pm\pm}(H)$: 1) $\mathbb{U}_r(H;A)$ is contained in $\mathbb{U}_r^{++}(H)$, 2) $\mathbb{U}_r(H;A)$ is contained in $\mathbb{U}_r^{++}(H) \cup \mathbb{U}_r^{+-}(H)$ but not in $\mathbb{U}_r^{++}(H)$, 3) $\mathbb{U}_r(H;A)$ is contained in $\mathbb{U}_r^{++}(H) \cup \mathbb{U}_r^{-+}(H)$ but not in $\mathbb{U}_r^{++}(H)$, 4) none of the cases (1), (2),or (3) holds.

Using the multiplication table (5.25), one can easily see that in the case (1) the set $US_H(A)$ has exactly 4 connected components, in the cases (2) and (3) this set has exactly 2 connected components, and in the case (4) $US_H(A)$ is connected.

In the case when H is either positive definite or negative definite Theorem 5.7 is evident in view of Theorem 5.8(ii). □

CHAPTER 6

FUNCTIONS OF H-SELFADJOINT AND H-UNITARY MATRICES

In Section 2.3 and Chapter 4 we have made use of special functions of matrices (Moebius, or Cayley, transformations) to examine relationships between H-selfadjoint and H-unitary matrices. In this chapter the objective is to present a more systematic investigation of functions mapping H-selfadjoint matrices to H-unitary matrices, H-unitary to H-unitary, and so on. In particular, we are to investigate how the sign characteristic is transformed.

6.1 Preliminaries

For any square matrix A with spectrum $\sigma(A)$, a function $f(\lambda)$ is said to be *defined on* $\sigma(A)$ if, for each point $\lambda_j \in \sigma(A)$, $f(\lambda_j)$ and the derivatives $f^{(1)}(\lambda_j),\cdots,f^{(m_j-1)}(\lambda_j)$ exist, where m_j is the *index* of λ_j, i.e. in a Jordan normal form for A, m_j is the size of the largest Jordan block with eigenvalue λ_j. The set of numbers of $f^{(r)}(\lambda_j)$, $\lambda_j \in \sigma(A)$, $r = 0,1,\cdots,m_j-1$ is known as the *values of* f *on* $\sigma(A)$. If $f(\lambda)$ is a function of the *complex* variable λ and is defined on $\sigma(A)$ then, of course, for each point $\lambda_j \in \sigma(A)$, either $f(\lambda_j)$ is defined but has no derivatives (as at a branch point) and $m_j = 1$, or $f(\lambda)$ is analytic in a neighbourhood of λ_j.

For any function $f(\lambda)$ defined on $\sigma(A)$, the matrix $f(A)$ is defined to be equal to $g(A)$ where $g(\lambda)$ is any polynomial which assumes the same values as f on $\sigma(A)$ (see [14], for example). It is well-known that this procedure defines $f(A)$ uniquely and for *any* scalar polynomial $p(\lambda) = \sum_j p_j\lambda^j$, $p(A) = \sum_j p_jA^j$.

For a function $f(\lambda)$ defined on $\sigma(A)$ it is also easily seen that if $A = SJS^{-1}$ and $J = J_1 \oplus \cdots \oplus J_k$ is a Jordan form for A with blocks $J_1, J_2, \cdots, J_k$, then

$$f(A) = S\,\mathrm{diag}[f(J_1), \cdots, f(J_k)]S^{-1} . \tag{6.1}$$

Furthermore, for $j = 1, 2, \cdots, k$

$$f(J_j) = \begin{bmatrix} f(\lambda_j) & f^{(1)}(\lambda_j) & \frac{1}{2!}f^{(2)}(\lambda_j) & \cdots & \frac{1}{(m_j-1)!}f^{(m_j-1)}(\lambda_j) \\ 0 & f(\lambda_j) & f^{(1)}(\lambda_j) & \ddots & \vdots \\ \vdots & & f(\lambda_j) & \ddots & \vdots \\ \vdots & & & f(\lambda_j) & f^{(1)}(\lambda_j) \\ 0 & \cdots & \cdots & 0 & f(\lambda_j) \end{bmatrix} \tag{6.2}$$

where $\{\lambda_j\} = \sigma(J_j)$ and the block J_j has size m_j.

If $f(\lambda)$ is analytic in a neighbourhood of $\sigma(A)$ it is, of course, defined on $\sigma(A)$ and an integral representation of $f(A)$ is available. Thus,

$$f(A) = \frac{1}{2\pi i}\int_\Gamma f(\lambda)(\lambda I - A)^{-1} d\lambda \tag{6.3}$$

where Γ is a composite contour consisting of a circle with sufficiently small radius around each distinct eigenvalue of A.

It is clear that in this chapter knowledge of the Jordan form of $f(A)$ given that for A will be required. The first proposition together with (6.1) and (6.2) clarifies this point.

PROPOSITION 6.1 *Let* X *be the* $m \times m$ *Jordan block with eigenvalue* λ_0 *and let* $f(\lambda)$ *be a function with* $m - 1$ *derivatives at* λ_0 *(i.e.* $f(\lambda)$ *is defined on* $\sigma(X)$*). Let* $f^{(r)}(\lambda_0)$ $(1 \leq r \leq m-1)$ *be the first non-vanishing derivative of* $f(\lambda)$ *at* λ_0 *and if* $f^{(j)}(\lambda_0) = 0$ *for* $j = 1, \cdots, m-1$, *put* $r = m$. *Then the sizes of the Jordan blocks of* $f(X)$ *are given by*

$$[\tfrac{m}{r}] \quad \textit{repeated} \quad r[\tfrac{m}{r}] - m + r \quad \textit{times},$$

and

$$[\tfrac{m}{r}] + 1 \quad \textit{repeated} \quad (m - r[\tfrac{m}{r}]) \quad \textit{times},$$

and $[x]$ *denotes the greatest integer less than or equal to* x.

PROOF. Without loss of generality, it can be assumed that $\lambda_0 = f(\lambda_0) = 0$. Then it is easily seen using (6.2) that $\dim \mathrm{Ker}\, f(X) = r$

and, more generally, if $t_j = \dim \mathrm{Ker}(f(X))^j$, then

$$t_j = \min(m, jr) \ . \tag{6.4}$$

Now the sizes of Jordan blocks of $f(X)$ are uniquely determined by the sequence $t_1, t_2, \cdots, t_m$. Indeed, the number of Jordan blocks of $f(X)$ with size not less than j is just $t_j - t_{j-1}$, where $j = 1,2,\cdots,m$ and $t_0 = 0$. This observation, together with (6.4) leads to the conclusion of the proposition. □

6.2 Exponential and Logarithmic Functions

We have seen in Section 2.3 that H-unitary matrices can be obtained from the H-selfadjoint ones by means of the Cayley transform, and this fact was exploited further in Chapter 4. Another method for the transformation of H-selfadjoint to H-unitary matrices depends on properties of the exponential and logarithmic functions and will be presented in this section. But it is necessary to take some care in the definition of the logarithm of an H-unitary matrix.

First define a neighbourhood Ω of $\sigma(U)$, where U is H-unitary, with the following properties:

(a) Ω is symmetric with respect to the unit circle, i.e. $\lambda \in \Omega$ implies $\bar{\lambda}^{-1} \in \Omega$.

(b) If $\sigma(U) = \{\lambda_1, \cdots, \lambda_k\}$ then $\Omega = \bigcup_{r=1}^{k} \Omega_r$ where $\Omega_1, \cdots, \Omega_k$ are disjoint neighbourhoods of $\lambda_1, \cdots, \lambda_k$, respectively.

(c) $0 \notin \Omega$.

Then define a function $\ln \lambda$ on Ω by assigning $\ln \lambda = \ln_r \lambda$ whenever $\lambda \in \Omega_r$ and $\ln_r \lambda$ denotes a branch of the logarithmic function, *and* if $\lambda_j \bar{\lambda}_k = 1$ then $\ln_j \lambda = \ln_k \lambda$, i.e. the same branch of the logarithm is used for domains Ω_j, Ω_k containing eigenvalues which are symmetric with respect to the unit circle. With these conventions, the function $\ln \lambda$ is defined on $\sigma(U)$.

THEOREM 6.2 (a) *If* A *is* H-*selfadjoint then* e^{iA} *is* H-*unitary*. (b) *If* U *is* H-*unitary and the function* $\ln \lambda$ *is defined in a neighbourhood* Ω *of* $\sigma(U)$ *as above, then* $V = -i \ln U$ *satisfies* $e^{iV} = U$ *and* V *is* H-*selfadjoint*.

PROOF. (a) Obviously $e^{i\lambda}$ is defined on the spectrum of any matrix, so e^{iA} is well-defined and, as is well known,

$$e^{iA} = I + iA + \frac{1}{2!} i^2A^2 + \cdots \quad .$$

Consequently, since $HA^r = A^{*r}H$ for $r = 1,2,\cdots$,

$$\begin{aligned} He^{iA} &= (I + iA^* + \frac{1}{2!} i^2A^{*2} + \cdots)H \\ &= e^{iA^*}H = [(e^{iA})^*]^{-1}H , \end{aligned}$$

which implies that e^{iA} is H-unitary.

(b) Let J be a Jordan form for U and $U = SJS^{-1}$. Then

$$e^{\ln U} = Sg(J)S^{-1}$$

where $g(\lambda) = e^{\ln_r\lambda}$ when $\lambda \in \Omega_r$, and in any case $g(\lambda) = \lambda$. Hence $e^{\ln U} = U$.

We also show that

$$\ln(U^{*-1}) = -(\ln U)^*. \tag{6.5}$$

Using the Jordan form for U this is seen to be the case if and only if $\ln(J_r^{*-1}) = -(\ln J_r)^*$ for each Jordan block of J. Using (6.2) we obtain

$$(\ln J_r)^* = \begin{bmatrix} \overline{\ln_r\lambda} & 0 & 0 & \cdot & \cdot & \cdot \\ \bar{\lambda}^{-1} & \overline{\ln_r\lambda} & 0 & \cdot & \cdot & \cdot \\ -\bar{\lambda}^{-2} & \bar{\lambda}^{-1} & \overline{\ln_r\lambda} & & & \\ \cdot & \cdot & \cdot & & & \\ & \cdot & \cdot & \cdot & & \\ & & \cdot & \cdot & \cdot & \end{bmatrix},$$

and a little calculation shows that

$$\ln(J_r^{*-1}) = \begin{bmatrix} \ln_r\bar{\lambda}^{-1} & 0 & 0 & \cdot & \cdot & \cdot \\ -\bar{\lambda}^{-1} & \ln_r\bar{\lambda}^{-1} & 0 & \cdot & \cdot & \cdot \\ \bar{\lambda}^{-2} & -\bar{\lambda}^{-1} & \ln_r\bar{\lambda}^{-1} & & & \\ \cdot & \cdot & \cdot & & & \\ & \cdot & \cdot & \cdot & & \\ & & \cdot & \cdot & \cdot & \end{bmatrix}.$$

But

$$\ln_r\bar{\lambda}^{-1} = -\ln|\lambda| + i(\arg\lambda + 2\pi m_r) = -\overline{\ln_r\lambda}$$

since $\ln_r\lambda$ and $\ln_r\bar{\lambda}^{-1}$ are obtained using the same branch of the logarithmic function. Consequently, $\ln(J_r^{*-1}) = -(\ln J_r)^*$ and (6.5) holds.

Finally, defining $V = -i \ln U$ we obtain $e^{iV} = U$ and also,

$$HV = -iH \ln(U) = -iH \ln(H^{-1}U^{*-1}H) = -i \ln(U^{*-1})H ,$$

so (6.5) yields

$$HV = i(\ln U)^* = V^*H . \quad \square$$

6.3 Functions of H-Selfadjoint Matrices

The main emphasis of this section and the next is on functions of H-selfadjoint and of H-unitary matrices, respectively. We start with a general result.

THEOREM 6.3 *If matrix* A *is either H-selfadjoint or H-unitary and* $f(\lambda)$ *is defined on* $\sigma(A)$, *then* $f(A)$ *is H-normal.*

PROOF. First consider the case in which A is H-selfadjoint. Since the property of H-normality is preserved under unitary similarity (Proposition 3.2), it can be assumed that (A,H) is in the canonical form $(J,P_{\varepsilon,J})$ of Theorem 3.3. Then recall that to a Jordan block of J with real eigenvalue there corresponds a block of $f(J)$ of Toeplitz form

$$B = \begin{bmatrix} \alpha_1 & \alpha_2 & \cdot & \cdot & \cdot & \alpha_k \\ 0 & \alpha_1 & \cdot & & & \cdot \\ \cdot & & \cdot & \cdot & & \cdot \\ \cdot & & & \cdot & \cdot & \cdot \\ \cdot & & & & & \alpha_2 \\ 0 & \cdot & \cdot & \cdot & 0 & \alpha_1 \end{bmatrix} \tag{6.6}$$

as indicated in (6.2). Also for a pair of complex conjugate eigenvalues of J there is a pair of blocks $B_1 \oplus B_2$ of $f(J)$ where each of B_1, B_2 has the form (6.6).

The proof of the theorem now reduces to verification of the facts that B of (6.6) is εP-normal (where $\varepsilon = 1$ or $\varepsilon = -1$) and that $B_1 \oplus B_2$ is $\begin{bmatrix} 0 & P \\ P & 0 \end{bmatrix}$-normal, and P is always the sip matrix of appropriate size. In both cases this reduces to verifying that $BB^{[*]} = B^{[*]}B$ where B has the form (6.6). Since $B^{[*]} = P^{-1}B^*P$ and $P^{-1}B^*P = \bar{B}$ it is to be shown that $B\bar{B} = \bar{B}B$. This is easily checked.

The case when A is H-unitary is easily transformed to the H-selfadjoint case already established. Let $g(\lambda)$ be the inverse of a Cayley transform (see Proposition 2.8), so that $g(A)$ is H-selfadjoint and let $h(\lambda) = f(g^{-1}(\lambda))$. Then $h(\lambda)$ is defined on the spectrum of $g(A)$. By the first part of the proof it follows that the matrix

$$h(g(A)) = f(g^{-1}(g(A))) = f(A)$$

is H-normal. □

To characterize those functions $f(\lambda)$ for which $f(A)$ is H-selfadjoint when A is H-selfadjoint, we need the following definition. The function $f(\lambda)$ defined on $\sigma(A)$ is said to be *real symmetric on* $\sigma(A)$ if for each point $\lambda_j \in \sigma(A)$ we have

$$f^{(k)}(\lambda_j) = \overline{f^{(k)}(\bar{\lambda}_j)} \;, \quad k = 0,1,\cdots,m_j-1 \;,$$

and m_j is the index of λ_j. For example, a scalar polynomial is symmetric on the spectrum of any H-selfadjoint matrix A if it has real coefficients. Also, if $f(\lambda) = \overline{f(\bar{\lambda})}$ and is analytic throughout a neighbourhood of $\sigma(A)$, then $f(\lambda)$ is real symmetric on $\sigma(A)$.

THEOREM 6.4 *Let* A *be H-selfadjoint and let* $f(\lambda)$ *be defined on* $\sigma(A)$. *Then* $f(A)$ *is H-selfadjoint if and only if* $f(\lambda)$ *is real symmetric on* $\sigma(A)$.

PROOF. Let $f(\lambda)$ be real symmetric on $\sigma(A)$. Suppose that $\lambda_j \in \sigma(A)$ has index m_j and let $m = \sum_{\lambda_j \in \sigma(A)} m_j$. It is clear that there is a *real* polynomial $p(\lambda)$ of degree $m - 1$ satisfying the m interpolating conditions

$$p^{(k)}(\lambda_j) = f^{(k)}(\lambda_j) \;, \quad \lambda_j \in \sigma(A) \;, \quad k = 0,1,\cdots,m_j-1 \;,$$

i.e. $p(\lambda)$ assumes the same values as $f(\lambda)$ on $\sigma(A)$. Since the definitions of $f(A)$ implies $f(A) = p(A)$ and $p(A)$ is obviously H-selfadjoint, it follows that $f(A)$ is H-selfadjoint.

Conversely, assume that $f(A)$ is H-selfadjoint. Use Theorem 3.3 to reduce (A,H) to a canonical pair so that $f(A)$ is given by (6.1) and (6.2). Then there are just two cases to consider:

(a) A is a single Jordan block with real eigenvalue λ_0 and size k and

$\pm H$ is the $k \times k$ sip matrix.

(b) $A = J \oplus \bar{J}$ where J is a $k \times k$ Jordan block with non-real eigenvalue λ_0 and H is the $2k \times 2k$ sip matrix.

In case (a) it follows from (6.2) that $f(A)$ is H-selfadjoint, i.e. $Hf(A) = f(A)^*H$ if and only if $f(\lambda_0), f^{(1)}(\lambda_0), \cdots, f^{(k-1)}(\lambda_0)$ are real numbers. This means that $f(\lambda)$ is real symmetric on $\sigma(A)$.

In case (b) it follows from (6.2) that $f(A)$ is H-selfadjoint if and only if $f^{(r)}(\lambda_0) = f^{(r)}(\bar{\lambda}_0)$ for $r = 0,1,\cdots,k-1$ and, again, this means that $f(\lambda)$ is real symmetric on $\sigma(A)$. □

It is clear that the conclusion of Theorem 6.4 can be para-phrased to give the statement that $f(A)$ is H-selfadjoint (when A is H-selfadjoint) if and only if $f(A) = p(A)$ for some *real* polynomial $p(\lambda)$.

For the more familiar situation in which H is positive definite A is H-selfadjoint implies that for each $\lambda_j \in \sigma(A)$, λ_j is real and $m_j = 1$. In this case $f(\lambda)$ is real symmetric on $\sigma(A)$ if and only if the numbers $f(\lambda_j)$ are real for each $\lambda_j \in \sigma(A)$.

We shall describe now an important class of functions of an H-selfadjoint matrix A for which $f(A)$ is a Riesz projector. Let Λ be a set of eigenvalues of A such that $\lambda \in \Lambda$ implies $\bar{\lambda} \in \Lambda$. Let $f_\Lambda(\lambda) \equiv 1$ in a neighbourhood of Λ, and $f_\Lambda(\lambda) \equiv 0$ in a neighbourhood of $\sigma(A) \setminus \Lambda$. Then $f_\Lambda(\lambda)$ is analytic in a neighbourhood of $\sigma(A)$ and is real symmetric on $\sigma(A)$. By Theorem 6.4, $f_\Lambda(A)$ is H-selfadjoint. Formula (6.3) shows that $f_\Lambda(A)$ is just the Riesz projector on the sum of the root subspaces of A corresponding to the eigenvalues in Λ. Now H-selfadjointness of $f_\Lambda(A)$ implies that $\operatorname{Ker} f_\Lambda(A)$ and $\operatorname{Im} f_\Lambda(A)$ are H-orthogonal. Indeed, for $x \in \operatorname{Im} f_\Lambda(A)$, $y \in \operatorname{Ker} f_\Lambda(A)$ we have

$$(Hx,y) = (Hf_\Lambda(A)x,y) = (f_\Lambda(A)^*Hx,y) = (Hx,f_\Lambda(A)y) = 0 .$$

In other words, the subspace $\operatorname{Ker} f_\Lambda(A)$, which is the sum of root subspaces of A corresponding to eigenvalues outside Λ, is an H-orthogonal complement to $\operatorname{Im} f_\Lambda(\lambda)$. We have observed this fact before (in another form) in Theorem 2.5.

The conditions under which a function of an H-selfadjoint matrix is H-unitary are more complicated than the counterparts of Theorems 6.3 and 6.4 in which the function is H-normal, or H-selfadjoint, respectively.

THEOREM 6.5 *Let* A *be H-selfadjoint and let* $f(\lambda)$ *be defined on* $\sigma(A)$. *Then* $f(A)$ *is H-unitary if and only if, for each* $\lambda \in \sigma(A)$,

$$\overline{f(\lambda)}f(\bar{\lambda}) = 1 \tag{6.7}$$

and

$$\sum_{j=0}^{k} \binom{k}{j} \overline{f^{(j)}(\lambda)}\, f^{(k-j)}(\bar{\lambda}) = 0 \ , \quad k = 1,2,\cdots,m \ , \tag{6.8}$$

and $m = m(\lambda)$ *is the index of* λ.

Note, in particular, that (6.7) implies $|f(\lambda)| = 1$ when $\lambda \in \sigma(A)$ and λ is real. Furthermore, if $f(\lambda)$ is analytic in a domain Ω containing $\sigma(A)$ and Ω is symmetric with respect to the real axis, then if (6.7) holds throughout Ω, (6.8) follows automatically. To see this, observe that for $\mu,\lambda \in \Omega$ with μ close enough to λ

$$f(\mu) = \sum_{j=0}^{\infty} \frac{(\mu-\lambda)^j}{j!} f^{(j)}(\lambda) \ , \qquad f(\bar{\mu}) = \sum_{k=0}^{\infty} \frac{(\bar{\mu}-\bar{\lambda})^k}{k!} f^{(k)}(\bar{\lambda}) \ .$$

Then comparing coefficients of powers of $(\bar{\mu}-\bar{\lambda})$ in the product $\overline{f(\mu)}f(\bar{\mu}) = 1$, (6.8) is obtained.

Note also that if $f(\lambda)$ and $R(\lambda)$ are analytic on a domain Ω as above and $f(\lambda) = e^{iR(\lambda)}$, then $f(\lambda)$ satisfies condition (6.7) on Ω, and hence (6.8), if $R(\lambda)$ is real symmetric on Ω. For

$$\overline{f(\lambda)}f(\bar{\lambda}) = e^{-i\overline{R(\lambda)}}\, e^{iR(\bar{\lambda})} = 1$$

provided $\overline{R(\lambda)} = R(\bar{\lambda})$, i.e. if $R(\lambda)$ is real symmetric on Ω. In particular, e^{iA} is H-unitary, as we have already observed in Theorem 6.2.

PROOF (of Theorem 6.5). Again, we can assume that (A,H) is in the canonical form of Theorem 3.3. So we have to consider the two cases (as in the proof of Theorem 6.4): (a) A is a Jordan block with real eigenvalue λ and size k, and $\pm H$ is the $k \times k$ sip matrix; (b) $A = J \oplus \bar{J}$, where J is a Jordan block with non-real eigenvalue λ and size k, and H is the $2k \times 2k$ sip matrix. In either case the equality $f(A)^* Hf(A) = H$ expressing the property that $f(A)$ is H-unitary holds if and only if $f(\lambda)$ satisfies (6.7) and (6.8), as one checks easily using equation (6.2). □

6.4 Functions of H-Unitary Matrices

It has been shown in Theorem 6.3 that if U is an H-unitary matrix and $f(\lambda)$ is defined on $\sigma(U)$ then $f(U)$ is necessarily H-normal. We now consider the conditions on $f(\lambda)$ under which $f(U)$ is H-selfadjoint or H-unitary. For this purpose, define integers c_{jk} for $j = 1,2,\cdots,k$ and $k = 1,2,\cdots$ by $c_{1k} = c_{kk} = 1$ for all k, and for $1 < j < k$,

$$c_{jk} = j\, c_{j,k-1} + c_{j-1,k-1} \,. \tag{6.9}$$

THEOREM 6.6 *Let* U *be H-unitary and* $f(\lambda)$ *be defined on* $\sigma(U)$.
(a) *The matrix* $f(U)$ *is H-selfadjoint if and only if, for any* $\mu \in \sigma(U)$,

$$\overline{f(\mu)} = f(\bar{\mu}^{-1}) \tag{6.10}$$

and for $k = 1,2,\cdots,m-1$

$$\overline{\sum_{j=1}^{k} c_{jk} f^{(j)}(\mu)\mu^{j}} = (-1)^k \sum_{j=1}^{k} c_{jk} f^{(j)}(\bar{\mu}^{-1})\bar{\mu}^{-j} \tag{6.11}$$

where m *is the index of* μ.
(b) *The matrix* $f(U)$ *is H-unitary if and only if, for any* $\mu \in \sigma(U)$,

$$\overline{f(\mu)}f(\bar{\mu}^{-1}) = 1 \tag{6.12}$$

and for $k = 1,2,\cdots,m-1$

$$\sum_{q=0}^{k} \binom{k}{q}\left(\overline{\sum_{j=0}^{q} c_{jq} f^{(j)}(\mu)\mu^{j}}\right)\left(\sum_{r=0}^{k-q} c_{r,k-q} f^{(j)}(\bar{\mu}^{-1})\bar{\mu}^{-j}\right) = 0 \tag{6.13}$$

(*and* c_{0q} *is defined to be* 1).

PROOF. It is easily seen that the integers c_{jk} defined in (6.9) appear as coefficients in the formula:

$$\frac{d^k(f(e^{i\lambda}))}{d\lambda^k} = i^k \sum_{j=1}^{k} c_{jk} f^{(j)} g^j \,, \quad k = 1,2,\cdots,m-1 \tag{6.14}$$

where $e^{i\lambda}$ belongs to the domain of definition of f and $f^{(j)} = f^{(j)}(e^{i\lambda})$, $g = e^{i\lambda}$. By Theorem 6.2 there exists an H-selfadjoint matrix A such that $U = e^{iA}$. Then $f(U) = f(g(A))$, where $g(\lambda) = e^{i\lambda}$. Using (6.14) the theorem follows from the necessary and sufficient conditions for function $f(g(A))$ of the H-selfadjoint matrix A to be H-selfadjoint or H-unitary (see Theorems 6.4 and 6.5). □

As in the case of functions of H-selfadjoint matrices, if equality

(6.10) holds for μ in a neighbourhood of $\sigma(U)$ (and not only for $\mu \in \sigma(U)$) and $f(\mu)$ is analytic in this neighbourhood, then equalities (6.11) follow automatically. Indeed, write $h(\lambda) = f(e^{i\lambda})$, where λ belongs to a suitable neighbourhood of $\sigma(A)$, and A is an H-selfadjoint matrix such that $U = e^{iA}$ (cf. the proof of Theorem 6.6). The equality (6.10) implies that $\overline{h(\lambda)} = h(\bar{\lambda})$ in a neighbourhood of $\sigma(A)$; by Theorem 6.4 and the remark preceding it, $h(A) = f(U)$ is H-selfadjoint. Now the equalities (6.11) follow from Theorem 6.6. In a similar way one checks that if (6.12) holds in a neighbourhood of $\sigma(U)$ and $f(\mu)$ is analytic in this neighbourhood, then equalities (6.13) follow.

6.5 The Canonical Form and Sign Characteristic for a Function of an H-Selfadjoint Matrix

Let A be H-selfadjoint and $f(\lambda)$ be defined on $\sigma(A)$. If $f(A)$ is also H-selfadjoint, i.e. $f(\lambda)$ is real symmetric on $\sigma(A)$ (ref. Theorem 6.4), the problem arises of defining the canonical form and sign characteristic for the pair $(f(A),H)$ in terms of corresponding properties of the pair (A,H). An intermediate step is, of course, the determination of the Jordan form for $f(A)$ in terms of that of A and this has already been examined in Proposition 6.1.

Let λ_0 be a real eigenvalue of $f(A)$. We first find the canonical form and sign characteristic of $(f(A),H)$ at λ_0. Let $\lambda_1,\cdots,\lambda_k,\lambda_{k+1},\cdots,\lambda_{k+\ell}$ be all the different eigenvalues of A such that $f(\lambda_i) = \lambda_0$, where $\lambda_1,\cdots,\lambda_k$ are real and $\lambda_{k+1},\cdots,\lambda_{k+\ell}$ are non-real. Note that the integer ℓ is necessarily even. The cases $k = 0$ or $\ell = 0$ are not excluded.

Let $s_{i1},\cdots,s_{i,p_i}$ $(1 \leq i \leq k+\ell)$ be the sizes of Jordan blocks with the eigenvalue λ_i in the Jordan form of A, and let $\varepsilon_{i1},\cdots,\varepsilon_{i,p_i}$ $(1 \leq i \leq k)$ be the corresponding signs in the sign characteristic of (A,H) (which exist, of course, only for real eigenvalues of A). Denote $m_i = \max\{s_{i1},\cdots,s_{i,p_i}\}$ for $1 \leq i \leq k+1$, i.e. the index of λ_i.

THEOREM 6.7 *For* $i = 1,\cdots,k+\ell$ *let* r_i *be the minimal integer* $(1 \leq r_i \leq m_i-1)$ *such that* $f^{(r_i)}(\lambda_i) \neq 0$ *(if* $f^{(j)}(\lambda_i) = 0$ *for* $j = 1,\cdots,m_i-1$, *put* $r_i = m_i$*). For a fixed positive integer* q, *put*

$$\gamma(q) = \sum_{i=1}^{k} \sum_{j=1}^{p_i} \varepsilon_{ij}\delta_{ij}(q) , \tag{6.15}$$

where

$$\delta_{ij}(q) = \begin{cases} 1 , & \text{if } r_i(q+1) - s_{ij} \text{ is odd and either} \\ & \qquad q = \left[\frac{s_{ij}}{r_i}\right] + 1 \text{ or } q = \left[\frac{s_{ij}}{r_i}\right] ; \\ 0 , & \text{otherwise} ; \end{cases}$$

put also

$$\eta(q) = \sum_{i=1}^{\ell} \sum_{j=1}^{p_i} \eta_{ij}(q) , \tag{6.16}$$

where

$$\eta_{ij}(q) = \begin{cases} r_i\left(\left[\frac{s_{ij}}{r_i}\right] + 1\right) - s_{ij} & \text{if } q = \left[\frac{s_{ij}}{r_i}\right] ; \\ s_{ij} - r_i\left[\frac{s_{ij}}{r_i}\right] & \text{if } q = \left[\frac{s_{ij}}{r_i}\right] + 1 ; \\ 0 & \text{otherwise} . \end{cases}$$

Then $\eta(q)$ *is the number of Jordan blocks of* $f(A)$ *with eigenvalue* λ_0 *and size* q*; and* $\frac{1}{2}(\eta(q)+\gamma(q))$ *(resp.* $\frac{1}{2}(\eta(q)-\gamma(q))$*) is the number of signs* $+1$ *(resp.* -1*) in the sign characteristic of* $(f(A),H)$ *which correspond to these Jordan blocks.*

From the proof of this theorem it will be seen that the numbers $\eta(q) \pm \gamma(q)$ are always even. In particular, Theorem 6.7 asserts that $\gamma(q)$ is the sum of signs corresponding to the Jordan blocks of $f(A)$ with eigenvalue λ_0 and size q; and by (6.15) this sum does not depend on the non-real eigenvalues λ_i $(k+1 \leq i \leq k+\ell)$.

PROOF. By Proposition 6.1 we see that $\eta(q)$ given by (6.16) is just the number of Jordan blocks of $f(A)$ with eigenvalue λ_0 and size q. So it remains to prove that $\gamma(q)$ is the sum of signs in the sign characteristic of $(f(A),H)$ corresponding to these blocks.

In this proof we shall use a perturbation argument based on stability of the sign characteristic of pairs (B,G), where B is G-selfadjoint, under certain perturbations of B and G. This result is proved in Part III (Theorem III.5.1).

We can assume that (A,H) is in the canonical form (Theorem 3.3). So we have to consider only two cases: (a) A is the Jordan block with real eigenvalue λ_1 and size s, and $\pm H$ is the $s \times s$ sip matrix; (b) $A = J \oplus \bar{J}$, where J is the Jordan block with non-real eigenvalue λ and size s, and H is the $2s \times 2s$ sip matrix.

Consider the case (a). Let r $(= r_1$ in the notation of the theorem) be the least integer $1 \leq r \leq s-1$ such that $f^{(r)}(\lambda_1) \neq 0$ (and $r = s$ if $f^{(i)}(\lambda_1) = 0$ for $1 \leq i \leq s-1$). Let $f_t(\lambda)$, $t \in [0,1]$ be a family of analytic functions in a neighbourhood of λ_1 such that $f_t(\lambda_1) = f(\lambda_1)$ $(= \lambda_0)$, $f_t^{(i)}(\lambda_1) = 0$ for $1 \leq i < r$; $f_t^{(r)}(\lambda_1) = (1-t)f^{(r)}(\lambda_1) + r!t$; $f_t^{(i)}(\lambda_1) = 0$ for $r < i \leq s-1$. Recall that $f(\lambda)$ is symmetric on $\sigma(A)$, so the numbers $f^{(i)}(\lambda_1)$, $i = 0,\cdots,s-1$ are real. Formula (6.2) shows that $f_t(A)$ is a continuous function of $t \in [0,1]$, and $f_0(A) = f(A)$. Furthermore, Proposition 6.1 shows that the eigenvalue of $f_t(A)$ and the sizes of Jordan blocks of $f_t(A)$ are independent of $t \in [0,1]$. As $f_t(A)$ is H-selfadjoint for $t \in [0,1]$, Theorem III.5.1 implies that the sign characteristic of $(f_t(A),H)$ is also independent of t. So the sign characteristics of $(f(A),H)$ and $(f_1(A),H)$ are the same, and we shall determine the latter.

As formula (6.2) shows,

$$f_1(A) = \begin{bmatrix} \lambda_0 & \cdots & 0 & 1 & 0 & \cdot & \cdot & \cdot & 0 \\ & \lambda_0 & \cdots & 0 & 1 & \ddots & & & \vdots \\ & & \lambda_0 & & \ddots & & \ddots & & 0 \\ & & & \ddots & & \ddots & & \ddots & 1 \\ & & & & & & \ddots & & 0 \\ & 0 & & & & & & \ddots & \vdots \\ & & & & & & & & \lambda_0 \end{bmatrix}, \tag{6.17}$$

where 1's appear in the entries $(1,r+1),(2,r+2),\cdots,(r+1,s)$. It is easily seen that, denoting $v = [\frac{s}{r}]$ and letting e_j be the vector $<0,\cdots,0,1,0,\cdots,0>$ with 1 in the j-th place, the chains of vectors

$$\begin{aligned} &e_i, e_{i+r}, \cdots, e_{i+vr} \quad ; \quad i = 1,2,\cdots,s-rv \\ &e_i, e_{i+r}, \cdots, e_{i+(v-1)r} \quad ; \quad i = s-rv+1,\cdots,r \end{aligned} \tag{6.18}$$

are Jordan chains of $f_1(A)$ and form a basis in $\mathbb{C}^s$. Now use the second description of the sign characteristic given in Section 3.9 (Theorem 3.16) to compute the sign characteristic of $(f_1(A),H)$. Indeed, in the notation of Theorem 3.16, the linear transformation G_{v+1} in the basis $e_1,\cdots,e_{s-rv}$ is given by the right upper $(s-rv) \times (s-rv)$ corner of H. The linear transformation G_v in the basis $e_1,\cdots,e_r$ is given by the submatrix of H formed by its first r rows and columns $r(v-1)+1,\cdots,r(v-1)+r$. (In case $v = 0$ this description should be properly modified). Taking into account that εH $(\varepsilon = \pm 1)$ is the sip matrix, with the help of Theorem 3.16 we find that indeed the sum of signs in the sign characteristic of $(f(A),H)$ which correspond to blocks of size q, is given by formula (6.15).

Consider the case (b). Applying a perturbation argument analogous to the one used in the proof of case (a), we can assume that $f(A)$ is a direct sum of two equal blocks (6.17). The chains of vectors

$$e_i,e_{i+r},\cdots,e_{i+vr} \quad ; \quad i = 1,2,\cdots,s-rv,s+1,\cdots,2s-rv \ ;$$

$$e_i,e_{i+r},\cdots,e_{i+(v-1)r} \ ; \quad i = s-rv+1,\cdots,r,2s-rv+1,2s-rv+2,\cdots,s+r \ ;$$

where $v = [\frac{s}{r}]$, are Jordan chains of $f(A)$ and form a basis in $\mathbb{C}^{2s}$. Now the sign characteristic of $(f(A),H)$ is computed by application of Theorem 3.16, as in the case (a). It turns out that the sum of signs in the sign characteristic of $(f(A),H)$ corresponding to equal Jordan blocks, is zero. □

EXAMPLE 6.1 Let

$$A = \begin{bmatrix} i & 1 \\ 0 & i \end{bmatrix} \oplus \begin{bmatrix} -i & 1 \\ 0 & -i \end{bmatrix}; \quad H = \begin{bmatrix} 0 & 0 & 0 & 1 \\ 0 & 0 & 1 & 0 \\ 0 & 1 & 0 & 0 \\ 1 & 0 & 0 & 0 \end{bmatrix}.$$

Then

$$A^2 = \begin{bmatrix} -1 & 2i & 0 & 0 \\ 0 & -1 & 0 & 0 \\ 0 & 0 & -1 & -2i \\ 0 & 0 & 0 & -1 \end{bmatrix}$$

is H-selfadjoint. According to Theorem 6.7 the sign characteristic of (A^2,H) at the eigenvalue -1 consists of $\{+1,-1\}$. Indeed, with

$$T = \begin{bmatrix} 2ic & 0 & 2i & 0 \\ 0 & c & 0 & 1 \\ -2i & 0 & -2ic & 0 \\ 0 & 1 & 0 & c \end{bmatrix}, \quad c = \frac{\sqrt{3}}{2} - \frac{i}{2}$$

we have

$$T^{-1}A^2T = \begin{bmatrix} -1 & 1 \\ 0 & -1 \end{bmatrix} \oplus \begin{bmatrix} -1 & 1 \\ 0 & -1 \end{bmatrix}; \quad T^*HT = \begin{bmatrix} 0 & 2 \\ 2 & 0 \end{bmatrix} \oplus \begin{bmatrix} 0 & -2 \\ -2 & 0 \end{bmatrix}. \quad \square$$

6.6 Functions of H-Selfadjoint Matrices which are Selfadjoint in another Indefinite Scalar Product

In Part II we shall need results on the sign characteristic of a function of an H-selfadjoint matrix, but with respect to another indefinite scalar product. To start with, let A be an invertible H-selfadjoint matrix, and let $\hat{H} = HA$. Obviously, $\hat{H}$ is invertible and hermitian. Moreover, A is $\hat{H}$-selfadjoint. Indeed

$$\hat{H}A = HA^2 = A^*HA = A^{*2}H = A^*\hat{H} .$$

We shall compute the sign characteristic of $(A,\hat{H})$ in terms of the sign characteristic of (A,H).

PROPOSITION 6.8 *Let* $\lambda_0 \in \sigma(A) \cap \mathbb{R}$, *and let* $\varepsilon_1,\cdots,\varepsilon_r$ *be the signs in the sign characteristic of* (A,H) *corresponding to the Jordan blocks of* A *with eigenvalue* λ_0. *Then the signs in the sign characteristic of* $(A,\hat{H})$ *corresponding to the same Jordan blocks are* $\varepsilon_1 \cdot \operatorname{sgn} \lambda_0,\cdots,\varepsilon_r \cdot \operatorname{sgn} \lambda_0$, *where* $\operatorname{sgn} \lambda_0$ *is* 1 *if* $\lambda_0 > 0$ *and* -1 *if* $\lambda_0 < 0$.

Recall that $\lambda_0 \neq 0$ since A is assumed invertible.

PROOF. We can assume that (A,H) is in the canonical form. Let J be a Jordan block in A with eigenvalue λ_0, and let $P = \varepsilon(\delta_{i+j,k+1})_{i,j=1}^k$, $\varepsilon = \pm 1$ be the corresponding part of H. Using the second description of the sign characteristic (Theorem 3.16), it is easily seen that the canonical form of (J,PJ) is $(J, \operatorname{sgn} \lambda \cdot P)$. $\square$

More generally, let A be an H-selfadjoint matrix, and let $f(\lambda)$ be a function defined on the spectrum of A such that $f(A)$ is H-selfadjoint and invertible (by Theorem 6.4 this means that $f(\lambda)$ is symmetric on $\sigma(A)$ relative to the real axis and $f(\lambda_0) \neq 0$ for every $\lambda_0 \in \sigma(A)$). Then $\hat{H} = Hf(A)$ is invertible and hermitian, and $f(A)$ is $\hat{H}$-selfadjoint. Combining Proposition 6.8 and the description of the sign characteristic of $(f(A),H)$ given in Theorem 6.7, we obtain the following description of the sign characteristic of $(f(A),\hat{H})$. The notation introduced before and in the statement of Theorem 6.7 will be used.

THEOREM 6.9 *Let* A *be* H-*selfadjoint, and let* f(A) *be* H-*selfadjoint and invertible. For a fixed positive integer* q, $\eta(q)$ *is the number of Jordan blocks of* f(A) *with eigenvalue* λ_0 *and size* q *and* $\frac{1}{2}(\eta(q)+(\operatorname{sgn}\lambda_0)\cdot\gamma(q))$ *(resp.* $\frac{1}{2}(\eta(q)-(\operatorname{sgn}\lambda_0)\cdot\gamma(q)))$ *is the number of signs* +1 *(resp.* -1*) in the sign characteristic of* (f(A),H) *which correspond to these blocks (here* $\operatorname{sgn}\lambda_0 = 1$ *if* $\lambda_0 > 0$ *and* $\operatorname{sgn}\lambda_0 = -1$ *if* $\lambda_0 < 0$*).*

In Part II the following fact will be used.

THEOREM 6.10 *Let* A *be* H-*selfadjoint and let* f(A) *be a function of* A *which is* H-*selfadjoint and invertible. Then an* f(A)-*invariant subspace* $M \subset \mathbb{C}^n$ *(where* n *is the size of* A*) is* H-*neutral if and only if it is* $\hat{H}$-*neutral, where* $\hat{H} = Hf(A)$.

PROOF. Let $\hat{A} = f(A)$ and observe that the subspace M is $\hat{A}$-invariant. Assuming M is H-neutral (so $(Hx,y) = 0$ for all $x,y \in M$), for every $x,y \in M$ we have

$$(\hat{H}x,y) = (H\hat{A}x,y) = 0 .$$

Conversely, if M is $\hat{H}$-neutral, then for every $x,y \in M$, we have (denoting $z = \hat{A}^{-1}x \in M$):

$$(Hx,y) = (\hat{H}z,y) = 0 . \quad \square$$

Note that Theorem 6.10 does not hold in general for A-invariant subspaces which are H-nonnegative or H-nonpositive.

NOTES TO PART I

The material of the first two chapters is well-known, and the greater part of it is known even in the infinite dimensional setting (see [6,2,23]). Theorem 2.11 appeared in [24].

The canonical form described in Chapter 3 was known at the end of the 19th century (see [46,28]), and it was developed by Weierstrass in the form of Theorem 3.18. The canonical form was rediscovered later by many authors [34,48a,44]. See [45b] for historical remarks. The sign characteristic as an important notion in its own right was introduced in [19a], where the equivalence of the three different descriptions of the sign characteristic was obtained in a more general setting. By systematic exposition of the material a few straightforward results are pointed out which may not have been noticed before (as in Sections 3.7,3.8,3.13). The results of Section 3.6 seem to be new. Theorem 3.21 (essentially) appeared in [31]. Theorem 3.22 is obtained in [40b].

A canonical form for real matrices which are selfadjoint in a skew symmetric scalar product is given in [48b,8]. Another approach to this canonical form via rational matrix functions is developed in [13b].

Chapter 4 contains material most of which probably did not appear before in such a form. It is obtained by a straightforward analysis based on the results of Chapter 3. H-normal matrices are introduced here probably for the first time.

The canonical form for real H-selfadjoint matrices is given in [45a]. Another approach to the canonical form based on analysis of rational matrix valued functions is developed in [13a]. Theorems 5.5 and 5.7 are probably new.

The main results of Chapter 6 are probably new.

A spectral theory for some classes of H-selfadjoint operators in an infinite dimensional space was developed in [27].

PART II

FIRST APPLICATIONS

The second phase of our analysis is concerned with matrix valued functions which take hermitian or unitary values. This general description applies to the solutions of the differential equations of Hamiltonian type considered here, to the hermitian matrix polynomials and rational functions evaluated on the real line, and to the symmetric matrix algebraic Riccati equation. In each case, the theory of H-selfadjoint and H-unitary matrices developed in Part I arises naturally and can be utilised in a fruitful and illuminating way. All of the problems discussed originate in systems theory or the theory of differential equations, with special reference to the analysis of transfer functions, optimal control, differential equations with periodic coefficients, and time-invariant systems of ordinary differential equations of higher order.

In general, the chapters of this part can be read independently of each other. As far as possible, each one is a self-contained unit and the two exceptions apply to the third chapter. Here a considerable body of knowledge on general (non-hermitian) rational matrix functions is postponed to an appendix and will cause no embarrassment to the reader who is already familiar with this material. Also, the section of this chapter on factorization of non-negative rational functions utilises a result from the second chapter.

To look ahead a little, we come back to the topics of Part II in Parts III and IV. This is done after extending the theory of Part I to include different aspects of perturbation theory.

CHAPTER 1

HAMILTONIAN AND SELFADJOINT DIFFERENTIAL EQUATIONS WITH PERIODIC COEFFICIENTS

Consider a mechanical system with n degrees of freedom. Using the generalized coordinates $q_1,\cdots,q_n$ and generalized momenta $p_1,\cdots,p_n$, Hamilton's canonical equations of motion for the system can be written in the form

$$\frac{dq_j}{dt} = \frac{\partial H}{\partial p_j} ; \quad \frac{dp_j}{dt} = -\frac{\partial H}{\partial q_j} ; \quad j = 1,\cdots,n ,$$

where $H = H(q_1,\cdots,q_n,p_1,\cdots,p_n,t)$ is the Hamiltonian function of the system (which is supposed to be known), $q_j = q_j(t)$, $p_j = p_j(t)$ are real functions of the real variable t, which usually represents the time.

The Hamiltonian generally has the form

$$H = \frac{1}{2} \sum_{j,k=1}^{2n} h_{jk}(t)x_j x_k + f(t) ,$$

where $x_j = q_j$ for $j = 1,2,\cdots,n$, $x_j = p_{j-n}$ for $j = n+1,\cdots,2n$, and $h_{jk}(t) = h_{kj}(t)$ $(j,k = 1,\cdots,2n)$ are real valued functions of t. Then the canonical system can be rewritten in vector-matrix notation as follows:

$$\frac{dx}{dt} = EH(t)x , \quad t \in \mathbb{R} \tag{1.1}$$

where $x = (x_1,\cdots,x_{2n})^T$ is a 2n-dimensional real vector function, $H(t) = [h_{jk}(t)]_{j,k=1}^{2n}$ is a symmetric $2n \times 2n$ real matrix function and $E = \begin{bmatrix} 0 & I_n \\ -I_n & 0 \end{bmatrix}$.

Physical phenomena which are periodic in t give rise to Hamiltonians H and matrices $H(t)$ which are also periodic in t. The purpose of this chapter is to study equation (1.1) and generalizations of it, under the assumption that $H(t)$ has this property. In particular, the connection between systems of the form (1.1) and families of matrices which are unitary in a certain indefinite scalar product will be established. In Parts III and IV of this book, deeper problems concerning stability of systems described by (1.1) will be investigated.

1.1 The Matrizant

Let E be an $n \times n$ (constant) invertible hermitian matrix, and let $H(t)$ be an $n \times n$ matrix-valued function which is hermitian, piecewise continuous, and periodic with period ω for $t \in \mathbb{R}$. Thus, for all real t, $H(t) \in \mathbb{C}^{n\times n}$ and $H(t)^* = H(t)$, $H(t+\omega) = H(t)$. Consider the system of differential equations

$$E\frac{dx}{dt} = iH(t)x \tag{1.2}$$

for $t \in \mathbb{R}$, where $x(t)$ is a function of t (which is to be found) with values in $\mathbb{C}^n$ and with a piecewise continuous derivative.

Consider also the closely related initial value problem for the matrix valued function $X(t)$:

$$E\frac{dX}{dt} = iH(t)X\ , \quad X(0) = I\ . \tag{1.3}$$

The unique solution $X(t)$ of (1.3) defined for $t \in [0,\omega]$ is called the *matrizant* of the system (1.2).

Knowledge of the matrizant leads directly to solution of the initial value problem (1.3) on the whole line. Indeed, if $X(t)$ is the solution of the initial value problem (1.3) then, in view of the periodicity of H, the matrix functions $X(t+\omega)$ and $X(t)X(\omega)$ are both solutions of the differential equation $E\frac{dX}{dt} = iH(t)X$. But also $X(t+\omega)|_{t=0} = [X(t)X(\omega)]|_{t=0}$, and by the uniqueness of solution of the initial value problem

$$E\frac{dX}{dt} = iH(t)X\ ; \quad X(0) = I\ ,$$

it follows that for all $t \in \mathbb{R}$,

$$X(t+\omega) = X(t)X(\omega)\ . \tag{1.4}$$

Repeating this argument, we obtain

$$X(t+n\omega) = X(t)(X(\omega))^n , \quad n = 1,2,\cdots , \tag{1.5}$$

and it is easily seen that (1.5) holds also for negative integers n. (Note that $X(t)$ is invertible for all real t. This follows from the first statement of the next theorem, for example.)

Since E is hermitian and invertible, it determines an indefinite scalar product in $\mathbb{C}^n$. The following theorem describes an important related property of the matrizant.

THEOREM 1.1 *The matrizant* $X(t)$ *is* E-*unitary for all* $t \in [0,\omega]$. *Conversely, any piecewise continuously differentiable curve* $X(t)$, $t \in [0,\omega]$ *which takes* E-*unitary values and for which* $X(0) = I$ *is a matrizant of a system* (1.2) *for a unique piecewise continuous hermitian matrix function* $H(t)$ *with period* ω.

PROOF. Let $X(t)$ be the matrizant of (1.2). Write:

$$\begin{aligned}\frac{d}{dt}(X^*EX) &= \frac{dX^*}{dt} EX + X^*E\frac{dX}{dt}\\ &= -iX^*H(t)E^{-1}EX + iX^*H(t)X = 0 ,\end{aligned}$$

where we have used the hermitian properties of $H(t)$ and E. Thus, X^*EX is independent of t, and evaluation at $t = 0$ gives $X^*EX = E$, i.e. $X = X(t)$ is E-unitary.

Assume now that $X(t)$, $t \in [0,\omega]$ is a piecewise continuously differentiable E-unitary curve such that $X(0) = I$. Put

$$H(t) = -iE\frac{dX(t)}{dt}X(t)^{-1} , \quad t \in [0,\omega]$$

and $H(t)$ is continued by periodicity to the whole real line. Then

$$\frac{dX(t)}{dt} = iE^{-1}H(t)X(t) , \quad t \in [0,\omega] .$$

Further, differentiate the relation

$$X(t)^*EX(t) = E$$

to obtain

$$\frac{dX(t)^*}{dt}EX(t) + X(t)^*E\frac{dX(t)}{dt} = 0 .$$

Premultiply by $(X(t)^*)^{-1}$ and postmultiply by $X(t)^{-1}$:

$$(X(t)^*)^{-1} \frac{dX(t)^*}{dt} E + E \frac{dX(t)}{dt} X(t)^{-1} = 0 .$$

This formula shows that the matrix $H(t)$ is hermitian for all $t \in [0,\omega]$.

Finally, the uniqueness of a piecewise continuous hermitian function $H(t)$ with a given matrizant $X(t)$ follows from the formula $H(t) = -iE \frac{dX}{dt} X^{-1}$. □

Denote by H_ω the (infinite dimensional) real linear space of all complex, $n \times n$, selfadjoint, piecewise continuous matrix functions of $t \in \mathbb{R}$ with period ω (this period is the same for every function from H_ω), and let G_ω be the set of all piecewise continuously differentiable paths $X(t)$, $t \in [0,\omega]$ such that $X(t)$ is E-unitary for all $t \in [0,\omega]$, and $X(0) = I$. Theorem 1.1 establishes a one-to-one correspondence:

$$\varphi : H_\omega \to G_\omega , \tag{1.6}$$

where, for given $H(t) \in H_\omega$, the image $\varphi(H(t))$ is the matrizant of (1.2). This correspondence means that many questions concerning systems of differential equations (1.2) can be reduced to more tractable questions concerning curves in the group of E-unitary matrices. This approach will be used extensively in Part IV.

It turns out that the correspondence φ given by (1.6) is not only one-to-one but also a homeomorphism. Let us make this statement precise. Introduce the norm in H_ω as follows:

$$|||H||| = \max_{0 \leq t \leq \omega} ||H(t)|| , \quad H(t) \in H_\omega .$$

It is easily seen that $|||\cdot|||$ satisfies the usual properties of a norm. In particular, H_ω is a metric space with the distance function induced by the norm: $\rho(H_1,H_2) = |||H_1-H_2|||$. The set G_ω is also a metric space with the distance function

$$\rho(X,Y) = \max_{t\in[0,\omega]} ||X(t)-Y(t)|| + \max_{t\in[0,\omega]} || \frac{dX}{dt} - \frac{dY}{dt} ||$$

for any $X,Y \in G_\omega$.

For any two metric spaces A and B with distance functions $\rho_A(.,.)$ and $\rho_B(.,.)$, respectively, a map $\alpha : A \to B$ is called *locally*

Lipschitz if for every $\alpha \in A$ there are positive constants ε and K such that

$$\rho_{\mathcal{B}}(\alpha(a'),\alpha(a)) \leqslant K\rho_A(a',a) \tag{1.7}$$

for every $a' \in A$ with $\rho_A(a',a) < \varepsilon$. Obviously, a locally Lipschitz map is continuous.

THEOREM 1.2 *The one-to-one map* $\varphi : H_\omega \to G_\omega$ *and its inverse are locally Lipschitz. In particular,* φ *is a homeomorphism.*

PROOF. The locally Lipschitz property of φ follows from the well-known results on dependence of the solutions of a differential equation on the parameters of the equation. Let us give the details. Fix $H_0(t) \in H_\omega$, and let $H(t) \in H_\omega$ be close enough to $H_0(t)$. Let $X_0(t)$ (resp. $X(t)$) be the matrizant of (1.2) with coefficient matrix $H_0(t)$ (resp. $H(t)$). Let $x_{i0}(t)$ and $x_i(t)$ be the i-th column of X_0 and X, respectively, $i = 1,\cdots,n$ so

$$E\frac{dx_{i0}}{dt} = iH_0(t)x_{i0}; \quad x_{i0}(0) = e_i ; \tag{1.8}$$

$$E\frac{dx_i}{dt} = iH(t)x_i \ ; \quad x_i(0) = e_i . \tag{1.9}$$

Here e_i is the vector $\langle 0,\cdots,0,1,\cdots,0\rangle$ with 1 in the i-th place.

Assuming $|||H-H_0||| \leqslant 1$, the solutions of (1.8) and (1.9) are easily estimated on the interval $(-2\omega,2\omega)$: we have

$$x_{i0}(t) = \int_0^t iE^{-1}H_0(s)x_{i0}(s)ds + e_i \ ; \quad t \geqslant 0 .$$

Consequently,

$$||x_{i0}(t)|| \leqslant ||E^{-1}|| \ |||H_0||| \int_0^t ||x_{i0}(s)||ds + ||e_i||$$

for $t \geqslant 0$. Now Gronwall's lemma (see, for example,[22], p. 19) gives

$$||x_{i0}(t)|| \leqslant \exp(K_1 t)$$

where $K_1 = ||E^{-1}|| \ |||H_0|||$ and consequently

$$\max_{t\in[0,2\omega]} ||x_{i0}(t)|| \leqslant \exp(2\omega K_1) .$$

In particular, $||X_0(\omega)|| \leqslant C \exp(2\omega K_1)$ where $C \geqslant 1$ is a constant depending only on n. Since $X_0(\omega)$ is E-unitary,

$$||X_0(\omega)^{-1}|| \leq ||E^{-1}|| \; ||X_0^*(\omega)|| \; ||E|| \leq K_2 \exp(2\omega K_1)$$

where $K_2 > 1$ depends only on E.

For the interval $[-2\omega,0]$ apply the same construction to the equation

$$\frac{dy_{i0}}{dt} = iE^{-1}H_0(t)y_{i0} \; , \quad y_{i0}(-2\omega) = f_i \; , \tag{1.10}$$

where f_i is the i-th column of $X_0(\omega)^{-2}$. Note that (1.10) and (1.8) have the same solution $x_{i0}(t)$. Indeed, (1.5) gives

$$X_0(t-2\omega) = X_0(t)X_0(\omega)^{-2} \; ,$$

so for $t = 0$ the i-th column of this equation has the form

$$x_{i0}(-2\omega) = X_0(\omega)^{-2}e_i = f_i$$

which coincides with the boundary condition in (1.10). We obtain

$$\max_{t\in[-2\omega,0]} ||x_{i0}(t)|| \leq ||f_i|| \exp(2\omega K_1) \leq ||X_0(\omega)^{-1}||^2 \exp(2\omega K_1)$$
$$\leq K_2^2(\exp(2\omega K_1))^3 \; .$$

Hence,

$$\max_{t\in[-2\omega,2\omega]} ||x_{i0}(t)|| \leq K_2^2(\exp(2\omega K_1))^3 \; ; \quad i = 1,\cdots,n \; .$$

The same estimation for (1.9) gives (using $|||H||| \leq |||H_0||| + 1$):

$$\max_{t\in[-2\omega,2\omega]} ||x_i(t)|| \leq K_2^2(\exp(2\omega(K_1+1)))^3 \; ; \quad i = 1,\cdots,n \; .$$

Denote $K_3 = K_2^2(\exp(2\omega(K_1+1)))^3$. Now apply Theorem 3.3.1 in [22] with $F(x,y) = iE^{-1}H_0(x)y$; $G(x,y) = iE^{-1}H(x)y$; $a = 2\omega$; $y_0 = e_i$; $b = K_3 + 1$; $M = ||E^{-1}|| \max_{|x|\leq 2\omega} ||H_0(x)||(K_3+1)$; $K = ||E^{-1}||(|||H_0|||+1)$ to get the following estimate (provided $|||H-H_0||| \leq 1$):

$$\max_{|t|\leq\omega} ||x_i(t)-x_{i0}(t)|| \leq \frac{|||H_0-H|||(K_3+1)}{||E^{-1}||(|||H_0|||+1)} \{\exp[(|||H_0|||+1)|t|] - 1\} \; ,$$

from which we obtain the inequality

$$\max_{t\in[0,\omega]} ||X(t)-X_0(t)|| \leq N_1 |||H-H_0||| \tag{1.11}$$

provided $|||H-H_0|||$ is small enough, where $N_1 > 0$ is a constant depending on H_0 only. Further, for $t \in [0,\omega]$:

$$||\frac{dX(t)}{dt} - \frac{dX_0(t)}{dt}|| \leq ||E^{-1}||\ ||H(t)||\ ||X(t)-X_0(t)|| + ||E^{-1}||\ ||H(t)-H_0(t)||\ ||X_0(t)|| , \tag{1.12}$$

and (1.11) and (1.12) ensure that φ is locally Lipschitz.

The formula

$$\varphi^{-1}(X(t)) = -iE \frac{dX(t)}{dt} X(t)^{-1} , \quad t \in [0,\omega]$$

(see the proof of Theorem 1.1) then implies easily that φ^{-1} is locally Lipschitz. □

1.2 The Monodromy Matrix

Let $X(t)$ be the matrizant of (1.2). As we shall see later, the stability properties of the system (1.2) largely depend on the structure of the *monodromy matrix* $X(\omega)$. In this section we shall study some elementary properties of the monodromy matrix and its dependence on $H(t)$. The following statement is in fact a corollary from Theorem 1.1.

THEOREM 1.3 *The monodromy matrix of* (1.2) *is* E-*unitary. Conversely, every* E-*unitary matrix is a monodromy matrix for some system of type* (1.2).

PROOF. The first statement is obvious in view of Theorem 1.1. To prove the second statement observe that the group of E-unitary matrices is arcwise connected (Lemma I.3.8), so there exists a continuous path from I to any given E-unitary matrix in this group. From the proof of Lemma I.3.8 it is seen that this path can be chosen to be piecewise continuously differentiable, and the second statement of Theorem 1.3 follows from Theorem 1.1. □

The group $\mathbb{U}(E)$ of all E-unitary matrices has the natural metric inherited from the space of all complex $n \times n$ matrices. By associating with each system (1.2) its monodromy matrix, we obtain a map

$$\varphi_0 : H_\omega \to \mathbb{U}(E) ,$$

which is continuous in view of Theorem 1.2. Moreover:

THEOREM 1.4 *The map* φ_0 *is open.*

PROOF. To prove the openess of φ it is sufficient to verify that for every sequence $X_m \to X$ with $X_m, X \in \mathbb{U}(E)$ there is a sequence $H_m(t) \in H_\omega$ and $H(t) \in H_\omega$ such that

$$\lim_{m\to\infty} |||H_m - H||| = 0 \quad \text{and} \quad \varphi_0(H_m) = X_m \ ; \quad \varphi_0(H) = X \ .$$

As the proof of Theorem 1.1 shows, it is sufficient to construct piecewise continuously differentiable paths $F_m(t)$, $m = 1,2,\cdots$ and $F(t)$ such that

$$F_m(0) = I \ ; \quad F_m(\omega) = X_m \ ; \quad F_m(t) \in \mathbb{U}(E) \ , \quad t \in [0,\omega] \ ; \tag{1.13}$$

$$F(0) = I \ ; \quad F(\omega) = X \ ; \quad F(t) \in \mathbb{U}(E) \ , \quad t \in [0,\omega] \ ; \tag{1.14}$$

and

$$\lim_{m\to\infty}\left\{ \max_{t\in[0,\omega]} ||F_m(t)-F(t)|| + \max_{t\in[0,\omega]} \left|\left|\frac{dF_m(t)}{dt} - \frac{dF(t)}{dt}\right|\right| \right\} = 0 \ . \tag{1.15}$$

We shall follow the idea of the proof of Lemma I.3.8. Let $\alpha \neq \pm 1$ be a unimodular complex number such that $X_m - \alpha I$, $m = 1,2,\cdots$ and $X - \alpha I$ are invertible. Put

$$A_m = (X_m - \alpha I)^{-1}(\zeta X_m - \bar{\zeta}\alpha I) \ ;$$

$$A = (X - \alpha I)^{-1}(\zeta X - \bar{\zeta}\alpha I) \quad ,$$

where $\zeta \in \mathbb{C} \setminus \mathbb{R}$ is a fixed number. We take $\zeta = \frac{|1-\alpha|^2}{\alpha - \bar{\alpha}}$. The matrices A_m and A are E-selfadjoint. Put

$$A(t) = \frac{\omega - t}{\omega} I + \frac{t}{\omega} A \ ; \quad t \in [0,\omega]$$

$$A_m(t) = \frac{\omega - t}{\omega} I + \frac{t}{\omega} A_m \ ; \quad t \in [0,\omega] \ .$$

Let $\rho(t) \in \mathbb{C} \setminus \mathbb{R}$ be a piecewise continuously differentiable arc such that $\rho(0) = \rho(\omega) = \zeta$;

$$\rho(t) \notin \sigma(A(t)) \ , \quad t \in [0,\omega] \ .$$

Since $A_m(t) \to A(t)$ uniformly on $t \in [0,\omega]$, also

$$\rho(t) \notin \sigma(A_m(t)) \quad \text{for} \quad m \geq m_0 .$$

Now put

$$\tilde{F}_m(t) = (A_m(t) - \overline{\rho(t)}I)(A_m(t) - \rho(t)I)^{-1} \; ; \quad m \geq m_0$$

$$\tilde{F}(t) = (A(t) - \overline{\rho(t)}I)(A(t) - \rho(t)I)^{-1} \quad .$$

Then

$$\tilde{F}_m(0) = I \; ; \quad \tilde{F}_m(\omega) = \alpha^{-1}X_m \; ; \quad \tilde{F}_m(t) \in \mathbb{U}(E) \; , \quad t \in [0,\omega] \; ;$$

$$\tilde{F}(0) = I \; ; \quad \tilde{F}(\omega) = \alpha^{-1}X \; ; \quad \tilde{F}(t) \in \mathbb{U}(E) \; , \quad t \in [0,\omega] \; .$$

So to satisfy (1.13) (for $m \geq m_0$), (1.14) and (1.15) it remains to put

$$F_m(t) = e^{i(\frac{t}{\omega} \arg \alpha)}\tilde{F}_m(t) \; ;$$

$$F(t) = e^{i(\frac{t}{\omega} \arg \alpha)}\tilde{F}(t) \quad . \quad \square$$

1.3 <u>The Floquet Theorem</u>

For the system of differential equations (1.2) there is a well-known result, attributed to Floquet, which establishes a strong connection between a system with periodic coefficients and a system of the same type with constant coefficients. This reduction can be achieved when the matrizant of the first system is known. A short account of this process is given in this section, but first we need a preliminary result concerning logarithms of E-unitary matrices.

LEMMA 1.5 *Let* U *be an E-unitary matrix, where* E *is an* $n \times n$ *invertible hermitian matrix. Then there exists an E-skew-adjoint logarithm* V *of* E, *i.e. an* $n \times n$ *matrix* V *for which* $e^V = U$ *and* $EV = -V^*E$.

This lemma is just a re-wording of Theorem I.6.2. The theorem of Floquet can now be stated as follows.

THEOREM 1.6 *Let* E *be an invertible hermitian matrix, and* $H(t)$ *a hermitian piecewise continuous periodic matrix-valued function with period* ω. *Let* $X(t)$ *be a solution of the system*

$$E\,\frac{dX}{dt} = iH(t)X\ , \tag{1.16}$$

for $t \in \mathbb{R}$ *and let* $Z(t)$ *be the matrizant of this system. If* V *is a (constant)* E-*skew-adjoint matrix, write* $P(t) = Z(t)e^{-Vt}$, *and define* $Y(t)$ *by* $X(t) = P(t)Y(t)$; *then* $Y(t)$ *satisfies the system with constant coefficients*

$$E\,\frac{dY}{dt} = iAY \tag{1.17}$$

where A *is the hermitian matrix* $-iEV$.

Furthermore, if ωV *is a logarithm of the monodromy matrix* $Z(\omega)$, *then the transforming matrix* $P(t)$ *is periodic with period* ω.

PROOF. It is clear that $X(t) = Z(t)X(0)$ and so $X(t) = P(t)Y(t)$ implies $X(0) = e^{-Vt}Y(t)$. But also $X(0) = Y(0)$, so $Y(t) = e^{Vt}Y(0)$ and hence $Y(t)$ is a solution of $\frac{dY}{dt} = VY$. Multiplying by E and using the fact that $EV = -V^*E$, equality (1.17) is obtained.

Since $Z(t+\omega) = Z(t)Z(\omega)$,

$$P(t+\omega) = Z(t)Z(\omega)e^{-V\omega}\cdot e^{-Vt}\ .$$

Since $Z(\omega)$ is E-unitary, Lemma 1.5 implies that there is an E-skew-adjoint matrix ωV such that $Z(\omega) = e^{\omega V}$. Consequently,

$$P(t+\omega) = Z(t)e^{-Vt} = P(t)\ . \quad \square$$

1.4 The Real Case

Consider the system of differential equations

$$E\,\frac{dx}{dt} = H(t)x\ , \quad t \in \mathbb{R}\ , \tag{1.18}$$

where E is a skew-symmetric ($E = -E^T$) real invertible $n \times n$ matrix and $H(t)$, $t \in \mathbb{R}$, is an $n \times n$ piecewise continuous real matrix function, which is symmetric and periodic:

$$H(t) = H(t)^T\ ; \quad H(t+\omega) = H(t)\ ; \quad t \in \mathbb{R}\ .$$

Note that invertibility of E demands that the size n of the matrices is even (indeed, all zeros of $\det(\lambda I - E)$ are pure imaginary).

The results of Section 1.1, together with their proofs, also hold for the real system (1.18). In this case, however, it is more convenient to speak in terms of the group $\mathbb{U}_r(E)$ of E-*orthogonal* matrices X, i.e. real matrices X such that $X^T E X = E$ (rather than in terms of real iE-unitary matrices, as in the preceding sections).

We shall need the following properties of the group $\mathbb{U}_r(E)$.

THEOREM 1.7 *Let* $X \in \mathbb{U}_r(E)$, *and let* $X = SU$ *be the polar decomposition of* X, *where* $S = (XX^T)^{\frac{1}{2}}$ *is real positive definite, and* U *is real unitary. Then* S *and* U *are also* E-*orthogonal. The set of all matrices which are unitary and* E-*orthogonal, is homeomorphic to the set of all* $k \times k$ *unitary matrices, where* $k = \frac{n}{2}$. *The set of all matrices which are positive definite and* E-*orthogonal is homeomorphic to* $\mathbb{R}^{k(k+1)}$. *So the set* $\mathbb{U}_r(E)$ *is homeomorphic to the direct product of* $\mathbb{R}^{k(k+1)}$ *and the set of all* $k \times k$ *unitary matrices. In particular,* $\mathbb{U}_r(E)$ *is connected.*

The proof of Theorem 1.7 will be given in the next section. It will be seen from the proof that the homeomorphisms in Theorem 1.7 as well as their inverses, are given by real analytic functions of the entries of appropriate matrices (so the homeomorphisms are in fact real analytic). This fact will be used in the proof of Theorem 1.9 below.

Using Theorem 1.7 we shall also prove the results of Section 1.2 in the real case. The next result is the real analogue of Theorem 1.3. But observe first that a "real version" of Theorem 1.1 is immediately obtained on making small adjustments to the wording of the proof.

THEOREM 1.8 *The monodromy matrix of* (1.18) *is* E-*orthogonal. Conversely, every* E-*orthogonal matrix is the monodromy matrix for some real system* (1.18).

PROOF. The first statement follows obviously from the real version of Theorem 1.1. Conversely, let $X \in \mathbb{U}_r(E)$. Since $\mathbb{U}_r(E)$ is connected, there is a continuous path from I to X in the group of E-orthogonal matrices. It follows that there is also a piecewise continuously differentiable path connecting I and X in $\mathbb{U}_r(E)$. This fact is intuitively clear; a rigorous proof can be given using the fact that $\mathbb{U}_r(E)$ is homeomorphic to the direct product of $\mathbb{R}^{\frac{n}{2}(\frac{n}{2}+1)}$ and the group of $\frac{n}{2} \times \frac{n}{2}$ unitary matrices (Theorem 1.7). Now apply the real version of Theorem 1.1 again. □

As in the complex case, we have a map

$$\varphi_0 : \mathcal{H}_{\omega,r} \to \mathbb{U}_r(E) ,$$

where $\mathcal{H}_{\omega,r}$ is the set of all real $n \times n$ matrix functions $H(t)$, $t \in \mathbb{R}$ which are symmetric and periodic with the period ω, and $\varphi_0(H(t))$ is the monodromy matrix of system (1.18).

THEOREM 1.9 *The map* φ_0 *is continuous and open.*

PROOF. The continuity of φ_0 follows from the real analogue of Theorem 1.2. To prove the openess of φ_0, for every sequence $X_m \to X_0$; $X_m, X_0 \in \mathbb{U}_r(E)$ we have to construct piecewise continuously differentiable paths $F_m(t)$, $m = 1,2,\cdots$ and $F(t)$ such that

$$F_m(0) = I , \quad F_m(\omega) = X_m ; \quad F_m(t) \in \mathbb{U}_r(E) , \quad t \in [0,\omega] \tag{1.19}$$

$$F(0) = I , \quad F(\omega) = X_0 ; \quad F(t) \in \mathbb{U}_r(E) , \quad t \in [0,\omega] \tag{1.20}$$

and

$$\lim_{m\to\infty}\left\{ \max_{t\in[0,\omega]} ||F_m(t)-F(t)|| + \max_{t\in[0,\omega]} ||\frac{dF_m(t)}{dt} - \frac{dF(t)}{dt}||\right\} = 0 . \tag{1.21}$$

In view of Theorem 1.7 it is sufficient to consider two cases separately: (1) the matrices X_m are E-orthogonal and unitary; (2) the matrices X_m are positive definite and E-orthogonal. The second case is trivial because the set of all positive definite E-orthogonal matrices is homeomorphic (with a real analytic homeomorphism) to $\mathbb{R}^{k(k+1)}$ (here $k = \frac{n}{2}$). The first can be dealt with as in the proof of Theorem 1.4, because the set of unitary E-orthogonal matrices is analytically homeomorphic to the group of $k \times k$ unitary matrices. □

The real analogues of the results of Section 1.3 hold also provided the monodromy matrix of (1.18) does not have real negative eigenvalues. We shall not prove these results, but refer the reader to [51], Section 4.5.

1.5 Proof of Theorem 1.7

We shall need the following auxilliary result.

LEMMA 1.10 *Let* X *be an* $n \times n$ *real positive definite matrix. Then there exists a unique real symmetric matrix* A *denoted* $A = \tau(X)$ *such that* $X = e^A$. *This correspondence* τ *is a homeomorphism between the set of all* $n \times n$ *real positive definite matrices and the set of all* $n \times n$ *real symmetric matrices. Moreover,* τ *and its inverse are analytic, i.e.* $\tau(X)$ *(resp.* $\tau^{-1}(A)$*) is given by analytic functions in the entries of* X *(resp. the entries of* A*).*

PROOF. Let $\lambda_1,\cdots,\lambda_n$ be the eigenvalues of X (all of them are real), and let S be a (real) orthogonal matrix such that $X = S\,\mathrm{diag}[\lambda_1,\cdots,\lambda_n]S^{-1}$. Any matrix Y (not necessarily real) such that $e^Y = X$ is given by the following formula (see [14], Ch. VIII):

$$Y = ST\,\mathrm{diag}[\ln\lambda_1+2i\pi m_1,\cdots,\ln\lambda_n+2i\pi m_n]T^{-1}S^{-1},$$

where T is an $n\times n$ matrix commuting with $\mathrm{diag}[\lambda_1,\cdots,\lambda_n]$, $m_1,\cdots,m_n$ are integers, and $\ln\lambda_j$ is understood as the real number such that $\exp(\ln\lambda_j) = \lambda_j$. Assume now that Y is hermitian. In particular, the eigenvalues of Y are real; then necessarily $m_1 = \cdots = m_n = 0$ and

$$Y = S\,\mathrm{diag}[\ln\lambda_1,\cdots,\ln\lambda_n]S^{-1} \tag{1.22}$$

is real. So there exists a unique real symmetric matrix A such that $X = e^A$. On the other hand, for every real symmetric matrix A the matrix $X = e^A$ is positive definite. Hence τ is one-to-one map of the set of all $n\times n$ real positive matrices onto the set of all $n\times n$ real symmetric matrices. Finally, continuity and analyticity of τ and τ^{-1} are established easily in view of the following formulas:

$$\tau^{-1}(A) = e^A;$$

$$\tau(X) = \frac{1}{2\pi i}\int_\Gamma(\lambda I-X)^{-1}\ln\lambda\, d\lambda. \tag{1.23}$$

Here Γ is a suitable contour such that $\sigma(X)$ is inside Γ and $\ln\lambda$ is the analytic branch of the logarithm defined by the property that $\ln\lambda$ is real for $\lambda > 0$. □

We pass now to the proof of Theorem 1.7 itself.

Without loss of generality, we can assume $E = \begin{bmatrix} 0 & I_k \\ -I_k & 0\end{bmatrix}$ (cf. Corollary III.1.2). Let $X \in \mathbb{U}_r(E)$, with the polar decomposition $X = SU$. We have:

$$X = E^{-1}(X^T)^{-1}E = E^{-1}(S^T)^{-1}(U^T)^{-1}E = E^{-1}(S^T)^{-1}E\cdot E^{-1}(U^T)^{-1}E.$$

The last product is again a polar decomposition of X, and in view of uniqueness of the polar decomposition we get $E^{-1}(S^T)^{-1}E = S$, $E^{-1}(U^T)^{-1}E = U$, i.e. $S \in \mathbb{U}_r(E)$ and $U \in \mathbb{U}_r(E)$.

Assume now that $X \in \mathbb{U}_r(E)$ and X is unitary. Write

$$X = \begin{bmatrix} A & B \\ C & D \end{bmatrix}$$

with $k \times k$ matrices A, B, C, D. Then conditions

$$X^T E X = E \ , \quad X^T X = I \tag{1.24}$$

imply that $XE = XE$ which means $A = D$, $C = -B$. Now it is easily seen that (1.24) holds if and only if $X = \begin{bmatrix} A & B \\ -B & A \end{bmatrix}$ with

$$A^T A + B^T B = I \ , \quad A^T B = B^T A \ . \tag{1.25}$$

But (1.25) means exactly that $A + iB$ is unitary.

Finally, let $X \in \mathbb{U}_r(E)$ be positive definite. By Lemma 1.10 there exists a unique real symmetric matrix A such that $X = e^A$. Now

$$e^A = E^{-1}(e^A)^{-1}E = E^{-1}e^{-A}E = e^{-E^{-1}AE} \ .$$

The matrix $-E^{-1}AE$ is real and symmetric; since the real symmetric matrix A with $e^A = X$ is unique, we obtain $A = -E^{-1}AE$. So the set of positive definite E-orthogonal matrices is homeomorphic (even analytically homeomorphic) to the set of real symmetric $n \times n$ matrices A with the property $A = -E^{-1}AE$. It is easily checked that all such matrices A have the form

$$A = \begin{bmatrix} A_1 & A_2 \\ A_2 & -A_1 \end{bmatrix} , \tag{1.26}$$

where $A_1 = A_1^T$ and $A_2 = A_2^T$ are real. Since the set of $k \times k$ real symmetric matrices is homeomorphic to $\mathbb{R}^{\frac{k(k+1)}{2}}$, we obtain the conclusion of Theorem 1.7 concerning the set of positive definite E-orthogonal matrices. □

1.6 Selfadjoint Equations With Periodic Coefficients

Consider a system of differential equations

$$iQ(t)\,\frac{dx}{dt} + \left(H(t) + \frac{i}{2}\,\frac{dQ(t)}{dt}\right)x = 0 \tag{1.27}$$

for $t \in \mathbb{R}$, where $Q(t)$ is an invertible, hermitian, complex, $n \times n$ matrix-valued function: $Q(t)^* = Q(t)$ for $t \in \mathbb{R}$, and $H(t)$ is an $n \times n$ hermitian matrix function: $H(t)^* = H(t)$ for $t \in \mathbb{R}$. It is assumed also that $Q(t)$ is piecewise continuously differentiable and $H(t)$ is piecewise continuous. Clearly, equation (1.2) is a specific example.

It will be shown that the system (1.27) is selfadjoint in an appropriate sense. For the purpose of definition, let Lx be a first order linear differential expression of the form

$$(Lx)(t) = L_0(t)\frac{dx}{dt} + L_1(t)x$$

for $t \in \mathbb{R}$, where $L_0(t)$ is a piecewise continuously differentiable $n \times n$ matrix function, $L_1(t)$ is a piecewise continuous $n \times n$ matrix function, and $x = x(t)$ is a piecewise continuously differentiable function with values in $\mathbb{C}^n$.

The *adjoint* differential expression, L^*x, of Lx is defined by

$$(L^*x)(t) = -(L_0(t)^*x)' + L_1(t)^*x$$

for all $t \in \mathbb{R}$, and has the characteristic property that for any real a and b, and any complex vector functions $x(t)$ and $y(t)$ with piecewise continuous derivatives the equality

$$\int_a^b (Lx,y)dt = P(\eta,\xi) + \int_a^b (x,L^*y)dt ,$$

holds, where $P(\eta,\xi)$ is a bilinear form in the variables $\eta = (x(a),x(b))$ and $\xi = (y(a),y(b))$. (See, e.g.,[36] for more details). One checks easily that the left-hand side of (1.27) is indeed selfadjoint; namely if

$$Lx = iQ(t)\frac{dx}{dt} + \left(H(t) + \frac{i}{2}\frac{dQ(t)}{dt}\right)x$$

and $Q(t)$, $H(t)$ have the properties mentioned in the beginning of this section, then $L = L^*$.

Conversely, every selfadjoint differential expression of the first order with invertible leading coefficient has the form (1.27). To see this, write $Lx = L_0\frac{dx}{dt} + L_1x$; then selfadjointness of L means $L_0(t) = -L_0(t)^*$ and

$$L_1(t) = -\frac{dL_0^*(t)}{dt} + L_1(t)^* = \frac{dL_0(t)}{dt} + L_1(t)^* .$$

Consequently,

$$L_1(t) - \frac{1}{2}\frac{dL_0(t)}{dt} = \left(L_1(t) - \frac{1}{2}\frac{dL_0(t)}{dt}\right)^* .$$

Putting $L_0 = iQ$ and $L_1 - \frac{1}{2}\frac{dL_0}{dt} = H$, it is found that L has the form (1.27).

In the rest of this section we consider the selfadjoint system (1.27) under the additional condition that $Q(t)$ and $H(t)$ have a common period ω: $Q(t+\omega) = Q(t)$; $H(t+\omega) = H(t)$. Evidently, (1.2) is still a particular case of (1.27) with $Q(t) \equiv E$, a constant matrix. As in Section 1.1 introduce the *matrizant* $X(t)$ of (1.27) which is the unique solution of the initial value problem

$$Q(t)\frac{dX}{dt} - \left(iH(t) - \frac{1}{2}\frac{dQ(t)}{dt}\right)X = 0 ; \quad X(0) = I . \tag{1.28}$$

for $t \in [0,\omega]$.

LEMMA 1.11 *If* $X(t)$ *is the matrizant of* (1.27), *then*

$$X(t)^*Q(t)X(t) = Q(0) . \tag{1.29}$$

PROOF. Differentiating the left hand side of (1.29) gives

$$\begin{aligned}\frac{d}{dt}(X^*QX) &= \frac{dX^*}{dt}QX + X^*\frac{dQ}{dt}X + X^*Q\frac{dX}{dt}\\ &= (Q\frac{dX}{dt})^*X + X^*\frac{dQ}{dt}X + X^*Q\frac{dX}{dt} ,\end{aligned}$$

where the hermitian property of Q was used. Since X is a solution of (1.28),

$$\frac{d}{dt}(X^*QX) = X^*(-iH^* - \frac{1}{2}\frac{dQ^*}{dt})X + X^*\frac{dQ}{dt}X + X^*(iH - \frac{1}{2}\frac{dQ}{dt})X ,$$

which reduces to zero since $Q = Q^*$ and $H = H^*$. So the left hand side of (1.29) is a constant and since $X(0) = I$, this constant must be $Q(0)$. □

Introduce the *monodromy matrix* $X(\omega)$ of (1.27), where $X(t)$ is the matrizant of (1.27). Since $Q(t)$ has period ω Lemma 1.11 shows, in particular, the monodromy matrix is $Q(0)$-unitary. This proves the first part of the next theorem.

THEOREM 1.12 *The matrizant* $X(t)$ *of* (1.27) *is a piecewise continuously differentiable curve of invertible matrices such that* $X(0) = I$ *and* $X(\omega)$ *is* $Q(0)$-*unitary. Conversely, any continuously differentiable curve* $X(t)$, $t \in [0,\omega]$ *of invertible matrices such that* $X(0) = I$ *and* $X(\omega)$ *is* G-*unitary for some invertible hermitian matrix* G, *is the matrizant of a system of type* (1.27), *where* $Q(0) = G$.

PROOF. It remains to prove the converse statement. Given a continuously differentiable curve $X(t)$, $t \in [0,\omega]$ as in Theorem 1.12 define

$$Q(t) = X(t)^{*-1} G X(t)^{-1}$$

for $t \in [0,\omega]$ and then define $Q(t)$ on $\mathbb{R}$ by periodicity. It is easily seen that $Q(t)$ is hermitian with piecewise continuous derivative and $Q(0) = Q(\omega)$. Also, $X(t)^* Q(t) X(t) = G$ and differentiation gives

$$\frac{dX^*}{dt} QX + X^* \frac{dQ}{dt} X + X^* Q \frac{dX}{dt} = 0 .$$

Hence

$$Q \frac{dX}{dt} - \left(-X^{*-1} \frac{dX^*}{dt} Q - \frac{dQ}{dt}\right) X = 0 .$$

An equation of the form (1.27) is obtained on defining

$$H(t) = iX^{*-1}(t) \frac{dX^*(t)}{dt} Q(t) + \frac{i}{2} \frac{dQ(t)}{dt} .$$

Let us prove that $H(t)$ is hermitian. Indeed,

$$\begin{aligned} H - H^* &= iX^{*-1} \frac{dX^*}{dt} Q + \frac{i}{2} \frac{dQ}{dt} + iQ^* \frac{dX}{dt} X^{-1} + \frac{i}{2} \frac{dQ^*}{dt} \\ &= iX^{*-1} \frac{dX^*}{dt} Q + iQ \frac{dX}{dt} X^{-1} + i \frac{dQ}{dt} \\ &= iX^{*-1} \left\{ \frac{dX^*}{dt} QX + X^* Q \frac{dX}{dt} + X^* \frac{dQ}{dt} X \right\} X^{-1} \equiv 0 . \quad \square \end{aligned}$$

In contrast with the hamiltonian system (1.2), the selfadjoint system (1.27) (with periodic coefficients) is not defined uniquely by its matrizant $X(t)$ and the hermitian invertible matrix G such that $X(\omega)$ is G-unitary. For instance, one can replace $Q(t)$ and $H(t)$ by $\alpha Q(t)$ and $\alpha H(t)$, respectively, where $\alpha \neq 0$ is a real number, without changing the matrizant.

Let $\mathcal{QH}_\omega$ be the set of all pairs $\{Q(t),H(t)\}$, where $Q(t)$ is an invertible, hermitian, piecewise continuously differentiable complex $n \times n$ matrix function with period ω, and $H(t)$ is an hermitian, piecewise continuous, complex $n \times n$ matrix function with period ω. Denote by JG_ω the set of all pairs $\{X(t),G\}$, where G is an invertible hermitian $n \times n$ matrix, and $X(t)$, $t \in [0,\omega]$, is a piecewise continuously differentiable curve of $n \times n$ invertible matrices such that $X(0) = I$ and $X(\omega)$ is G-unitary. There is a natural map

$$\psi : QH_\omega \to JG_\omega , \tag{1.30}$$

which is defined as follows: $\psi\{Q(t),H(t)\} = \{X(t),G\}$, where $G = Q(0)$ and $X(t)$ is the matrizant of system (1.27). Theorem 1.12 says that ψ is onto.

Extend the definition of Section 1.1 to introduce distance functions ρ_1 and ρ_2 in QH_ω and JG_ω, respectively, thereby making QH_ω and JG_ω into metric spaces. For $\{Q_i(t),H_i(t)\} \in QH_\omega$, $i = 1,2$, put

$$\begin{aligned}
&\rho_1(\{Q_1(t),H_1(t)\} , \{Q_2(t),H_2(t)\}) \\
&= \sup_{t\in[0,\omega]} ||Q_1(t)-Q_2(t)|| + \sup_{t\in[0,\omega]} \left|\left|\frac{dQ_1(t)}{dt} - \frac{dQ_2(t)}{dt}\right|\right| \\
&\quad + \sup_{t\in[0,\omega]} ||H_1(t)-H_2(t)|| ,
\end{aligned}$$

and for $\{X_i(t),G_i\} \in JG_\omega$, $i = 1,2$, put

$$\begin{aligned}
\rho_2(\{X_1(t),G_1\} , \{X_2(t),G_2\}) &= \sup_{t\in[0,\omega]} ||X_1(t)-X_2(t)|| \\
&+ \sup_{t\in[0,\omega]} \left|\left|\frac{dX_1(t)}{dt} - \frac{dX_2(t)}{dt}\right|\right| + ||G_1-G_2|| .
\end{aligned}$$

With these definitions of metric topologies in QH_ω and JG_ω, the following result holds.

THEOREM 1.13 *The map* ψ *defined by* (1.30) *is locally Lipschitz (in particular, continuous) and open.*

PROOF. The local Lipschitz property of ψ can be proved as in the proof of Theorem 1.2. Let us prove the openess of ψ. Let $\{X_m(t),G_m\} \in JG_\omega$, $m = 1,2,\cdots$ be a sequence such that

$$\{X_m(t),G_m\} \to \{X_0(t),G_0\} , \quad \text{and} \quad \{X_0(t),G_0\} \in JG_\omega .$$

We shall verify that there exists a sequence $\{Q_m(t),H_m(t)\} \in \mathcal{QH}_\omega$, $m = 1,2,\cdots$ and a pair $\{Q_0(t),H_0(t)\} \in \mathcal{QH}_\omega$ such that $\{Q_m(t),H_m(t)\} \to \{Q_0(t),H_0(t)\}$ and

$$\psi\{Q_j(t),H_j(t)\} = \{X_j(t),G_j\} , \quad j = 0,1,\cdots .$$

The proof of Theorem 1.12 shows that one can take

$$Q_j(t) = X_j(t)^{*-1}G_jX_j(t)^{-1} , \quad t \in [0,\omega] , \quad j = 0,1,\cdots ,$$

and

$$H_j(t) = iX_j^{*-1}\frac{dX_j^*}{dt}Q_j + \frac{i}{2}\frac{dQ_j}{dt} , \quad t \in [0,\omega] , \quad j = 0,1,\cdots ,$$

and then continue $Q_j(t)$ and $H_j(t)$ by periodicity to the whole real line. □

We pass now to analysis of the monodromy matrix. Consider the set $\mathcal{U}$ of all pairs (A,G), where G is an invertible, hermitian, $n \times n$ matrix, and A is G-unitary. The set $\mathcal{U}$ has a natural metric topology. Define the map

$$\psi_0 : \mathcal{QH}_\omega \to \mathcal{U} , \tag{1.31}$$

which maps $\{Q(t),H(t)\} \in \mathcal{QH}_\omega$ to the pair $(X(\omega),Q(0))$, where $X(\omega)$ is the monodromy matrix of system (1.27).

THEOREM 1.14 *The map* ψ_0 *is locally Lipschitz (in particular, continuous) and open.*

PROOF. Theorem 1.12 ensures that ψ_0 is locally Lipschitz. Let $(A_m,G_m) \in \mathcal{U}$, $m = 1,2,\cdots$, and $(A_m,G_m) \to (A_0,G_0) \in \mathcal{U}$. We can assume that the signature of G_m, $m = 1,2,\cdots$ coincides with the signature of G_0. Then there exist invertible matrices S_m, $m = 1,2,\cdots$ such that $G_0 = S_m^*G_mS_m$. Since $G_m \to G_0$, we can assume, moreover, that $S_m \to I$. (Indeed, let T_m, $m = 1,2,\cdots$ and T_0 be invertible matrices such that

$$T_m^*G_mT_m = T_0^*G_0T_0 = \begin{bmatrix} I_p & 0 \\ 0 & -I_q \end{bmatrix} .$$

Lagrange's algorithm (see [14] or [30b]) shows that one can choose T_m in such a way that $T_m \to T_0$. Now put $S_m = T_mT_0^{-1}$.)

If $B_m = S_m^{-1}A_mS_m$, then B_m is G_0-unitary and $B_m \to A_0$. By Theorem 1.4 there are piecewise differentiable paths $F_j(t)$, $j = 0,1,\cdots$, $t \in [0,\omega]$ with G_0-unitary values such that $F_j(0) = I$, $F_0(\omega) = A_0$, $F_m(\omega) = B_m$,

$m = 1,2,\cdots$, and

$$\lim_{m\to\infty}\left\{\max_{t\in[0,\omega]} ||F_m(t)-F_0(t)|| + \max_{t\in[0,\omega]} ||\frac{dF_m}{dt} - \frac{dF_0}{dt}||\right\} = 0 .$$

Put $X_m(t) = S_m F_m(t) S_m^{-1}$, $t \in [0,\omega]$, $m = 1,2,\cdots$, and $X_0(t) = F_0(t)$. For $m = 0,1,2,\cdots$ and any $t \in [0,\omega]$ define

$$Q_m(t) = X_m(t)^{*-1} G_m X_m(t)^{-1} ;$$

$$H_m(t) = iX_m(t)^{*-1} \frac{dX_m^*(t)}{dt} Q_m(t) + \frac{i}{2}\frac{dQ_m(t)}{dt} .$$

Continue $Q_m(t)$ and $H_m(t)$ by periodicity to the whole real line; then $\{Q_m(t),H_m(t)\} \in \mathcal{QH}_\omega$, and $\{Q_m(t),H_m(t)\} \to \{Q_0(t),H_0(t)\}$. Moreover,

$$\psi_0\{Q_m(t),H_m(t)\} = (A_m,G_m) , \quad m = 0,1,2,\cdots ,$$

which proves the openess of ψ_0. □

1.7 The Real Case for Selfadjoint Equations

Consider the system of differential equations

$$Q(t)\frac{dx}{dt} - \left(H(t) - \frac{1}{2}\frac{dQ(t)}{dt}\right)x = 0 , \quad t \in \mathbb{R} ,$$

where $Q(t)$ and $H(t)$ are real $n \times n$ matrix functions with $Q(t)^T = -Q(t)$, $H(t)^T = H(t)$ and smoothness conditions as in the preceding section (in this case n is necessarily even; cf. Section 1.5). Lemma 1.11 and Theorem 1.12 and their proofs are valid in the real case as well. The map ψ (which is the analogue of that given by (1.30)) is defined by

$$\psi : (\mathcal{QH}_\omega)_r \to (\mathcal{JG}_\omega)_r ,$$

where $(\mathcal{QH}_\omega)_r$ consists of all $\{iQ(t),H(t)\} \in \mathcal{QH}_\omega$ with real $Q(t)$ and $H(t)$, and $(\mathcal{JG}_\omega)_r$ consists of all pairs $\{X(t),G\} \in \mathcal{JG}_\omega$ with real $X(t)$ and every entry in G is pure imaginary. This map is onto, locally Lipschitz and open (the proof is similar to the proof of Theorem 1.13). The real version of Theorem 1.14 holds also (instead of Lagrange's algorithm which was used in the proof of Theorem 1.14 one can use the algorithm for reduction of a skew-symmetric form (see [35], Section 91)).

1.8 Boundedness of Solutions of Selfadjoint Equations

Consider the selfadjoint equation (1.27) with periodic coefficients with period ω. The notion of stability appears naturally in physical systems governed by (1.27); namely, the system should not "blow up" for any choice of initial data. Mathematically, this means boundedness of all solutions of (1.27) on the half-line $t \geq 0$. Evidently, the solutions of (1.27) are bounded in this sense if and only if

$$\sup_{0\leq t\leq\infty} ||X(t)|| < \infty , \tag{1.32}$$

where $X(t)$ is the matrizant of (1.27). It turns out that (1.32) holds if and only if

$$\sup_n ||X(\omega)^n|| < \infty \tag{1.33}$$

where the supremum is taken over all nonnegative integers n. Indeed, assuming (1.33) holds, from the equality (1.5) (which holds also for the selfadjoint equation) we find that

$$\sup_{0\leq t\leq\infty} ||X(t)|| \leq \sup_{0\leq t\leq\omega} ||X(t)|| \cdot \sup_n ||X(\omega)^n|| < \infty .$$

Conversely, if (1.32) holds, then, in particular,

$$\sup_n ||X(n\omega)|| = \sup_n ||X(\omega)^n|| < \infty .$$

A necessary and sufficient condition for (1.33) can be easily given in terms of the Jordan structure of the monodromy matrix $X(\omega)$. Namely, (1.33) holds if and only if all the eigenvalues of $X(\omega)$ are inside or on the unit circle, and the elementary divisors of $I\lambda - X(\omega)$ corresponding to the eigenvalues on the unit circle (if any) are all linear. But by Theorem 1.3 $X(\omega)$ is G-unitary (where $G = Q(0)$) and the spectrum of a G-unitary matrix is symmetric relative to the unit circle (Proposition I.2.7); therefore we obtain the following result.

THEOREM 1.15 *All solutions of the system* (1.27) *are bounded if and only if its monodromy matrix is similar to a diagonal matrix with unimodular eigenvalues.*

In particular, Theorem 1.15 gives the necessary and sufficient condition for boundedness of the solutions of the hamiltonian system (1.2) with complex, as well as with real, coefficients.

We remark that (again by Theorem 1.15) the solutions of (1.27) are bounded on the whole real line if and only if they are bounded on the half-line $t \geq 0$.

In Parts III and IV of this book we shall study deeper stability properties of hamiltonian and selfadjoint systems with periodic coefficients.

CHAPTER 2

HERMITIAN MATRIX POLYNOMIALS

As we have seen in the preceding chapter, the solutions of a hamiltonian system with periodic coefficients are intimately connected with those of a system with constant coefficients. This leads naturally to consideration of matrix polynomials of type $E + \lambda H$, where E and H are hermitian matrices. Matrix polynomials with hermitian coefficients of second degree occur in other applications, e.g. in the theory of vibrating systems (see [30a]).

In this chapter we consider matrix polynomials with hermitian coefficients of arbitrary degree. Let $L(\lambda) = \sum_{j=0}^{\ell} A_j\lambda^j$, $A_j = A_j^*$, $j = 0,\cdots,\ell$, be such a polynomial. Assuming A_ℓ is invertible, one forms the *companion matrix*

$$C_L = \begin{bmatrix} 0 & I & 0 & \cdot & \cdot & \cdot & 0 \\ 0 & 0 & I & & & & 0 \\ \cdot & & & \cdot & & & \cdot \\ \cdot & & & & \cdot & & \cdot \\ \cdot & & & & & \cdot & I \\ -\tilde{A}_0 & -\tilde{A}_1 & \cdot & \cdot & \cdot & \cdot & -\tilde{A}_{\ell-1} \end{bmatrix}, \tag{2.1}$$

where $\tilde{A}_j = A_\ell^{-1}A_j$, $j = 0,\cdots,\ell-1$. It is well-known that the spectral structure of $L(\lambda)$ (i.e. the eigenvalues, eigenvectors and generalized eigenvectors) and that of C_L are intimately related; for example, if x_0 is an eigenvector of $L(\lambda)$ corresponding to λ_0 (so that $L(\lambda_0)x_0 = 0$, $x_0 \neq 0$), then the vector

$$<x_0,\lambda_0 x_0,\dots,\lambda_0^{\ell-1}x_0> \overset{\text{def}}{=} \begin{bmatrix} x_0 \\ \lambda_0 x_0 \\ \cdot \\ \cdot \\ \cdot \\ \lambda_0^{\ell-1}x_0 \end{bmatrix}$$

is an eigenvector of C_L corresponding to its eigenvalue λ_0. This observation applies generally to matrix polynomials with invertible leading coefficient (not only with hermitian coefficients). A property peculiar to a matrix polynomial with hermitian coefficients is the fact that its companion matrix is selfadjoint with respect to an indefinite scalar product:

$$B_L C_L = C_L^* B_L ,$$

where

$$B_L = \begin{bmatrix} A_1 & A_2 & \cdot & \cdot & \cdot & A_\ell \\ A_2 & & & & \cdot & \\ \cdot & & & \cdot & & \\ \cdot & & \cdot & & & \\ \cdot & A_\ell & & & 0 & \\ A_\ell & & & & & \end{bmatrix}$$

is hermitian and invertible (note that B_L is never positive (or negative) definite except for the trivial case when $\ell = 1$ and A_ℓ is positive (or negative) definite). Hence the properties of selfadjoint matrices in indefinite scalar products play a key role in the investigation of hermitian matrix polynomials.

In this chapter we shall focus mainly on the immediate applications of Part I to hermitian matrix polynomials. Later (in Parts III and IV) we shall study deeper properties of these polynomials.

The results presented in this chapter are also based on the general theory of matrix polynomials, where a considerable amount of knowledge is available now (see, for example, the authors' monograph [19b]). For the reader's convenience, we shall give here a short account of some basic facts from this theory.

2.1 Preliminaries

In this section attention is confined to matrix polynomials $L(\lambda) = \sum_{j=0}^{\ell} A_j\lambda^j$ $(\lambda \in \mathbb{C})$ where the coefficients $A_0, A_1, \cdots, A_\ell$ are $n \times n$ complex matrices and the leading coefficient A_ℓ is invertible. The *degree* of such a polynomial is ℓ. The degree of the scalar polynomial $\det L(\lambda)$ will obviously be $n\ell$, so the *spectrum* $\sigma(L)$ of L, defined by

$$\sigma(L) = \{\lambda \in \mathbb{C} \mid \det L(\lambda) = 0\} ,$$

is finite and consists of not more than $n\ell$ different complex numbers; these are called the *eigenvalues* of L.

We now define an $n\ell \times n\ell$ matrix whose spectral properties are intimately related to those of L. Let $\hat{A}_j = A_\ell^{-1}A_j$, $j = 0,1,\cdots,\ell-1$ and define the *companion matrix* of L by (2.1). The relationship referred to can be concentrated in the statement that $I\lambda - C_L$ and $L(\lambda) \oplus I_{n(\ell-1)}$ are *equivalent*, where $I_{n(\ell-1)}$ is the unit matrix of size $n(\ell-1)$. That is, there exist $n\ell \times n\ell$ matrix polynomials $E(\lambda)$ and $F(\lambda)$, whose inverses are also matrix polynomials, for which

$$L(\lambda) \oplus I_{\ell-1} = E(\lambda)(I\lambda - C_L)F(\lambda) . \tag{2.2}$$

This fact can be demonstrated by writing out $E(\lambda)$ and $F(\lambda)$ explicitly. In fact, we may take

$$F(\lambda) = \begin{bmatrix} I & 0 & \cdots & 0 \\ I\lambda & I & & \cdot \\ \cdot & & \cdot & \cdot \\ \cdot & & \cdot & \cdot \\ \cdot & & \cdot & \cdot \\ I\lambda^{\ell-1} & I\lambda^{\ell-2} & \cdots & I \end{bmatrix}, \quad E(\lambda) = \begin{bmatrix} K_{\ell-1}(\lambda) & K_{\ell-2}(\lambda) & \cdots & K_1(\lambda) & K_0(\lambda) \\ I & 0 & & & \cdot \\ 0 & I & & & \cdot \\ \cdot & & \cdot & & \cdot \\ \cdot & & & \cdot & \cdot \\ 0 & 0 & \cdots & I & 0 \end{bmatrix}$$

where $K_0(\lambda) = A_\ell$ and $K_{r+1}(\lambda) = \lambda K_r(\lambda) + A_{\ell-r-1}$ for $r = 0,1,\cdots,\ell-2$. Clearly, $\det F(\lambda) \equiv 1$, $\det E(\lambda) \equiv \pm\det A_\ell \neq 0$, and hence $E^{-1}(\lambda)$ and $F^{-1}(\lambda)$ are polynomials. A direct multiplication shows that (2.2) is satisfied.

It follows from this equivalence that the elementary divisors of $L(\lambda)$ and $I\lambda - C_L$ coincide. In particular, the eigenvalues of L coincide with those of C_L, and, furthermore, their partial multiplicities agree.

More generally, any $\ell n \times \ell n$ matrix T for which $L(\lambda) \oplus I_{n(\ell-1)}$ and $I\lambda - T$ are equivalent is called a *linearization* of L. It follows that *all linearizations* of L *are similar* to one another and, in particular, to the companion matrix C_L.

To study the matrix polynomial $L(\lambda)$ it is convenient to introduce pairs (X,T) of matrices as follows: X is an $n \times n\ell$ matrix and T is an $n\ell \times n\ell$ matrix such that

$$X = [I\ 0\cdots 0]S\ , \quad T = S^{-1}C_L S$$

for some $n\ell \times n\ell$ invertible matrix S, where C_L is the companion matrix of $L(\lambda)$. In particular, T is also a linearization of $L(\lambda)$. Such a pair (X,T) will be called a (right) *standard pair* of $L(\lambda)$. Note that the $n\ell \times n\ell$ matrix

$$\begin{bmatrix} X \\ XT \\ \cdot \\ \cdot \\ \cdot \\ XT^{\ell-1} \end{bmatrix} \tag{2.3}$$

is invertible for every standard pair (X,T) of a matrix polynomial of degree ℓ with invertible leading coefficient. Indeed

$$\begin{bmatrix} X \\ XT \\ \cdot \\ \cdot \\ \cdot \\ XT^{\ell-1} \end{bmatrix} = \begin{bmatrix} [I\ 0\cdots 0] \\ [I\ 0\cdots 0]C_L \\ \cdot \\ \cdot \\ \cdot \\ [I\ 0\cdots 0]C_L^{\ell-1} \end{bmatrix} S = S\ .$$

When the matrix polynomial $L(\lambda)$ is monic (i.e. with leading coefficient $A_\ell = I$), we have the following representation theorem (see [19b]):

THEOREM 2.1 *Let* (X,T) *be a standard pair of a monic matrix polynomial* $L(\lambda)$. *Then*

$$L(\lambda) = \lambda^\ell I - XT^\ell(V_1+\cdots+V_\ell\lambda^{\ell-1})\ , \tag{2.4}$$

where V_i, $i = 1,\cdots,\ell$ *are* $n\ell \times n$ *matrices such that*

$$[\,V_1 \ \cdot\ \cdot\ \cdot\ V_\ell\,] = \begin{bmatrix} X \\ XT \\ \cdot \\ \cdot \\ \cdot \\ XT^{\ell-1} \end{bmatrix}^{-1} .$$

PROOF. Let $L(\lambda) = \lambda^\ell I + \sum_{j=0}^{\ell-1} \lambda^j A_j$. It is sufficient to prove Theorem 2.1 for $X = [I\ 0\cdots 0]$, $T = C_L$. In this case we have

$$[I\ 0\cdots 0]C_L^\ell = [-A_0,-A_1,\cdots,-A_{\ell-1}]$$

and

$$\begin{bmatrix} [I\ 0\cdots 0] \\ [I\ 0\cdots 0]C_L \\ \cdot \\ \cdot \\ \cdot \\ [I\ 0\cdots 0]C_L^{\ell-1} \end{bmatrix} = I ,$$

so the formula (2.4) follows. □

Consider now the system of differential equations with constant coefficients

$$\sum_{j=0}^{\ell} A_j \frac{d^j x}{dt^j} = 0 , \quad t \in \mathbb{R} \tag{2.5}$$

where A_j are $n \times n$ (complex) matrices and A_ℓ is invertible. It turns out that the general solution $x(t)$ of (2.5) can be conveniently expressed via a standard pair (X,T) of the monic matrix polynomial $L(\lambda)$:

$$x(t) = Xe^{tT}x_0 , \quad t \in \mathbb{R} , \tag{2.6}$$

where $x_0 \in \mathbb{C}^{n\ell}$ is an arbitrary vector. Again, it is sufficient to prove (2.6) for $(X,T) = ([I\ 0\cdots 0],C_L)$. We have $x^{(j)}(t) = [I\ 0\cdots 0]C_L^j e^{tC_L}x_0$, and therefore

$$\sum_{j=0}^{\ell} A_j \frac{d^j x}{dt^j} = \left(\sum_{j=0}^{\ell} A_j [I\ 0\cdots 0]C_L^j\right)e^{tC_L}x_0 . \tag{2.7}$$

But $[I\ 0\cdots 0]C_L^j = [0\cdots 0\ I\ 0\cdots 0]$ with I in the $(j+1)$-th place, $j = 0,\cdots,\ell-1$; and $[I\ 0\cdots 0]C_L^\ell = [-A_\ell^{-1}A_0,-A_\ell^{-1}A_1,\cdots,-A_\ell^{-1}A_{\ell-1}]$. Substitution in (2.7) yields $\sum_{j=0}^{\ell} A_j \frac{d^jx}{dt^j} = 0$. To see that formula (2.6) gives *all* solutions of (2.5), observe first that the space of the solutions of (2.5) is $n\ell$-dimensional, as A_ℓ is invertible. So it is sufficient to check that the formula (2.6) gives the zero solution only for $x_0 = 0$. Assume $Xe^{tT}x_0 \equiv 0$; differentiation gives $XT^ie^{tT}x_0 \equiv 0$. Now

$$\begin{bmatrix} X \\ XT \\ \cdot \\ \cdot \\ \cdot \\ XT^{\ell-1} \end{bmatrix} e^{tT}x_0 \equiv 0 ,$$

which in view of the invertibility of (2.3) and invertibility of e^{tT} implies $x_0 = 0$.

See [19b] for more details concerning standard pairs and triples of matrix polynomials.

2.2 Matrix Polynomials With Hermitian Coefficients

Consider the matrix polynomial $L(\lambda) = \sum_{j=0}^{\ell} \lambda^j A_j$, where A_ℓ is invertible, and assume that all the coefficients A_j are hermitian: $A_j = A_j^*$. In this case the polynomial $L(\lambda)$ is said to be *hermitian*.

In the study of the hermitian matrix polynomial $L(\lambda)$ important roles are played by the companion matrix C_L and standard pairs (X,T) as discussed above, and by the $\ell n \times \ell n$ matrix B_L defined by

$$B_L = \begin{bmatrix} A_1 & A_2 & \cdot & \cdot & \cdot & A_{\ell-1} & A_\ell \\ A_2 & & & & & A_\ell & 0 \\ \cdot & & & & \cdot & & \cdot \\ \cdot & & & \cdot & & & \cdot \\ \cdot & & \cdot & & & & \cdot \\ A_{\ell-1} & A_\ell & & & & & \cdot \\ A_\ell & 0 & \cdot & \cdot & \cdot & \cdot & 0 \end{bmatrix} .$$

It is clear that B_L is hermitian and invertible and so it may be used to form an indefinite scalar product in $\mathbb{C}^{n\ell}$. By a straightforward calculation one checks that $B_L C_L = C_L^* B_L$; so C_L *is* B_L*-selfadjoint*. In particular, $\sigma(C_L)$ (and therefore also $\sigma(L)$) is symmetric relative to the real axis, i.e. if $\lambda_0 \in \sigma(C_L)$ then $\bar{\lambda}_0 \in \sigma(C_L)$. Moreover, the degrees of elementary divisors of $L(\lambda)$ corresponding to λ_0 and to $\bar{\lambda}_0$ are the same.

Observe that the signature of B_L is given by the following formula:

$$\operatorname{sig} B_L = \begin{cases} 0 & \text{if } \ell \text{ is even} \\ \operatorname{sig} A_\ell & \text{if } \ell \text{ is odd} . \end{cases} \tag{2.8}$$

To see this, consider the continuous family of hermitian matrices:

$$B(\varepsilon) = \begin{bmatrix} \varepsilon A_1 & \varepsilon A_2 & \cdots & \varepsilon A_{\ell-1} & A_\ell \\ \varepsilon A_2 & & & A_\ell & 0 \\ \cdot & & \cdot & & \cdot \\ \cdot & & \cdot & & \cdot \\ \cdot & \cdot & & & \cdot \\ \varepsilon A_{\ell-1} & A_\ell & & & \cdot \\ A_\ell & 0 & \cdots & \cdots & 0 \end{bmatrix}, \quad \varepsilon \in [0,1] .$$

Clearly, $B(1) = B_L$ and $B(\varepsilon)$ is invertible for all $\varepsilon \in [0,1]$. Hence $\operatorname{sig} B(\varepsilon)$ is independent of ε on this interval; so we have $\operatorname{sig} B_L =$ $= \operatorname{sig} B(0)$, and the latter is easily calculated to yield (2.8).

In later chapters real eigenvalues having only linear elementary divisors will be of special interest. The next proposition gives a geometrical characterization of a real eigenvalue with this property.

A non-zero vector $x_0 \in \mathbb{C}^n$ is an *eigenvector* of $L(\lambda)$ corresponding to $\lambda_0 \in \sigma(L)$ if $L(\lambda_0)x_0 = 0$. When this is the case it is easily seen that the vector

$$\hat{x}_0 = \langle x_0, \lambda_0 x_0, \cdots, \lambda_0^{\ell-1} x_0 \rangle \tag{2.9}$$

is an eigenvector of C_L. In fact, $C_L \hat{x}_0 = \lambda_0 \hat{x}_0$. Conversely, the structure of C_L implies that every eigenvector $\hat{x}_0$ of C_L corresponding the eigenvalue λ_0 of C_L is of the form (2.9) where x_0 is an eigenvector of $L(\lambda)$.

It is not difficult to see that an eigenvalue λ_0 has only linear elementary divisors if and only if the dimension of $\operatorname{Ker} L(\lambda_0)$ coincides with the multiplicity of λ_0 as a zero of $\det L(\lambda)$. Or, equivalently, if and only if the dimension of $\mathrm{Ker}(I\lambda_0 - C_L)$ coincides with the multiplicity of λ_0 as a zero of $\det(I\lambda - C_L)$.

PROPOSITION 2.2 *Let* $L(\lambda)$ *be a hermitian matrix polynomial with invertible leading coefficient, and let* $\lambda_0 \in \sigma(L)$ *be real. Assume* $L(\lambda)$ *has only linear elementary divisors corresponding to* λ_0, *with* p_+ *(resp.* p_-*) associated positive (resp. negative) signs in the sign characteristic of* (C_L, B_L). *Then the quadratic form defined on the* $(p_+ + p_-)$*-dimensional subspace* $\mathrm{Ker}\, L(\lambda_0)$ *by* $(x, L^{(1)}(\lambda_0)x)$ *is non-singular and has* p_+ *(resp.* p_-*) positive (resp. negative) squares in its canonical form. Conversely, if the quadratic form* $(x, L^{(1)}(\lambda_0)x)$, $x \in \mathrm{Ker}\, L(\lambda_0)$ *is non-singular, then* $L(\lambda)$ *has only linear elementary divisors corresponding to* λ_0, *and the number of positive (resp. negative) signs associated with* λ_0 *in the sign characteristic of* (C_L, B_L) *coincides with the number of positive (resp. negative) squares in the canonical form of* $(x, L^{(1)}(\lambda_0)x), x \in \mathrm{Ker}\, L(\lambda_0)$.

In this statement $L^{(1)}(\lambda)$ is the derivative of $L(\lambda)$ with respect to λ. Also, the quadratic form is said to be nonsingular if for a fixed $x_0 \in \mathrm{Ker}\, L(\lambda_0)$, $(x_0, L^{(1)}(\lambda_0)y) = 0$ for all $y \in \mathrm{Ker}\, L(\lambda_0)$ implies $x_0 = 0$.

PROOF. Let $x, y \in \mathrm{Ker}\, L(\lambda_0)$. A direct computation shows that

$$(x, L^{(1)}(\lambda_0)y) = (\hat{x}, B_L \hat{y}) \tag{2.10}$$

where $\hat{x}$, $\hat{y}$ are related to x, y as in equation (2.9) and are therefore in $\mathrm{Ker}(I\lambda_0 - C_L)$. Now the proposition follows from Theorem I.3.16. □

Note that the last quoted theorem could be used to generalize this proposition to any real eigenvalues, i.e. with no hypotheses on the elementary divisors. (See Theorem 10.14 in [19b]).

A more general class of matrix polynomials are matrix polynomials with H-selfadjoint coefficients: $L(\lambda) = \sum_{j=0}^{\ell} \lambda^j A_j$, where all coefficients A_j are H-selfadjoint with fixed H ($HA_j = A_j^* H$, $j = 0, \cdots, \ell$) and A_ℓ is invertible. However, such polynomials do not exhibit essentially new properties comparing with polynomials with hermitian coefficients, as the companion matrix C_L is selfadjoint in the indefinite scalar product determined by the matrix

$$\begin{bmatrix} H & & & \\ & H & & 0 \\ & & \ddots & \\ & 0 & & \\ & & & H \end{bmatrix} \begin{bmatrix} A_1 & A_2 & \cdots & A_{\ell-1} & A_\ell \\ A_2 & & \cdot & A_\ell & \\ \vdots & & \cdot\,\cdot & & \\ \vdots & \cdot\,\cdot & & & \\ \vdots & \cdot & & & \\ A_{\ell-1} & A_\ell & & 0 & \\ A_\ell & & & & \end{bmatrix}.$$

Consider now a system of differential equations with constant coefficients of the form

$$\sum_{j=0}^{\ell} i^j A_j \frac{d^j x}{dt^j} = 0 \tag{2.11}$$

where $A_0, A_1, \cdots, A_\ell$ are hermitian and A_ℓ is invertible. Such systems also arise in applications, and form a natural generalization of the Hamiltonian equations studied in Chapter 1 (i.e. the case $\ell = 1$ of (2.11)). The associated matrix polynomial is again hermitian:

$$L(\lambda) = \sum_{j=0}^{\ell} A_j \lambda^j . \tag{2.12}$$

We shall be interested in cases when all solutions of (2.11) are bounded. To describe these cases in terms of $L(\lambda)$ it is convenient to introduce the following definition. A hermitian matrix polynomial has *simple structure* if all its eigenvalues are real and all elementary divisors are linear.

THEOREM 2.3 *The solutions of* (2.11) *are bounded on the whole real line if and only if* $L(\lambda)$ *has simple structure.*

PROOF. As we mentioned in the preceding section, the general solution of the system $L(\frac{d}{dt})x(t) = 0$ is given by the formula $x(t) = Pe^{tC_L}x_0$, where $P = [I\ 0\cdots0]$ and $x \in \mathbb{C}^{n\ell}$ is arbitrary. It is easily seen then that the general solution of (2.11) is given by the formula $x(t) = Pe^{itC_L}x_0$. So if all eigenvalues of C_L (or, what is the same, the eigenvalues of $L(\lambda)$) are real with linear elementary divisors, then $\sup_{-\infty<t<\infty} ||x(t)|| < \infty$ for all x_0. Conversely, assume that the solution of (2.11) are bounded $(t \in \mathbb{R})$. Since $x^{(k)}(t)$, $k = 1,2,\cdots$ are also solutions of (2.11) and therefore are bounded, we find that the vector

$$\begin{bmatrix} x(t) \\ x'(t) \\ \cdot \\ \cdot \\ \cdot \\ x^{(\ell-1)}(t) \end{bmatrix} = \begin{bmatrix} P \\ iPC_L \\ \cdot \\ \cdot \\ \cdot \\ i^{\ell-1}PC_L^{\ell-1} \end{bmatrix} e^{itC_L}x_0$$

is bounded on the real line, for all x_0. But $PC_L^{k-1} = [0\cdots I\ 0\cdots 0]$ with I in the k-th place, so in fact $\sup_{-\infty<t<\infty} ||e^{itC_L}x_0|| < \infty$ for all x_0. This happens only if all eigenvalues of C_L are real with linear elementary divisors. □

Finally, we remark that when (2.11) has only bounded solutions, Proposition 2.2 can be applied to each eigenvalue of L and will, in fact, play an important part in the characterization of these systems (2.11) for which all neighbouring systems of the same kind will have only bounded solutions.

2.3 Factorization of Hermitian Matrix Polynomials

We start with the following description of divisibility of matrix polynomials (not necessarily hermitian) in terms of their standard pairs (see [19b] for the proof and more details).

We say that the matrix polynomial $L_1(\lambda)$ is a right divisor of a matrix polynomial $L(\lambda)$ if $L(\lambda) = L_2(\lambda)L_1(\lambda)$ for some matrix polynomial $L_2(\lambda)$.

THEOREM 2.4 *Let* $L(\lambda)$ *be a matrix polynomial of degree* ℓ *with invertible leading coefficient and with standard pair* (X,T). *Let* L *be a* T-*invariant subspace such that the linear map*

$$Q_k(L) \overset{\text{def}}{=} \begin{bmatrix} X \\ XT \\ \cdot \\ \cdot \\ \cdot \\ XT^{k-1} \end{bmatrix} \Bigg|_L : L \to \mathbb{C}^{nk}$$

is invertible (in particular, $\dim L = nk$). *Then the monic matrix polynomial of degree* k

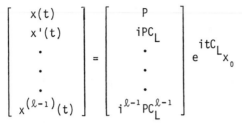

$$L_1(\lambda) = \lambda^k I - XT^k|_{L}(V_1+V_2\lambda+\cdots+V_k\lambda^{k-1}) , \qquad (2.13)$$

where $[V_1V_2\cdots V_k] = Q_k(L)^{-1}$, $V_i:\mathbb{C}^n \to L$, *is a right divisor of* $L(\lambda)$. *Conversely, if a monic matrix polynomial* $L_1(\lambda)$ *of degree* k *is a right divisor of* $L(\lambda)$, *then there exists a unique* T-*invariant subspace* L *such that* $Q_k(L)$ *is invertible and formula* (2.13) *holds.*

So there is a one-to-one correspondence between the set of T-invariant subspaces L for which $Q_k(L)$ is invertible, and the set of monic right divisors of degree k of $L(\lambda)$. We say that L is the *supporting subspace* (*with respect to* T) of the right monic divisor $L_1(\lambda)$ of $L(\lambda)$.

Note that, in the notation of Theorem 2.4, $(X|_L,T|_L)$ is a standard pair for $L_1(\lambda)$ (we choose a basis in L so that $(X|_L,T|_L)$ are represented by matrices).

We assume now that the matrix polynomial $L(\lambda)$ is hermitian. In this case we know that C_L is B_L-selfadjoint. So there are C_L-invariant maximal B_L-nonnegative (or B_L-nonpositive) subspaces (Theorem I.3.20). It turns out that such subspaces are supporting subspaces for certain monic divisors of hermitian matrix polynomials, provided the leading coefficient is positive definite:

THEOREM 2.5 *Let* $L(\lambda)$ *be an hermitian matrix polynomial of degree* ℓ *with positive definite leading coefficient* A_ℓ. *Let* L_+ (*resp.* L_-) *be a* C_L-*invariant maximal* B_L-*nonnegative* (*resp. maximal* B_L-*nonpositive*) *subspace. Then:*

(i) $\dim L_\pm = \frac{\ell n}{2}$ *if* ℓ *is even;* $\dim L_\pm = \frac{\ell\pm1}{2} n$ *if* ℓ *is odd;*

(ii) *the linear map*

$$Q_k(L_\pm) = \begin{bmatrix} P \\ PC_L \\ \cdot \\ \cdot \\ \cdot \\ PC_L^{k-1} \end{bmatrix} \Big|_{L_\pm} : L_\pm \to \mathbb{C}^{nk}$$

is invertible, where $P = [I\ 0\cdots 0]$, *and* $k = \dim L_\pm$;

(iii) *the monic matrix polynomial*

$$L_1(\lambda) = \lambda^k I - P(C_L|_{L_\pm})^k(V_1+V_2\lambda+\cdots+V_{k-1}\lambda^{k-1}) ,$$

where $[V_1 V_2 \cdots V_{k-1}] = (Q_k(L_\pm))^{-1}$, *is a right divisor of* $L(\lambda)$, *with the supporting subspace* $L_\pm$.

PROOF. The statement (i) follows from (2.8) and Theorem I.1.3. The statement (iii) follows from (ii) in view of Theorem 2.4. So it remains to prove (ii). It is sufficient to check that $Q_k(L_\pm)$ is one-to-one. Assume first that ℓ is even, and let $x = \langle x_1, \cdots, x_\ell \rangle \in \operatorname{Ker} Q_k(L_\pm)$, where $x_i \in \mathbb{C}^n$. Since

$$\begin{bmatrix} P \\ PC_L \\ \cdot \\ \cdot \\ \cdot \\ PC_L^{k-1} \end{bmatrix} = [I_{nk}, 0] ,$$

we have $x_1 = \cdots = x_k = 0$. This implies $(B_L x, x) = 0$. But since $L_\pm$ is B-nonpositive (or B-nonnegative) and $C_L x \in L_\pm$, we have by Schwartz' inequality,

$$|(B_L C_L x, x)| \leqslant (B_L C_L x, C_L x)(B_L x, x) \tag{2.14}$$

(note that Schwartz inequality holds if restricted to a nonpositive (or non-negative) subspace, see Section I.1.3). Now (2.14) yields $(B_L C_L x, x) = 0$. But

$$(B_L C_L x, x) = (B_L \langle x_2, \cdots, x_\ell, y \rangle , \langle x_1, \cdots, x_\ell \rangle)$$

for some $y \in \mathbb{C}^n$, and using $x_1 = \cdots = x_k = 0$ and the definition of B_L, we conclude that $(A_\ell x_{k+1}, x_{k+1}) = 0$. In view of the positive definiteness of A_ℓ, $x_{k+1} = 0$. Now $(B_L C_L x, C_L x) = 0$. Using again Schwartz inequality, we obtain that $(B_L C_L^2 x, C_L x) = 0$ which amounts to $x_{k+2} = 0$, and so on.

Assume now that ℓ is odd, and consider L_+ (so that $k = \frac{\ell+1}{2}$). Let $x = \langle x_1, \cdots, x_\ell \rangle \in \operatorname{Ker} Q_k(L_+)$; then, in particular, $x_1 = \cdots = x_k = 0$. As in the case of even ℓ, we have $(B_L x, x) = 0$ and by Schwartz inequality, $(B_L C_L x, C_L x) = (B_L C_L^2 x, x) = 0$. But $(B_L C_L x, C_L x) = (A_\ell x_{k+1}, x_{k+1})$, so $x_{k+1} = 0$. Applying an analogous argument using $(B_L C_L x, C_L x) = 0$ we obtain $x_{k+2} = 0$, and so on. For L_- we apply a similar argument starting as follows: given $x = \langle x_1, \cdots, x_\ell \rangle \in \operatorname{Ker} Q_{\frac{\ell-1}{2}}(L_-)$, we have

$$0 \geq (B_L x,x) = (A_\ell x_{\frac{\ell+1}{2}}, x_{\frac{\ell+1}{2}}),$$

which implies $x_{\frac{\ell+1}{2}} = 0$ and $(B_L x,x) = 0$. □

For more information about factorizations of hermitian matrix polynomials see [19b]. Here, we shall only state the following result (without proof), which describes a factorization of nonnegative matrix polynomials. An $n \times n$ matrix polynomial $L(\lambda)$ is *nonnegative* if $(L(\lambda)x,x) \geq 0$ for all $x \in \mathbb{C}^n$ and $\lambda \in \mathbb{R}$. Clearly, a nonnegative matrix polynomial is hermitian.

THEOREM 2.6 *The following statements are equivalent for a monic* $n \times n$ *matrix polynomial* $L(\lambda)$ *with hermitian coefficients and even degree* $\ell = 2k$:

(i) $L(\lambda)$ *is non-negative;*

(ii) *all elementary divisors of* $L(\lambda)$ *corresponding to the real eigenvalues (if any) have even degrees;*

(iii) *all signs in the sign characteristic of* (C_L, B_L) *are* $+1$*'s;*

(iv) $L(\lambda)$ *admits the factorization*

$$L(\lambda) = (M(\bar{\lambda}))^* M(\lambda) \tag{2.15}$$

for some monic $n \times n$ *matrix polynomial* $M(\lambda)$ *of degree* k.

In fact, there exists a one-to-one correspondence between C_L-invariant nk-dimensional B_L-neutral subspaces in $\mathbb{C}^{n\ell}$ and factorizations (2.15). Indeed, we know from Theorem 2.5 that each such subspace L is a supporting subspace for a monic divisor $L_1(\lambda)$ of $L(\lambda)$. It turns out that the B_L-neutrality of L implies that the quotient $L(\lambda)L_1(\lambda)^{-1}$ is just $(L_1(\bar{\lambda}))^*$. Conversely, by Theorem 2.4, every factorization (2.15) is generated by a C_L-invariant supporting subspace, and the special form of the factorization (2.15) (namely, the quotient $L(\lambda)M(\lambda)^{-1}$ is equal to $(M(\bar{\lambda}))^*$) implies that this subspace is B_L-neutral.

We remark also that if (i)-(iv) hold, then the matrix polynomial $M(\lambda)$ in (2.15) can be chosen in such a way that $\sigma(M)$ lies in the closed upper halfplane (or $\sigma(M)$ lies in the closed lower halfplane). More generally, let S be a set of non-real eigenvalues of $L(\lambda)$ such that $\lambda \in S$ implies $\bar{\lambda} \notin S$, and S is maximal with respect to this property. Then there exists a monic matrix polynomial $M(\lambda)$ satisfying (2.15) for which $\sigma(M) \setminus \mathbb{R} = S$.

2.4 Difference Equations and Hermitian Matrix Polynomials on the Unit Circle

Consider the following difference equation:

$$A_0 x_i + A_1 x_{i+1} + \cdots + A_\ell x_{i+\ell} = 0 , \quad i = 0,1,\cdots , \tag{2.16}$$

where $\{x_i\}_{i=0}^{\infty}$ is a sequence of n-dimensional complex vectors to be found, and $A_0,\cdots,A_\ell$ are given complex $n \times n$ matrices. We shall assume throughout this section that A_ℓ is invertible.

As in the case of the system of differential equations (2.5), a general solution of (2.16) can be found in terms of a standard pair (X,T) of the matrix polynomial $L(\lambda) = \sum_{j=0}^{\ell} \lambda^j A_j$: Namely, the general solution of (2.16) (2.16) is given by the formula

$$x_i = XT^i z , \quad i = 0,1,\cdots , \tag{2.17}$$

where z is an $n\ell$-dimensional vector. To see this we can assume $X = [I\ 0\cdots0]$, $T = C_L$; then $\sum_{j=0}^{\ell} A_j XT^j = 0$ (cf. the proof of (2.6)), so the formula (2.17) indeed gives solutions for (2.16). Further, invertibility of A_ℓ implies that

$$x_{\ell+i} = -A_\ell^{-1}(A_0 x_i + \cdots + A_{\ell-1} x_{i+\ell-1}) , \quad i = 0,1,\cdots ,$$

and therefore the solution of (2.16) is determined by the values of $x_0,\cdots,x_{\ell-1}$; in other words, the dimension of the space of solutions of (2.16) is $n\ell$. So in order to prove that (2.17) gives all solutions of (2.16) we have to check that $x_i = 0$, $i = 0,1,2,\cdots$ implies $z = 0$. But this follows from the invertibility of the matrix

$$\begin{bmatrix} X \\ XT \\ \cdot \\ \cdot \\ \cdot \\ XT^{\ell-1} \end{bmatrix} .$$

Consider now the system (2.16) with the additional assumptions that ℓ is even, say $\ell = 2k$, and

$$A_i^* = A_{\ell-i}\,, \quad i = 0,\cdots,\ell\,. \tag{2.18}$$

Note, in particular, that $A_0^* = A_\ell$ and is therefore invertible. This case occurs, for instance, if one approximates the differential equation

$$\sum_{j=0}^{\ell} i^j \tilde{A}_j \frac{d^j x}{dt^j} = 0$$

with hermitian coefficients $\tilde{A}_j$ by symmetric finite differences (i.e. replacing the derivative $\frac{dx}{dt}$ at the point kh $(k = 0,\pm1,\pm2,\cdots)$ by $\frac{1}{2h}[x((k+1)h) - x((k-1)h)]$; here h is a fixed positive number).

Let $L(\lambda) = \sum_{j=0}^{\ell} \lambda^j A_j$ be the associated matrix polynomial. Condition (2.18) means that the rational function $\hat{L}(\lambda) = \lambda^{-k} L(\lambda)$ is hermitian on the unit circle: $(\hat{L}(\lambda))^* = \hat{L}(\bar{\lambda}^{-1})$, $\lambda \in ¢$. So $(\hat{L}(\lambda))^* = \hat{L}(\lambda)$ for $|\lambda| = 1$. This means that the matrix polynomial $L(\lambda)$ is *hermitian* with respect to the unit circle. By analogy with the case of hermitian matrix polynomials we expect that the companion matrix C_L is unitary in a suitable indefinite scalar product. This is indeed so:

$$C_L^* \hat{B}_L C_L = \hat{B}_L\,, \tag{2.19}$$

where

$$\hat{B}_L = i\left[\begin{array}{cccc|cccc} & & & & A_\ell & & & 0 \\ & 0 & & & \vdots & \cdot & & \\ & & & & & & \cdot & \\ & & & & A_{k+1} & \cdot & \cdot & A_\ell \\ \hline -A_0 & \cdot & \cdot & -A_{k-1} & & & & \\ & \cdot & & \vdots & & & 0 & \\ & & \cdot & & & & & \\ 0 & & & -A_0 & & & & \end{array}\right]$$

is hermitian and invertible. To check (2.19), observe that

$$C_L^* = \begin{bmatrix} 0 & \cdot & \cdot & \cdot & 0 & -A_\ell A_0^{-1} \\ I & \cdot & & & 0 & \cdot \\ \cdot & & \cdot & & & \cdot \\ \cdot & & & \cdot & & \cdot \\ 0 & \cdot & \cdot & \cdot & I & -A_1 A_0^{-1} \end{bmatrix}$$

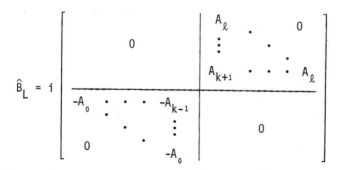

$$C_L^{*-1} = \begin{bmatrix} -A_{\ell-1}A_\ell^{-1} & I & 0 & \cdot & \cdot & \cdot & 0 \\ -A_{\ell-2}A_\ell^{-1} & 0 & I & \cdot & \cdot & \cdot & 0 \\ \cdot & \cdot & & \cdot & \cdot & \cdot & \cdot \\ \cdot & \cdot & & & & & I \\ -A_0A_\ell^{-1} & 0 & \cdot & \cdot & \cdot & \cdot & 0 \end{bmatrix} .$$

Now the equality $\hat{B}_L C_L = C_L^{*-1}\hat{B}_L$ is verified by a direct computation.

Let λ_0 be a unimodular (i.e. $|\lambda_0| = 1$) eigenvalue of $L(\lambda)$. For each $x \in \operatorname{Ker} L(\lambda_0)$ denote $\hat{x} = \langle x, \lambda_0 x, \cdots, \lambda_0^{\ell-1}x\rangle$. Then for $x,y \in \operatorname{Ker} L(\lambda_0)$ it follows that

$$(x, i\lambda_0 \hat{L}^{(1)}(\lambda_0)y) = (\hat{x}, \hat{B}_L \hat{y}) , \tag{2.20}$$

where $\hat{L}^{(1)}(\lambda)$ is the derivative of $\hat{L}(\lambda) = \lambda^{-k}L(\lambda)$ with respect to λ. The equality (2.20) can be checked by a direct computation of $(\hat{x}, \hat{B}_L \hat{y})$ using the property $\bar{\lambda}_0 = \lambda_0^{-1}$. Using the definition of the sign characteristic of a unitary matrix in an indefinite scalar product (Section I.4.2) we obtain the following statement which is an analogue of Proposition 2.2.

PROPOSITION 2.7 *Let* $L(\lambda)$ *be a matrix polynomial of even degree* ℓ, *which is hermitian on the unit circle, and with invertible leading coefficient. Let* λ_0 *be a unimodular eigenvalue of* $L(\lambda)$. *Then the quadratic form* $(x, i\lambda_0(\lambda^{-\frac{\ell}{2}}L(\lambda))^{(1)}(\lambda_0)x)$, $x \in \operatorname{Ker} L(\lambda_0)$ *is non-singular if and only if all the elementary divisors of* $L(\lambda)$ *corresponding to* λ_0 *are linear. In this case the number of positive (resp. negative) squares in the canonical form of* $(x, i\lambda_0(\lambda^{-\frac{\ell}{2}}L(\lambda))^{(1)}(\lambda_0)x)$, $x \in \operatorname{Ker} L(\lambda_0)$ *coincides with the number of signs* +1 *(resp.* -1*) in the sign characteristic of* $(C_L, \hat{B}_L)$ *(here the* $\hat{B}_L$*-unitary matrix* C_L *is the companion matrix of* $L(\lambda)$*).*

Going back to the system (2.16), formula (2.17) shows that all solutions of (2.16) are bounded if and only if the spectrum of T is in the closed unit disc, and the elementary divisors of $\lambda I - T$ corresponding to the unimodular eigenvalues, are all linear. If, in addition, (2.18) holds, then C_L is $\hat{B}_L$-unitary and therefore the spectrum of C_L is symmetric relative to the unit circle. We obtain the following theorem (an analogue of Theorem 2.3). We say that a matrix polynomial $L(\lambda)$ has *simple structure* with respect to the unit circle if the spectrum $\sigma(L)$ lies on the unit circle and all elementary divisors of $L(\lambda)$ are linear.

THEOREM 2.8 *All solutions of the system*

$$A_0 x_i + A_1 x_{i+1} + \cdots + A_\ell x_{i+\ell} = 0 , \quad i = 0,1,\cdots ,$$

where $A_j^* = A_{\ell-j}$ *for* $j = 0,\cdots,\ell$ *and* A_ℓ *is invertible, are bounded if and only if the matrix polynomial* $L(\lambda) = \sum_{j=0}^{\ell} \lambda^j A_j$ *has simple structure with respect to the unit circle.*

In concluding this section we note that one can obtain factorization results for hermitian matrix polynomials on the unit circle from corresponding results for hermitian matrix polynomials on the real line. To illustrate this approach let us prove the following result, which is an analogue of Theorem 2.6.

THEOREM 2.9 *Let* $R(\lambda) = \sum_{j=-k}^{k} \lambda^j R_j$ *be a rational* $n \times n$ *matrix function such that* $(R(\lambda)x,x) \geq 0$ *for all* $|\lambda| = 1$ *and* $x \in \mathbb{C}^n$, *and* $\det R(\lambda) \not\equiv 0$. *Then* $R(\lambda)$ *admits a factorization*

$$R(\lambda) = (A(\bar{\lambda}^{-1}))^* A(\lambda) , \quad \lambda \in \mathbb{C} \tag{2.21}$$

where $A(\lambda)$ *is a matrix polynomial.*

PROOF. Let

$$L(\lambda) = \sum_{j=0}^{2k} (1+i\lambda)^j (1-i\lambda)^{2k-j} R_{j-k} .$$

It is easily checked that $L(\lambda)$ is a matrix polynomial which is nonnegative on the real line, and $\det L(\lambda) \not\equiv 0$. Let $a \in \mathbb{R}$ be such that $L(a)$ is invertible. Then the matrix polynomial

$$M(\lambda) = \lambda^{2k} L(a)^{-\frac{1}{2}} L(\lambda^{-1}+a) L(a)^{-\frac{1}{2}}$$

is monic and nonnegative on the real line. By Theorem 2.6, $M(\lambda)$ admits a factorization $M(\lambda) = (M_1(\bar{\lambda}))^* M_1(\lambda)$ for some monic matrix polynomial $M_1(\lambda)$ of degree k. This factorization leads to the factorization $L(\lambda) = (L_1(\bar{\lambda}))^* L_1(\lambda)$ with matrix polynomial $L_1(\lambda) = (\lambda-a)^k M_1((\lambda-a)^{-1}) L(a)^{\frac{1}{2}}$. Now

$$R\left(\frac{1+i\lambda}{1-i\lambda}\right) = \frac{L(\lambda)}{(1+\lambda^2)^k} = \left[\frac{L_1(\bar{\lambda})}{(1-i\bar{\lambda})^k}\right]^* \frac{L_1(\lambda)}{(1-i\lambda)^k}$$

and denoting $\mu = \frac{1+i\lambda}{1-i\lambda}$, we obtain

$$R(\mu) = \left[\frac{(1+\bar{\mu})^k}{(2\bar{\mu})^k} L_1\left(\frac{-i(1-\bar{\mu})}{1+\bar{\mu}}\right)\right]^* \frac{(1+\mu)^k}{2^k} L_1\left(\frac{i(1-\mu)}{1+\mu}\right),$$

which is a factorization of type (2.21) with

$$A(\lambda) = \frac{(1+\lambda)^k}{2^k} L_1\left(\frac{i(1-\lambda)}{1+\lambda}\right). \quad \square$$

As with Theorem 2.6, the matrix polynomial $A(\lambda)$ in (2.21) can be chosen with an additional spectral property. Let S be a set of non-unimodular eigenvalues of $R(\lambda)$ such that $\lambda \in S$ implies $\bar{\lambda}^{-1} \notin S$, and S is maximal with respect to this property. Then there exists $A(\lambda)$ satisfying (2.21) for which the non-unimodular spectrum of $A(\lambda)$ coincides with S.

CHAPTER 3

HERMITIAN RATIONAL MATRIX FUNCTIONS

The subject of this chapter is rational matrix valued functions $W(\lambda)$ for which $W(\lambda) = W(\bar{\lambda})^*$ for all $\lambda \in \mathbb{C}$ at which $W(\lambda)$ is defined and, in particular, $W(\lambda)$ is hermitian whenever $\lambda \in \mathbb{R}$. For the purpose of illustration, consider a function $W(\lambda)$ with a representation

$$W(\lambda) = I + C(\lambda I - A)^{-1}B \tag{3.1}$$

for all $\lambda \notin \sigma(A)$. If $HA = A^*H$ and $C^* = HB$ for some nonsingular hermitian matrix H, it is easily verified that $W(\lambda) = W(\bar{\lambda})^*$ as required. It turns out that there is a converse statement: Every rational matrix-valued function $W(\lambda)$ for which $W(\lambda) = W(\bar{\lambda})^*$ and $W(\lambda) \to I$ as $|\lambda| \to \infty$ has a representation (3.1) and $HA = A^*H$, $C^* = HB$ for some H with $\det H \neq 0$ and $H^* = H$. This is the way in which H-selfadjoint matrices enter the analysis of hermitian rational matrix functions.

A new feature arises here due to the presence of poles as well as zeros (or eigenvalues) for rational matrix functions. Indeed, there may be points which are simultaneously zeros and poles of $W(\lambda)$. The first part of the chapter is devoted to generalization to rational matrix functions of concepts and results introduced for polynomial matrix functions in the preceding chapter. This structure plays an important role in the factorization problem for rational matrix functions, which arises in a natural way in systems theory. For example, if a function $W(\lambda)$ is nonnegative in the sense that $(W(\lambda)x,x) \geq 0$ for $x \in \mathbb{C}^n$ and all $\lambda \in \mathbb{R}$ at which $W(\lambda)$ is defined, the problem of finding symmetric factorizations $W(\lambda) = L(\bar{\lambda})^*L(\lambda)$ is important.

Rational matrix functions (hermitian and otherwise) play a central role in systems theory and appear as "transfer functions" in this context.

In order to make this presentation more self-contained, an appendix is provided for the chapter in which an introduction to the general theory of rational matrix functions (without the symmetry property) is presented.

3.1 Minimal Nodes

A rational $n \times n$ matrix function $W(\lambda)$ is called *hermitian* if $W(\lambda) = W(\bar{\lambda})^*$ for every $\lambda \in \mathbb{C}$ which is not a pole of W (in particular, $W(\lambda)$ is hermitian for every $\lambda \in \mathbb{R}$ which is not a pole of W). The first result characterizes hermitian rational matrix functions in terms of their minimal nodes.

THEOREM 3.1 *Let* $W(\lambda))$ *be a rational matrix function with* $W(\infty) = I$, *and let* $\theta = (A,B,C,I;\mathbb{C}^m,\mathbb{C}^n)$ *be a minimal node for* $W(\lambda)$. *Then the following statements are equivalent:*

(i) $W(\lambda)$ *is hermitian;*

(ii) *the node* θ *is similar to the minimal node* $\theta^* \overset{\text{def}}{=} (A^*,C^*,B^*,I;\mathbb{C}^m,\mathbb{C}^n)$, *i.e.*

$$A^* = HAH^{-1}, \quad C^* = HB, \quad B^* = CH^{-1} \tag{3.2}$$

for some invertible matrix H;

(iii) θ *is similar to* θ^* *with a hermitian* $m \times m$ *matrix* H.

Note that if two minimal modes are similar then the similarity matrix is uniquely defined (see Appendix). In particular, the matrix H which satisfies (3.2) is unique (if such an H exists).

PROOF. (i)⇒(iii). Let us check first that θ^* is minimal. We have $\bigcap_{i=0}^{\infty} \operatorname{Ker}(CA^i) = \{0\}$ and $\sum_{i=0}^{\infty} \operatorname{Im}(A^iB) = \mathbb{C}^m$. Therefore

$$\bigcap_{j=0}^{\infty} \operatorname{Ker} B^*A^{*j} = \bigcap_{j=0}^{\infty} \operatorname{Ker}(A^jB)^* = \bigcap_{j=0}^{\infty} [\operatorname{Im}(A^jB)]^{\perp},$$

which is equal to the orthogonal complement of $\sum_{j=0}^{\infty} \operatorname{Im} A^jB = \mathbb{C}^m$; so $\bigcap_{j=0}^{\infty} \operatorname{Ker} B^*A^{*j} = \{0\}$. Also

$$\sum_{i=0}^{\infty} \operatorname{Im} A^{*j}C^* = \sum_{i=0}^{\infty} [\operatorname{Ker} CA^j]^{\perp} = [\bigcap_{j=0}^{\infty} \operatorname{Ker} CA^j]^{\perp} = \mathbb{C}^m,$$

so θ^* is indeed a minimal node. The transfer function of θ^* is

$$I + B^*(\lambda I-A^*)^{-1}C^* = [I + C(\bar{\lambda}I-A)^{-1}B]^* = (W(\bar{\lambda}))^*,$$

and is equal to $W(\lambda)$ becuase $W(\lambda)$ is hermitian. So θ and θ^* are similar, as minimal nodes with the same transfer function, i.e. (3.2) holds for some invertible H. We claim that H is hermitian. Indeed, taking adjoints in (3.2) we obtain

$$A = H^{*-1}A^*H^* , \quad C = B^*H^* , \quad B = H^{*-1}C^* ,$$

i.e. (3.1) holds with H replaced by H^*. By the uniqueness of the similarity of minimal nodes (Proposition A.1 in the Appendix), $H = H^*$.
(iii)⇒(ii) is trivial.
(ii)⇒(i) follows from the fact that the transfer functions $W(\lambda)$ and $(W(\bar{\lambda}))^*$ of the similar nodes θ and θ^* coincide. □

Consider a hermitian rational matrix function $W(\lambda)$ with a minimal node $\theta = (A,B,C,I;\mathbb{C}^m,\mathbb{C}^n)$. As we know from Theorem 3.1, there exists a unique invertible hermitian $m \times m$ matrix H such that (3.2) holds. The first of these relations shows that A is H-selfadjoint. Also, $A^\times = A - BC$ is H-selfadjoint. Indeed,

$$H(A-BC) = HA - HBC = A^*H - C^*B^*H = (A-BC)^*H .$$

Let $(J,P_{\varepsilon,J})$ be the canonical form of (A,H) so that

$$J = S^{-1}AS ; \quad P_{\varepsilon,J} = S^*HS , \tag{3.3}$$

where S is an $m \times m$ invertible matrix. So using (3.2) and (3.3), we obtain:

$$\begin{aligned} W(\lambda) &= I + C(\lambda I-A)^{-1}B = I + CS(\lambda I-S^{-1}AS)^{-1}S^{-1}B \\ &= I + B^*HS(\lambda I-J)^{-1}S^{-1}B = I + B^*S^{*-1}S^*HS(\lambda I-J)^{-1}S^{-1}B \\ &= I + B_0^*P_{\varepsilon,J}(\lambda I-J)^{-1}B_0 , \end{aligned}$$

where $B_0 = S^{-1}B$. We arrive at the following result:

THEOREM 3.2 *Every hermitian rational matrix function* $W(\lambda)$ *such that* $W(\infty) = I$ *admits representations*

$$W(\lambda) = I + B_0^*P_{\varepsilon,J}(\lambda I-J)^{-1}B_0 = I + C_0^*(\lambda I-J)^{-1}P_{\varepsilon,J}C_0 . \tag{3.4}$$

Here, the Jordan matrix J *is a direct sum of Jordan blocks with real eigenvalues and blocks of type* $J_i \oplus \bar{J}_i$, *where* J_i *is a Jordan block with non-*

real eigenvalues. The matrix $P_{\varepsilon,J}$ *is the direct sum of corresponding sip matrices in which those corresponding to real Jordan blocks are multiplied by* +1 *or* -1, *and* ε *denotes this set of signs. Conversely, every rational matrix function of the form* (3.4) *is hermitian.*

The representation (3.4) will be called the *canonical form* of the hermitian rational function $W(\lambda)$.

PROOF. The first equality in (3.4) has already been proved. The second is obtained similarly:

$$W(\lambda) = I + CS(\lambda I-J)S^{-1}B = I + CS(\lambda I-J)^{-1}S^{-1}H^{-1}C^{*}$$
$$= I + CS(\lambda I-J)^{-1}P_{\varepsilon,J}^{-1}S^{*}C^{*} = I + C_0(\lambda I-J)^{-1}P_{\varepsilon,J}C_0 ,$$

with $C_0 = CS$. The converse statement of Theorem 3.2 is trivial. □

3.2 The Sign Characteristic: Definition and Main Result

This section, as well as the next two sections are devoted to the sign characteristic of a hermitian rational matrix function.

Let $W(\lambda)$ be a rational $n \times n$ matrix function with determinant not identically zero (and not necessarily with the value I at infinity). Then $W(\lambda)$ and $W(\lambda)^{-1}$ admit the following minimal realizations (see Appendix):

$$\left.\begin{aligned} W^{-1}(\lambda) &= Q_N(\lambda I-A_N)^{-1}R_N + \{\text{polynomial in } \lambda\} ; \\ W(\lambda) &= Q_P(\lambda I-A_P)^{-1}R_P + \{\text{polynomial in } \lambda\} . \end{aligned}\right\} \qquad (3.5)$$

The triple (Q_N, A_N, R_N) will be called the *null* triple of $W(\lambda)$, and (Q_P, A_P, R_P) is its *pole triple*. Assume now, in addition, that $W(\lambda)$ is hermitian and apply Theorem 3.1 to the rational hermitian matrix functions $Q_N(\lambda I-A_N)^{-1}R_N + I$ and $Q_P(\lambda I-A_P)^{-1}R_P + I$ defined by the null triple (Q_N, A_N, R_N) and the pole triple (Q_P, A_P, R_P) of $W(\lambda)$. It is found that there exist unique invertible matrices H_N and H_P such that

$$\left.\begin{aligned} H_N A_N &= A_N^{*} H_N ; \quad H_N R_N = Q_N^{*} ; \quad Q_N = R_N^{*} H_N ; \\ H_P A_P &= A_P^{*} H_P ; \quad H_P R_P = Q_P^{*} ; \quad Q_P = R_P^{*} H_P ; \end{aligned}\right\} \qquad (3.6)$$

and the matrices H_N and H_P are hermitian, A_N is H_N-selfadjoint and A_P is H_P-selfadjoint. The sign characteristics of (A_N, H_N) and (A_P, H_P) will be called the *zero*, and the *pole, sign characteristic* of $W(\lambda)$, respectively.

These properties of $W(\lambda)$ do not depend on the choice of the minimal realization (3.6). To see this, let

$$W^{-1}(\lambda) = Q_N'(\lambda I - A_N')^{-1}R_N' + \{\text{polynomial in } \lambda\}$$

be another minimal realization of $W^{-1}(\lambda)$. Since any two minimal realizations are similar, we have

$$Q_N' = Q_N S \ , \quad A_N' = S^{-1}A_N S \ , \quad R_N' = S^{-1}R_N \tag{3.7}$$

for some invertible matrix S. Let $H_N' = H_N'^{*}$ be invertible and such that

$$H_N'A_N' = A_N'^{*}H_N' \ , \quad H_N'R_N' = Q_N'^{*} \ , \quad Q_N' = R_N'^{*}H_N' \ .$$

Substituting from (3.7) for A_N', Q_N', R_N':

$$(S^{-1*}H_N'S^{-1})A_N = A_N^{*}(S^{-1*}H_N'S^{-1}) \ , \quad (S^{-1*}H_N'S^{-1})R_N = Q_N^{*} \ ,$$

$$Q_N = R_N^{*}(S^{-1*}H_N'S^{-1}) \ .$$

Comparing with (3.6) and using the uniqueness of H_N we find that $H_N = S^{-1*}H_N'S^{-1}$. So the pairs (A_N, H_N) and (A_N', H_N') are unitarily similar, and therefore they have the same sign characteristic.

It turns out (see Theorem 3.4) that these sign characteristics can also be described in terms of the eigenvalues of $\mu I - W(\lambda)$, i.e. considering $W(\lambda)$ as a hermitian matrix for every fixed real λ (which is not a pole of $W(\lambda)$). For this purpose some ideas and results of analytic perturbation theory are required.

With $W(\lambda)$ as above, i.e. a hermitian rational $n \times n$ matrix function with $\det W(\lambda) \not\equiv 0$, a meromorphic real valued function $\mu(\lambda)$ of the real variable λ is called a *proper value* of $W(\lambda)$ if there exists an analytic (generally complex-valued) vector function $\varphi(\lambda)$ of the real variable λ such that $\varphi(\lambda) \neq 0$ and $W(\lambda)\varphi(\lambda) = \mu(\lambda)\varphi(\lambda)$ for all $\lambda \in \mathbb{R}$ which are not poles of $\mu(\lambda)$. In this case the function $\varphi(\lambda)$ is called a *proper vector* corresponding to the proper value $\mu(\lambda)$.

THEOREM 3.3 *Let* $W(\lambda)$ *be a hermitian rational matrix function with determinant not identically zero. Then there exist proper values* $\mu_1(\lambda), \cdots, \mu_n(\lambda)$ *such that the corresponding proper vectors* $\varphi_1(\lambda), \cdots, \varphi_n(\lambda)$

form an orthonormal basis in $\mathbb{C}^n$ *for every real* λ. *These proper values are uniquely defined up to permutation.*

The proper values $\mu_1(\lambda),\cdots,\mu_n(\lambda)$ of $W(\lambda)$ with the properties described in Theorem 3.3 will be called *standard*.

The result of Theorem 3.3 holds under the more general hypothesis that $W(\lambda)$ is a hermitian meromorphic function defined on an interval $(a,b) \subset \mathbb{R}$ (here $-\infty \leq a < b \leq \infty$). See Chapter S6 of [19b] for the proof in the case of analytic $W(\lambda)$; if $W(\lambda)$ is meromorphic, the result is proved by applying the "analytic version" to an analytic selfadjoint function of the form $p(\lambda)W(\lambda)$, where $p(\lambda)$ is a suitable real-valued analytic function of a real variable λ. The uniqueness of the standard proper values follows from the fact that, for each fixed λ, they are the eigenvalues of $W(\lambda)$.

The conclusions of Theorem 3.3 can be re-phrased to assert the existence of an analytic matrix-valued function $U(\lambda)$ in a real variable λ, and of real meromorphic functions $\mu_1(\lambda),\cdots,\mu_n(\lambda)$ on $\mathbb{R}$ such that $U(\lambda)U(\lambda)^* = I$ and

$$W(\lambda) = U(\lambda)^{-1}\text{diag}[\mu_1(\lambda),\cdots,\mu_n(\lambda)]U(\lambda)$$

for all $\lambda \in \mathbb{R}$. Comparing this result with the Smith form (see the Appendix) it is clear that the partial multiplicities of $W(\lambda)$ at λ_0 are just the unique integers ν_j such that $\mu_j(\lambda)(\lambda-\lambda_0)^{-\nu_j}$ is analytic and non-zero at λ_0 for $j = 1,2,\cdots,n$.

Recall (see the Appendix), that given a null triple (Q_N,A_N,R_N) and a pole triple (Q_P,A_P,R_P) of $W(\lambda)$, for every $\lambda_0 \in \mathbb{C}$ the sizes of Jordan blocks in the Jordan form of A_N (resp. A_P) with eigenvalue λ_0 coincide with the positive (resp. absolute values of the negative) partial multiplicities of $W(\lambda)$ at λ_0. In particular, λ_0 is an eigenvalue (resp. a pole) of $W(\lambda)$ if and only if $\lambda_0 \in \sigma(A_N)$ (resp. $\lambda_0 \in \sigma(A_P)$). So the sign characteristic of $W(\lambda)$ can be understood also as a sign (+1 or -1) attached to each non-zero partial multiplicity of $W(\lambda)$ corresponding to a real eigenvalue or a real pole of $W(\lambda)$.

The following result, which is the main result of this section, describes the sign characteristic in terms of the standard proper values.

THEOREM 3.4 *Let* $W(\lambda)$ *be a hermitian rational matrix function with* $\det W(\lambda) \not\equiv 0$. *Let* $\mu_1(\lambda),\cdots,\mu_n(\lambda)$ *be the standard proper values of*

$W(\lambda)$, *and for real* λ_0 *which is an eigenvalue or a pole of* $W(\lambda)$, *let* $\nu_1,\nu_2,\cdots,\nu_n$ *be the corresponding partial multiplicities of* $W(\lambda)$ *at* λ_0. *Then the sign in the sign characteristic of* $W(\lambda)$ *at* λ_0 *associated with a non-zero partial multiplicity* ν_j *is the sign of the real number* $[\mu_j(\lambda)(\lambda-\lambda_0)^{-\nu_j}]_{\lambda=\lambda_0}$.

The proof of Theorem 3.4 will be given in Section 3.4. In the next section we present the background results on null functions which are necessary for the proof of Theorem 3.4.

3.3 Null Functions and Jordan Chains

In this section we present the results in the context of rational matrix functions (however, they also hold for meromorphic matrix functions in a neighbourhood of a given point, see [19d]).

Let $M(\lambda)$ be an $n \times n$ rational matrix function with determinant not identically zero, and let $\lambda_0 \in \mathbb{C}$. An analytic vector function

$$\psi(\lambda) = \sum_{j=0}^{\infty} (\lambda-\lambda_0)^j \psi_j$$

is said to be a *null function* of $M(\lambda)$ at λ_0 if $\psi_0 \neq 0$, $M(\lambda)\psi(\lambda)$ is analytic in a neighbourhood of λ_0, and $[M(\lambda)\psi(\lambda)]_{\lambda=\lambda_0} = 0$. The multiplicity of λ_0 as a zero of the vector function $M(\lambda)\psi(\lambda)$ is the *order* of $\psi(\lambda)$, and ψ_0 is the *null vector* of $\psi(\lambda)$. From this definition it follows immediately that for rational and invertible $n \times n$ matrix-valued functions $U(\lambda)$ and $V(\lambda)$ in a neighbourhood of λ_0, $\psi(\lambda)$ is a null function of $V(\lambda)M(\lambda)U(\lambda)$ at λ_0 or order k if and only if $U(\lambda)\psi(\lambda)$ is a null function of $M(\lambda)$ at λ_0 of order k.

A set of null functions $\psi_1(\lambda),\cdots,\psi_p(\lambda)$ of $M(\lambda)$ at λ_0 with orders $k_1,\cdots,k_p$, respectively, is said to be *canonical* if the null vectors $\psi_1(\lambda_0),\cdots,\psi_p(\lambda_0)$ are linearly independent and the sum $k_1 + k_2 + \cdots + k_p$ is maximal.

PROPOSITION 3.5 *Let* $M(\lambda)$ *be as defined above and let* $\psi_1(\lambda),\cdots,\psi_p(\lambda)$ *be a canonical set of null functions of* $M(\lambda)$ *(resp.* $M(\lambda)^{-1}$*) at* λ_0. *Then the number* p *is the number of positive (resp. negative) partial multiplicities of* $M(\lambda)$ *at* λ_0, *and the corresponding orders* $k_1,\cdots,k_p$ *are the positive (resp. absolute values of the negative) partial multiplicities of* $M(\lambda)$ *at* λ_0.

PROOF. Briefly, reduce $M(\lambda)$ to Smith form as described in the Appendix and apply the observation made in the second paragraph of this section. □

Now let (Q_P, A_P, R_P) be a pole triple of $M(\lambda)$. We shall establish a relationship between Jordan chains of A_P at λ_0 and null functions of $M(\lambda)^{-1}$ at λ_0, as follows. We denote by $P = P(\lambda_0)$ the Riesz projector on the root subspace of A_P corresponding to λ_0:

$$P = \frac{1}{2\pi i} \int_\Gamma (\lambda I - A_P)^{-1} d\lambda \ ,$$

where $\Gamma = \Gamma(\lambda_0)$ is a circle of sufficiently small radius around λ_0.

PROPOSITION 3.6 *Let* $\psi(\lambda)$ *be a null function of* $M(\lambda)^{-1}$ *at* λ_0 *of order* k. *Then*

$$x_j = \sum_{\nu=k}^{\infty} P(A_P - \lambda_0 I)^{\nu-j-1} R_P \varphi_\nu \ , \quad j = 0, \cdots, k-1 \tag{3.8}$$

where φ_ν *are the coefficients of* $\varphi(\lambda) \overset{\text{def}}{=} M(\lambda)^{-1}\psi(\lambda)$,

$$\varphi(\lambda) = \sum_{j=k}^{\infty} (\lambda - \lambda_0)^j \varphi_j \ , \tag{3.9}$$

is a Jordan chain for A_P *at* λ_0. *Conversely, if* $x_0, \cdots, x_{k-1}$ *is a Jordan chain of* A_P *at* λ_0, *then there is a null function* $\psi(\lambda)$ *of* $M(\lambda)^{-1}$ *at* λ_0 *for which* (3.8) *holds.*

Recall that $A_P P = P A_P$; as both subspaces Im P and Ker P are A_P-invariant and $\sigma(A_P|_{\operatorname{Im} P}) = \{\lambda_0\}$, the series in (3.8) is actually finite.

PROOF. Vectors (3.8) form a Jordan chain for A_P at λ_0 if, by definition, the following equalities hold:

$$\begin{cases} (A_P - \lambda_0 I) x_0 = 0 \ , & x_0 \neq 0 \\ (A_P - \lambda_0 I) x_j = x_{j-1} \ , & j = 1, 2, \cdots, k-1 \ . \end{cases}$$

The last $k - 1$ statements follow immediately from (3.8). Also

$$(A_P - \lambda_0 I) x_0 = \sum_{\nu=k}^{\infty} (A_P - \lambda_0 I)^\nu R_P \varphi_\nu \ .$$

Now the Laurent series for $M(\lambda)$ at λ_0, say $M(\lambda) = \sum_{j=-q}^{\infty} (\lambda - \lambda_0)^j M_j$, has the following coefficients of negative powers of $(\lambda - \lambda_0)$ (compare with (3.5)): for $j = 1, 2, \cdots, q$

$$M_{-j} = Q_P P(A_P-\lambda_0 I)^{j-1}R_P$$

and it is easily seen that q is the least positive integer for which $P(A_P-\lambda_0 I)^q = 0$. Now recall that $\psi(\lambda) = M(\lambda)\varphi(\lambda)$ is analytic near λ_0; so equating coefficients of negative powers of $(\lambda-\lambda_0)$ to zero and using the fact that $P(A_P-\lambda_0 I)^q = 0$, we obtain for $j = 1,2,\cdots$:

$$0 = \sum_{\nu=k}^{q-j} M_{-\nu-j}\varphi_\nu = \sum_{\nu=k}^{q-j} Q_P P(A_P-\lambda_0 I)^{\nu+j-1}R_P\varphi_\nu = \sum_{\nu=k}^{\infty} Q_P P(A_P-\lambda_0 I)^{\nu+j-1}R_P\varphi_\nu$$
$$= Q_P P(A_P-\lambda_0 I)^{j-1}(A_P-\lambda_0 I)x_0 = Q_P(A_P-\lambda_0 I)^{j-1}(A_P-\lambda_0 I)x_0$$

(the last equality follows from the facts that $x_0 \in \operatorname{Im} P$ and P commutes with A_P). Consequently,

$$Q_P A_P^{j-1}(A_P-\lambda_0 I)x_0 = 0 \,, \quad j = 1,2,\cdots . \tag{3.10}$$

As the node $(A_P,R_P,Q_P;\mathbb{C}^m,\mathbb{C}^n)$ is minimal, the matrix

$$\operatorname{col}[Q_P A_P^j]_{j=0}^{m-1} \overset{\text{def}}{=} \begin{bmatrix} Q_P \\ Q_P A_P \\ \cdot \\ \cdot \\ \cdot \\ Q_P A_P^{m-1} \end{bmatrix}$$

is left invertible, and (3.10) implies $(A_P-\lambda_0 I)x_0 = 0$ as required. Finally, since $\psi(\lambda_0) = x_0$ it is also true that $x_0 \neq 0$. Thus, as asserted, formula (3.8) does associate a Jordan chain for A_P corresponding to λ_0 with the null function $\psi(\lambda)$.

Conversely, let $x_0,x_1,\cdots,x_{k-1}$ be a Jordan chain of A_P at λ_0. From the definition of a pole triple it follows that the matrix

$$P[R_P,A_P R_P,\cdots,A_P^{m-1}R_P]$$

(as a linear transformation from $\mathbb{C}^{nm}$ to $\operatorname{Im} P$) is right invertible (where m is the size of A_P) and, consequently, so is the matrix

$$P[R_P,(A_P-\lambda_0 I)R_P,\cdots,(A_P-\lambda_0 I)^{m-1}R_P] =$$

$$=P[R_p, A_pR_p, \cdots, A_p^{m-1}R_p] \begin{bmatrix} I & \zeta & \binom{2}{0}\zeta^2 & \cdots & \binom{m-1}{0}\zeta^{m-1} \\ 0 & I & \binom{2}{1}\zeta & \cdots & \binom{m-1}{1}\zeta^{m-2} \\ 0 & 0 & I & & \vdots \\ \vdots & \vdots & \vdots & \ddots & \vdots \\ 0 & 0 & 0 & \cdots & I \end{bmatrix},$$

where we write $\zeta = -\lambda_0$. Hence, taking into account that $x_{k-1} \in \operatorname{Im} P$, there exist vectors $\varphi_k, \varphi_{k+1}, \cdots$, with only finitely many non-zero, such that

$$x_{k-1} = \sum_{j=k}^{\infty} P(A_p-\lambda_0 I)^{j-k} R_p\varphi_j \ . \tag{3.11}$$

The definition of a Jordan chain includes $(A_p-\lambda_0 I)x_j = x_{j-1}$ for $j = 1,2,\cdots,k-1$ and so equations (3.8) follow immediately from (3.11). It only remains to check that $M(\lambda)\varphi(\lambda)$, where $\varphi(\lambda) = \sum_{j=k}^{\infty} (\lambda-\lambda_0)^j\varphi_j$, is now a null function of $M(\lambda)^{-1}$ at λ_0.

Observe first that $x_0 \neq 0$ and that $Q_pA_p^jx_0 = \lambda_0^jQ_px_0$ for $j = 0,1,2,\cdots$. As the matrix $\operatorname{col}[Q_p(A_p-\lambda_0 I)^j]_{j=0}^{m-1}$ is left invertible, so is $\operatorname{col}[Q_pA_p^j]_{j=0}^{m-1}$, and it follows that $Q_px_0 \neq 0$. But using (3.8),

$$0 \neq Q_px_0 = \sum_{\nu=k}^{\infty} Q_pP(A_p-\lambda_0 I)^{\nu-1}R_p\varphi_\nu = \lim_{\lambda\to\lambda_0} M(\lambda)\varphi(\lambda) \ . \quad \square$$

If the Jordan chain $x_0,\cdots,x_{k-1}$ of A_p at λ_0 cannot be prolonged, then $x_{k-1} \notin \operatorname{Im}(A_p-\lambda_0 I)$, and it follows from (3.11) that $\varphi_k \neq 0$. Thus, a maximal Jordan chain of length k determines, by means of (3.9), an *associated* null function $\psi(\lambda)$ of $M(\lambda)^{-1}$ of order k.

3.4 Proof of Theorem 3.4

We shall prove explicitly only the statement concerning the negative partial multiplicities; the statement concerning positive partial multiplicities can then be obtained co considering $W(\lambda)^{-1}$ in place of $W(\lambda)$.

Let $W(\lambda)$ be as in Theorem 3.4 with a pole triple (Q_p, A_p, R_p). Recall the existence of a unique invertible hermitian matrix H_p (introduced in Section 3.2) for which

$$Q_p = R_p^*H_p \ ; \quad A_p = H_p^{-1}A_p^*H_p \ . \tag{3.12}$$

Let λ_0 be a pole of $W(\lambda)$, i.e. an eigenvalue of A_P, and assume that λ_0 is real. As in the preceding section, denote by P the Riesz projector or the root subspace of A_P corresponding to λ_0. Then (cf. Section I.6.3) P is H_P-selfadjoint:

$$H_P P = P^* H_P \ . \tag{3.13}$$

The proof is divided into three steps.

STEP 1. Let $\psi(\lambda)$ be a null function of order k of $W^{-1}(\lambda)$ at λ_0 and let $x_0, x_1, \cdots, x_{k-1}$ be the associated Jordan chain of A_P. Then the following equality holds:

$$(W^{-1}(\lambda)\psi(\lambda),\psi(\lambda)) = \sum_{j=0}^{k-1} (H_P x_j, x_{k-1})(\lambda-\lambda_0)^{k+j} + O(|\lambda-\lambda_0|^{2k}) , \tag{3.14}$$

for λ from a real neighbourhood of λ_0. Indeed, put $m_j = (H_P x_j, x_{k-1})$, $j = 0,\cdots,k-1$; then, using (3.8) and denoting $\varphi(\lambda) = \sum_{j=k}^{\infty} (\lambda-\lambda_0)^j \varphi_j = W^{-1}(\lambda)\psi(\lambda)$, we obtain:

$$\begin{aligned} m_j = (x_{k-1}, H_P x_j) &= \Big(x_{k-1}, H_P \sum_{\nu=k}^{\infty} P(A_P-\lambda_0 I)^{\nu-j-1} R_P \varphi_\nu\Big) \\ &= \sum_{\nu=k}^{\infty} \Big(R_P^*(A_P^*-\lambda_0 I)^{\nu-j-1} P^* H_P x_{k-1}, \varphi_\nu\Big) \\ &= \sum_{\nu=k}^{\infty} \Big(Q_P(A_P-\lambda_0 I)^{\nu-j-1} P x_{k-1}, \varphi_\nu\Big) = \sum_{\nu=k}^{k+j} (Q_P x_{k-1+j}, \varphi_\nu) \ . \end{aligned}$$

Here (3.12) and (3.13) were used. In view of (3.11), one easily checks that $Q_P x_{k-\nu+j}$ is the coefficient (denoted $\psi_{k-\nu+j}$) of $(\lambda-\lambda_0)^{k-\nu+j}$ in $\psi(\lambda)$. Hence,

$$m_j = \sum_{\nu=k}^{k+j} (\psi_{k-\nu+j}, \varphi_\nu) \ .$$

On the other hand,

$$(W^{-1}(\lambda)\psi(\lambda),\psi(\lambda)) = \sum_{\ell=k}^{\infty}\Big(\sum_{\nu=k}^{\infty} (\varphi_\nu, \psi_{k-\nu})\Big)(\lambda-\lambda_0)^{\ell} \ .$$

Setting $\ell = k + j$ and comparing these two relations it is found that the coefficient of $(\lambda-\lambda_0)^{k+j}$ in $(W^{-1}(\lambda)\psi(\lambda),\psi(\lambda))$ is $(H_P x_j, x_{k-1})$.

The calculations used in the proof of step 1 can be extended to show that for $i,j \geqslant 0$,

$$(H_P x_i, x_j) = \begin{cases} (H_P x_0, x_{k-1}) & \text{if} \quad i + j = k - 1 \\ 0 & \text{if} \quad i + j < k - 1 \end{cases} . \tag{3.15}$$

Also,

$$(H_P x_i, x_j) = (H_P x_\ell, x_m) \quad \text{if} \quad i + j = \ell + m .$$

STEP 2. There exists a canonical set of null functions $\psi_1(\lambda),\cdots,\psi_p(\lambda)$ of $W^{-1}(\lambda)$ at λ_0 with orders $k_1 \geq k_2 \geq \cdots \geq k_p$, respectively, such that for all λ in a real neighbourhood of λ_0,

$$(W(\lambda)^{-1}\varphi_j(\lambda),\varphi_k(\lambda)) = \begin{cases} 0 & \text{if} \quad j \neq k \\ (\lambda-\lambda_0)^{k_j} c_j & \text{if} \quad j = k \end{cases} , \tag{3.16}$$

where $c_j = \pm 1$.

Indeed, if $W(\lambda)$ is diagonal, say $W(\lambda) = \mathrm{diag}[\mu_1(\lambda),\cdots,\mu_n(\lambda)]$, the proof is easy. The canonical set is then of the form

$$\psi_j(\lambda) = |\mu_{m_j}(\lambda)(\lambda-\lambda_0)^{k_j}|^{\frac{1}{2}} e_{m_j}$$

where e_{m_j} is the m_j'th coordinate unit vector, and $\mu_{m_j}(\lambda)$ has a pole of order k_j at λ_0. When $W(\lambda)$ is not diagonal use Theorem 3.3 to reduce the problem to the diagonal case.

STEP 3. We will show that k_j in (3.16) can be equated to $-\nu_j$ (where ν_j is a negative partial multiplicity of $W(\lambda)$ at λ_0) so that c_{k_j} is the sign of $[\mu_j(\lambda)(\lambda-\lambda_0)^{-\nu_j}]_{\lambda=\lambda_0}$ which is to be identified with the sign characteristic of $W(\lambda)$ at λ_0 corresponding to ν_j.

Let $\psi_1(\lambda),\cdots,\psi_p(\lambda)$ be a canonical set of null functions for $W(\lambda)^{-1}$ at $\lambda_0 \in \mathbb{R}$ with respective orders $k_1,\cdots,k_p$ for which (3.16) holds. Proposition 3.5 asserts that $k_1,\cdots,k_p$ are also the absolute values of the negative partial multiplicities of $W(\lambda)$ at λ_0; they are also the sizes of the Jordan blocks in the Jordan form for A_P with eigenvalue λ_0, where A_P belong to a pole triple (Q_P,A_P,R_P) for $W(\lambda)$. Let

$$\left\{x_0^{(1)},\cdots,x_{k_1-1}^{(1)}\right\} , \left\{x_0^{(2)}, \quad ,x_{k_2-1}^{(2)}\right\} , \cdots , \left\{x_0^{(p)},\cdots,x_{k_p-1}^{(p)}\right\} \tag{3.17}$$

be the Jordan chains of A_P corresponding to λ_0 associated with the respective null functions $\psi_1(\lambda),\cdots,\psi_p(\lambda)$ as described in Proposition 3.6. Then the definition (3.16) together with (3.14) and (3.15) give:

$$(H_P x_\alpha^{(i)}, x_\beta^{(j)}) = \begin{cases} 0 & \text{if} \quad i \neq j \\ 0 & \text{if} \quad i = j \quad \text{and} \quad \alpha + \beta \neq k_i - 1 \\ c_{k_i} & \text{if} \quad i = j \quad \text{and} \quad \alpha + \beta = k_i - 1 \end{cases} \tag{3.18}$$

where $c_{k_i} = \pm 1$, as in (3.16). As the set (3.17) of Jordan chains of A_P corresponding to λ_0 is canonical (i.e. the vectors (3.17) are linearly independent and span the root subspace of A corresponding to λ_0), the relations (3.18) imply that c_{k_i} is the sign in the sign characteristic of (A_P, H_P) corresponding to the partial multiplicity k_i (see the second description of the sign characteristic of (A,H), Theorem I.3.16). This completes the proof of Theorem 3.4.

3.5 Factorization of Hermitian Rational Matrix Functions

Let $W(\lambda)$ be a non-constant hermitian rational matrix function with value I at infinity, and let $\theta = (A,B,C,I;\mathbb{C}^m,\mathbb{C}^n)$ be a minimal node for $W(\lambda)$. As we have seen in Section 3.1, there exists a unique hermitian invertible H such that $HA = A^*H$, $HB = C^*$, $C = B^*H$. In particular, the matrices A and $A^\times = A - BC$ are H-selfadjoint. As we know (Section I.3.12), there exists an A-invariant (resp. $A^\times$-invariant) subspace which is maximal H-nonpositive (H-nonnegative). The following result, which is the main result of this section, shows that these subspaces are always direct complements to each other, and therefore (see Theorem A.2 in the Appendix) give rise to a minimal factorization of $W(\lambda)$.

THEOREM 3.7 *Let* $W(\lambda)$ *be a hermitian rational matrix function with minimal node* $\theta = (A,B,C,I;\mathbb{C}^m,\mathbb{C}^n)$, *and let* H *be the unique hermitian* $m \times m$ *matrix such that* $HA = A^*H$, $HB = C^*$, $C = B^*H$. *Let* M_1 *be an* $A^\times$*-invariant maximal H-nonnegative subspace, and let* M_2 *be an A-invariant maximal H-nonpositive subspace. Then* M_1 *and* M_2 *are direct complements of each other in* $\mathbb{C}^m$, *and* $W(\lambda)$ *admits the minimal factorization* $W(\lambda) = W_1(\lambda)W_2(\lambda)$ *with respect to* M_1 *and* M_2; *i.e.,* $W_1(\lambda)$ *is the transfer function of* $\mathrm{pr}_{I-\pi}(\theta)$, $W_2(\lambda)$ *is the transfer function of* $\mathrm{pr}_\pi(\theta)$, *and* π *is the projector on* M_1 *along* M_2.

The definition of the nodes $\mathrm{pr}_{I-\pi}(\theta)$ and $\mathrm{pr}_\pi(\theta)$ is given in the Appendix. In particular, the main matrix of $\mathrm{pr}_{I-\pi}(\theta)$ is $A|_{M_2}$, and the main matrix of $(pr_\pi(\theta))^\times$ is $A^\times|_{M_1}$.

PROOF. Let us prove first that $M_1 \cap M_2 = \{0\}$. Choose $x \in M_1 \cap M_2$. As M_1 is H-nonnegative and M_2 is H-nonpositive, we have $(Hx,x) = 0$. Also, $A^\times x \in M_1$, and applying inequalities (I.1.8), we have,

$$|(HA^\times x,x)| \leq (HA^\times x,A^\times x)(Hx,x) = 0 ,$$

and $(HA^\times x,x) = 0$. Similarly, we have $(HAx,x) = 0$. Now

$$(HBCx,x) = (HAx,x) - (HA^\times x,x) = 0 ;$$

on the other hand, in view of (3.2),

$$(HBCx,x) = (C^*Cx,x) = ||Cx||^2 .$$

Thus, $Cx = 0$, and we have proved that $M_1 \cap M_2 \subset \operatorname{Ker} C$. Then for all $x \in M_1 \cap M_2$ we have $A^\times x = Ax - BCx = Ax$. So $M_1 \cap M_2$ is A-invariant. Hence for every $x \in M_1 \cap M_2$ we have $CA^j x = 0$, $j = 0,1,\cdots$, and

$$M_1 \cap M_2 \subset \bigcap_{i=0}^{\infty} \operatorname{Ker}(CA^i) = \{0\} .$$

Now Theorem I.1.3 shows that $\dim M_1 + \dim M_2 = m$; so M_1 and M_2 are direct complements to each other. It remains to apply Theorem A.2 of the Appendix. □

By Theorem I.3.20, there are subspaces M_1 and M_2 as in Theorem 3.7 with $\sigma((A-BC)|_{M_1})$ and $\sigma(A|_{M_2})$ in the closed lower half-plane. This choice of M_1 and M_2 leads to the next conclusion.

COROLLARY 3.8 *Let* $W(\lambda)$ *be a hermitian rational matrix function with* $W(\infty) = I$. *Then* $W(\lambda)$ *admits a minimal factorization*

$$W(\lambda) = W_1(\lambda)W_2(\lambda) \tag{3.19}$$

where $W_i(\lambda)$ *are rational matrix functions with* $W_i(\infty) = I$, $i = 1,2$, *and such that* $W_1(\lambda)$ *and* $W_2(\lambda)^{-1}$ *have no poles in the open upper half-plane.*

PROOF. Let $\theta = (A,B,C,I;\mathbb{C}^m,\mathbb{C}^n)$, be a minimal node for $W(\lambda)$. Let M_1 be an (A-BC)-invariant maximal H-nonnegative subspace such that $\operatorname{Im} \sigma((A-BC)|_{M_1}) \leq 0$. Let M_2 be an A-invariant maximal H-nonpositive subspace such that $\operatorname{Im} \sigma(A|_{M_2}) \leq 0$. By Theorem 3.7, $W(\lambda) = W_1(\lambda)W_2(\lambda)$, where

$W_1(\lambda)$ is the transfer function of $pr_{I-\pi}(\theta)$, and $W_2(\theta)$ is the transfer function of $pr_\pi(\theta)$; here π is the projector on M_1 along M_2. Since the main matrix of $pr_{I-\pi}(\theta)$ is $A|_{M_2}$, it follows that $W_1(\lambda)$ has no poles in the open upper half-plane. Also $(A-BC)|_{M_1}$ is the main matrix in the node $(pr_\pi(\theta))^\times$; so $W_2(\lambda)^{-1}$ has no poles outside $\sigma((A-BC)|_{M_1})$. □

Similarly, one can obtain minimal factorizations (3.19), where $W_1(\lambda)$ has no poles in the open upper/lower half-plane, while $W_2(\lambda)^{-1}$ has no poles in the open upper/lower half-plane, and all 4 possibilities are included.

Recall that our analysis has been confined to hermitian rational matrix functions which have the value I at infinity. We remark that the results of this section can be extended by the use of a suitable Moebius transformation to hermitian rational matrix function $W(\lambda)$ with the property that $W(\lambda_0)$ is positive definite for some real λ_0. Indeed, the rational hermitian matrix function

$$\tilde{W}(\lambda) = [W(\lambda_0)]^{-\frac{1}{2}} W\left(\frac{\lambda_0\lambda}{\lambda+1}\right)[W(\lambda_0)]^{-\frac{1}{2}}$$

then has the value I at infinity.

3.6 Symmetric Factorizations of Hermitian Rational Matrix Functions

In the preceding section we studied minimal factorizations of a hermitian rational matrix function $W(\lambda)$ in general. However, minimal factorizations of the following type, in which the hermitian property of $W(\lambda)$ is reflected, are of particular interest:

$$W(\lambda) = L(\lambda)_* N(\lambda) L(\lambda) \ . \tag{3.20}$$

Here, $N(\lambda)$ and $L(\lambda)$ are rational matrix functions and the rational matrix function $L(\lambda)_*$ is defined as follows: $L(\lambda)_* = (L(\bar{\lambda}))^*$. As in the preceding sections we shall assume that all rational matrix functions encountered have value I at infinity. If (3.20) holds, the function $N(\lambda)$ is necessarily hermitian. We call factorizations of type (3.20) *symmetric*.

By Theorem A.2 in the Appendix, a minimal factorization of $W(\lambda)$ is determined by a supporting projector π of a minimal node $\theta = (A,B,C,I;\mathbb{C}^m,\mathbb{C}^n)$ with transfer function W. Again, let H be the unique, $m \times m$, invertible, hermitian matrix such that $HA = A^*H$, $HB = C^*$, and $C = B^*H$. It turns out that symmetric factorizations (3.20) can be characterized in terms of the supporting projector π as follows:

THEOREM 3.9 *Let* π *be a supporting projector of a minimal node* $\theta = (A,B,C,I;\mathbb{C}^m,\mathbb{C}^n)$ *with hermitian transfer function* $W(\lambda)$, *and let* $W(\lambda) = K(\lambda)L(\lambda)$ *be the minimal factorization of* $W(\lambda)$ *(with* $K(\infty) = L(\infty) = I$*) corresponding to* π. *Then there exists a hermitian rational matrix function* $N(\lambda)$ *with* $N(\infty) = I$ *such that*

$$W(\lambda) = L_*(\lambda)N(\lambda)L(\lambda) \tag{3.21}$$

is a minimal factorization of $W(\lambda)$ *if and only if the subspace* $\operatorname{Im}\pi$ *is* H-*neutral and the subspace* $(\operatorname{Ker}\pi)^\perp$ *is* H^{-1}-*neutral.*

Thus, there is a one-to-one correspondence between symmetric minimal factorizations (3.20) and supporting projectors π of the minimal node with transfer function $W(\lambda)$, such that $\operatorname{Im}\pi$ is H-neutral and $(\operatorname{Ker}\pi)^\perp$ is H^{-1}-neutral.

Some preliminary discussion and a lemma are required before proving Theorem 3.9. First, observe that because $W(\lambda)$ is hermitian, a minimal factorization

$$W(\lambda) = K(\lambda)L(\lambda) \tag{3.22}$$

of $W(\lambda)$ with $K(\infty) = L(\infty) = I$ gives rise to the factorization

$$W(\lambda) = L_*(\lambda)K_*(\lambda) , \tag{3.23}$$

where $L_*(\lambda) = (L(\bar{\lambda}))^*$ and $K_*(\lambda) = (K(\bar{\lambda}))^*$ are rational matrix functions. The factorization (3.23) is again minimal. To see this, let $L(\lambda) = \sum_{j=-q}^{\infty} (\lambda-\lambda_0)^j L_j$ be the Laurent series of $L(\lambda)$ in a neighbourhood of λ_0, so that $L_*(\lambda) = \sum_{j=-\infty}^{\infty} (\lambda-\bar{\lambda}_0)^j L_j^*$. If $\operatorname{Ker}(A;\lambda_0)$ denotes the subspace of $\mathbb{C}^n$ spanned by all the eigenvectors of a rational matrix function $A(\lambda)$ at λ_0 (eigenvectors are defined in the Appendix), then

$$\operatorname{Ker}(L_*;\bar{\lambda}_0) = \overline{\operatorname{Ker}(L^T;\lambda_0)} \tag{3.24}$$

where the bar denotes complex conjugation. Similarly,

$$\operatorname{Ker}(L_*^{T-1};\bar{\lambda}_0) = \overline{\operatorname{Ker}(L^{-1};\lambda_0)} , \tag{3.25}$$

$$\mathrm{Ker}(K_*^{-1};\bar\lambda_0) = \overline{\mathrm{Ker}((K^T)^{-1};\lambda_0)} \ ; \quad \mathrm{Ker}(K_*^T;\bar\lambda_0) = \overline{\mathrm{Ker}(K;\lambda_0)}$$

By definition, minimality of factorization (3.22) means

$$\mathrm{Ker}(K;\lambda_0) \cap \mathrm{Ker}(L^{-1};\lambda_0) = \mathrm{Ker}((K^T)^{-1};\lambda_0) \cap \mathrm{Ker}(L^T;\lambda_0) = \{0\}$$

for all $\lambda_0 \in \mathbb{C}$. Taking complex conjugates in this equality, we find, using (3.24) and (3.25) that

$$\mathrm{Ker}(K_*^T;\bar\lambda_0) \cap \mathrm{Ker}((L_*^T)^{-1});\bar\lambda_0) = \mathrm{Ker}(K_*^{-1};\bar\lambda_0) \cap \mathrm{Ker}(L_*;\bar\lambda_0) = 0$$

for all $\lambda_0 \in \mathbb{C}$, which means exactly the minimality of factorization (3.23).

From Theorem A.2 of the Appendix we know that the minimal factorizations (3.22) and (3.23) correspond to a supporting projector $\pi\colon\ \mathbb{C}^m \to \mathbb{C}^m$ of the node θ. It turns out that the supporting projector corresponding to the minimal factorization (3.23) is just $(I-\pi)^{[*]}$, i.e. the adjoint of $I - \pi$ with respect to the indefinite scalar product defined by H, as the following lemma shows.

LEMMA 3.10 *Let* π *be the supporting projector of* $\theta = (A,B,C,I;\mathbb{C}^m,\mathbb{C}^n)$ *corresponding to the minimal factorization* $W(\lambda) = K(\lambda)L(\lambda)$ *of the transfer function* $W(\lambda)$ *of* θ. *Then* $(I-\pi)^{[*]} = H^{-1}(I-\pi^*)\ H$ *is the supporting projector of* θ *corresponding to the minimal factorization* $W(\lambda) = L_*(\lambda)K_*(\lambda)$ *of* $W(\lambda)$, *where* H *is the unique hermitian invertible matrix satisfying* $HA = A^*H$, $HB = C^*$, $C = B^*H$.

PROOF. Denote $\tau = H^{-1}(I-\pi^*)H$. Let us check first that τ is indeed a supporting projector of θ. Since π is a projector, so is $I - \pi^*$ and, consequently, so is τ. Further, observe that

$$H(\mathrm{Ker}\ \tau) = \mathrm{Ker}(I-\pi^*) = \mathrm{Im}\ \pi^* = [\mathrm{Ker}\ \pi]^\perp ,$$

where "$\perp$" denotes the orthogonal complement with respect to the usual scalar product in $\mathbb{C}^m$. Pick $x \in \mathrm{Ker}\ \tau$; then $Hx \in \mathrm{Im}\ \pi^*$, say $Hx = \pi^*y$. Now

$$\begin{aligned}\tau Ax &= Ax - H^{-1}\pi^*HAx = H^{-1}(I-\pi^*)A^*Hx = H^{-1}(I-\pi^*)A^*\pi^*y \\ &= H^{-1}[\pi A(I-\pi)]^*y ,\end{aligned}$$

and since $\mathrm{Ker}\ \pi$ is A-invariant, we have $A(I-\pi) = (I-\pi)A(I-\pi)$, and therefore $\pi A(I-\pi) = 0$. So $\tau Ax = 0$, and $Ax \in \mathrm{Ker}\ \tau$. Hence $\mathrm{Ker}\ \tau$ is A-invariant.

We check now that

$$A^{\times}(\mathrm{Im}\ \tau) \subset \mathrm{Im}\ \tau\ . \tag{3.26}$$

Choose $x \in \mathrm{Im}\ \tau$; as $H(\mathrm{Im}\ \tau) = \mathrm{Im}(I-\pi^*)$, we have $Hx = (I-\pi^*)y$. Now $\tau A^{\times}x = H^{-1}(A^{\times}(I-\pi))^* Hx = H^{-1}(A^{\times}(I-\pi))^*(I-\pi^*)y = H^{-1}((I-\pi)A^{\times}(I-\pi))^* y$, and since $\mathrm{Im}\ \pi$ is $A^{\times}$-invariant, i.e. $A^{\times}\pi = \pi A^{\times}\pi$, we get

$$\tau A^{\times}x = H^{-1}((I-\pi)A^{\times})^* y = H^{-1}(A^{\times})^* Hx = A^{\times}x\ ,$$

in view of the H-selfadjointness of $A^{\times}$. So $A^{\times}x \in \mathrm{Im}\ \tau$, (3.26) is checked, and it follows that τ is a supporting projector of θ.

It remains to show that the transfer function of $pr_{\tau}(\theta)$ is K_*. We have:

$$pr_{\tau}(\theta) = (\tau A|_{\mathrm{Im}\ \tau}, \tau B, C|_{\mathrm{Im}\ \tau}, I; \mathrm{Im}\tau, \mathbb{C}^n)$$

$$= (H^{-1}(I-\pi^*)HA|_{\mathrm{Im}\ \tau}, H^{-1}(I-\pi^*)HB, C|_{\mathrm{Im}\ \tau}, I; \mathrm{Im}\ \tau, \mathbb{C}^n)$$

$$= (H^{-1}(I-\pi^*)A^*H|_{H^{-1}(\mathrm{Im}\ \pi)^{\perp}}, H^{-1}(I-\pi^*)C^*, B^*H|_{H^{-1}(\mathrm{Im}\ \pi)^{\perp}}, I, H^{-1}(\mathrm{Im}\ \pi)^{\perp}, \mathbb{C}^n)\ .$$

A calculation shows that the transfer function of $pr_{\tau}(\theta)$ is

$$U(\lambda) = I + B^*(\lambda I - (I-\pi^*)A^*)^{-1}(I-\pi^*)C^*\ .$$

Now

$$U_*(\lambda) = I + C(I-\pi)(\lambda I - A(I-\pi))^{-1}B\ ,$$

and since $(I-\pi)(\lambda I-A(I-\pi))^{-1} = (\lambda I-(I-\pi)A)^{-1}(I-\pi)$,

$$U_*(\lambda) = I + C(\lambda I - (I-\pi)A)^{-1}(I-\pi)B\ ,$$

which coincides with the transfer function of $pr_{\pi}(\theta)$, i.e. with $K(\lambda)$. So $U(\lambda) = K_*(\lambda)$ as desired. □

We are ready now to prove Theorem 3.9.

Proof of Theorem 3.9. Assume that $W(\lambda) = L(\lambda)_* N(\lambda)L(\lambda) = K(\lambda)L(\lambda)$ is a minimal factorization. By Lemma 3.10 the minimal factorization $W(\lambda) = L_*(\lambda)K_*(\lambda)$ corresponds to the supporting projector $\tau = H^{-1}(I-\pi^*)H$

of θ. On the other hand, there exists a supporting projector ν of $pr_\tau(\theta)$ corresponding to the minimal factorization $K_*(\lambda) = N(\lambda)L(\lambda)$, and the supporting projector of θ corresponding to $L(\lambda)$ as a right factor of $W(\lambda)$ is just $\nu\tau$ (see Appendix). So

$$\nu\tau = \pi , \tag{3.27}$$

and

$$\text{Im } \pi \subset \text{Im } \tau = \text{Im}(H^{-1}(I-\pi^*)H) = H^{-1}(\text{Im}(I-\pi^*)) = H^{-1}(\text{Ker } \pi^*)$$
$$= H^{-1}((\text{Im } \pi)^\perp) ,$$

or $H(\text{Im } \pi) \subset (\text{Im } \pi)^\perp$. Hence $\text{Im } \pi$ is H-neutral. Also

$$\text{Ker } \pi \supset \text{Ker } \tau = \text{Ker}(H^{-1}(I-\pi^*)H) = H^{-1}(\text{Ker}(I-\pi^*)) = H^{-1}((\text{Ker } \pi)^\perp) ,$$

which means $(\text{Ker } \pi)^\perp$ is H^{-1}-neutral.

Conversely, assume $\text{Im } \pi$ is H-neutral and $(\text{Ker } \pi)^\perp$ is H^{-1}-neutral, i.e.

$$\text{Im } \pi \subset \text{Im } \tau ; \quad \text{Ker } \pi \supset \text{Ker } \tau . \tag{3.28}$$

Then there exists a projector ν on $\text{Im } \tau$ such that $\text{Im } \nu = \text{Im } \pi$ and $\text{Ker } \nu = \text{Ker } \pi \cap \text{Im } \tau$. One checks easily that ν is a supporting projector of $pr_\tau(\theta)$; further, inclusions (3.28) imply (3.27). Since (3.27) holds, $L(\lambda)$ is a right factor in the minimal factorization of $K_*(\lambda)$ determining by the supporting projector ν of $pr_\tau(\theta)$. So $W(\lambda) = L_*(\lambda)N(\lambda)L(\lambda)$ for some rational matrix function $N(\lambda)$ with $N(\infty) = I$, which is clearly self-adjoint. □

The proof of Theorem 3.9 shows that the middle factor $N(\lambda)$ in the minimal factorization $W = L_*NL$ is equal to I if and only if $\nu = I$, which means

$$H(\text{Im } \pi) = (\text{Im } \pi)^\perp \quad \text{and} \quad H(\text{Ker } \pi) = (\text{Ker } \pi)^\perp .$$

3.7 Non-Negative Definite Rational Matrix Functions

A rational $n \times n$ matrix function $W(\lambda)$ is called *non-negative definite* if $(W(\lambda)x,x) \geq 0$ for every $x \in \mathbb{C}^n$ and every $\lambda \in \mathbb{R}$ which is not a pole of $W(\lambda)$. Clearly, a non-negative definite matrix function is hermitian. For instance, any function of the form

$$W(\lambda) = L_*(\lambda)L(\lambda) \tag{3.29}$$

where $L(\lambda)$ is a rational matrix function with $L(\infty) = I$, is non-negative definite. Note that factorization (3.29) is a particular form of (3.21) with $N(\lambda) \equiv I$. The following result shows, in particular, that every non-negative definite rational matrix function has the form (3.29).

THEOREM 3.11 *Let* $W(\lambda)$ *be a hermitian rational matrix function with* $W(\infty) = I$, *which is the transfer function of minimal node* $\theta = (A,B,C,I;\mathbb{C}^m,\mathbb{C}^n)$, *and let* $H = H^*$ *be invertible and such that* $HA = A^*H$, $HB = C^*$, $C = B^*H$. *Then the following conditions are equivalent:*

(i) $W(\lambda)$ *admits a minimal symmetric factorization*

$$W(\lambda) = L_*(\lambda)L(\lambda) \ ; \tag{3.30}$$

(ii) *there exists a maximal* H-*nonnegative subspace which is* H-*neutral and* $A^\times$-*invariant (where* $A^\times = A - BC$) *and there exists a maximal* H-*nonpositive subspace, which is* H-*neutral and* A-*invariant;*

(iii) *the partial multiplicities of* $W(\lambda)$ *corresponding to real eigenvalues and poles are all even, and the sign characteristic of* $W(\lambda)$ *consists only of* +1'*s*.

(iv) $W(\lambda)$ *is nonnegative definite for each real* λ *which is not a pole of* $W(\lambda)$.

PROOF. (i)⇒(iv) is clear (and was observed above).

(ii)⇒(i). Let M (resp. N) be the $A^\times$- invariant (resp. A-invariant) subspace whose existence is assumed in (ii). By Theorem 3.7 there is a minimal factorization $W(\lambda) = W_1(\lambda)W_2(\lambda)$ with a supporting projector π such that $\operatorname{Im}\pi = M$, $\operatorname{Ker}\pi = N$. Since M is H-neutral, we have $HM \subset M^\perp$, and since $\dim M = \dim N = \dim M^\perp$ we have, in fact, $HM = M^\perp$. Also, N is H-neutral, which implies $HN = N^\perp$. Now Theorem 3.9 (see also the remark after its proof) shows that $W_1(\lambda) = W_{2*}(\lambda)$, so (i) holds.

(iii)⇒(ii). Existence of subspaces as in (ii) follows from Theorem I.3.22 (see formula (I.3.26)).

(iv)⇒(iii). By definition, the sign characteristic of $W(\lambda)$ at $\lambda_0 \in \mathbb{R}$ consists of the signs in the sign characteristic of (A,H) (resp. $(A^\times,H)$)

attached to the negative (resp. positive) partial multiplicities of $W(\lambda)$ at λ_0.

Let $x_0,\cdots,x_{k-1}$ be a Jordan chain of A, corresponding to its real eigenvalue λ_0, which cannot be prolonged. As we have seen in Section 3.3, there is a null function $\psi(\lambda)$ of order k of $W^{-1}(\lambda)$ at λ_0 associated with $x_0,\cdots x_{k-1}$. Equality (3.14) implies that

$$(W^{-1}(\lambda)\psi(\lambda),\psi(\lambda)) = (\lambda-\lambda_0)^k(Hx_0,x_{k-1}) + O(|\lambda-\lambda_0|^{k+1})$$

in a neighbourhood of λ_0. Since $W^{-1}(\lambda)$ is nonnegative definite for real λ which are not poles of $W^{-1}(\lambda)$, we have

$$(\lambda-\lambda_0)^k(Hx_0,x_{k-1}) + O(|\lambda-\lambda_0|^{k+1}) \geqslant 0 \tag{3.31}$$

for real λ close enough to λ_0. This implies $(Hx_0,x_{k-1}) \geqslant 0$. In view of the second description of the sign characteristic (Theorem I.3.16) it follows that all the signs in the sign characteristic of (A,H) are +1's; moreover, $(Hx_0,x_{k-1}) > 0$ for any Jordan chain $x_0,\cdots,x_{k-1}$ of A at λ_0 which cannot be prolonged. Now in view of (3.31) the integer k is even. Thus, the blocks of the Jordan form for A corresponding to real eigenvalues are all of even size. As these sizes are equal to the negative partial multiplicities of $W(\lambda)$ at the same points (see Appendix), we obtain the conclusion of (iii) concerning negative partial multiplicities of $W(\lambda)$. For the positive partial multiplicities of $W(\lambda)$ apply the arguments above to $A^\times$ instead of A. □

In fact, Theorem 3.11 establishes a one-to-one correspondence between pairs of subspaces as in (ii) and factorizations (3.31). More exactly, given a maximal H-nonnegative subspace M which is H-neutral and $A^\times$-invariant, and given a maximal H-nonnegative subspace N which H-neutral and A-invariant, a factorization (3.31) holds where $L_*(\lambda)$ is the transfer function of $pr_{I-\pi}(\theta)$, $L(\lambda)$ is the transfer function of $pr_\pi(\theta)$, and π is the projector with $\operatorname{Im}\pi = M$, $\operatorname{Ker}\pi = N$. In particular, since the signature of H is zero (Theorem 3.11(iii)) it follows that the dimension m from the minimal node $\theta = (A,B,C,I;\mathbb{C}^m,\mathbb{C}^n)$ of the non-negative definite rational matrix function $W(\lambda)$, is even, say $m = 2k$, and $\dim M = \dim N = k$. Using the uniqueness of the subspaces M and N (Theorems I.3.21 and I.3.22) we obtain the next result (recall that a c-set of an H-selfadjoint matrix A is a maximal set of non-real eigenvalues which does not contain conjugate pairs of complex numbers).

THEOREM 3.12 *Let* $W(\lambda)$ *and* $\theta = (A,B,C,I;\mathbb{C}^m,\mathbb{C}^n)$ *be as in Theorem 3.11. Then for every pair of c-sets* C_A *and* $C_{A^\times}$ *of* A *and* $A^\times$ *respectively, there exists a unique minimal factorization* $W(\lambda) = L_*(\lambda)L(\lambda)$ *with* $L(\infty) = I$, *where the rational matrix function* $L(\lambda)$ *has no poles outside* $C_A \cup \mathbb{R}$, *while the rational matrix function* $L(\lambda)^{-1}$ *has no poles outside* $C_{A^\times} \cup \mathbb{R}$. *In particular, there exists a unique minimal factorization* $W(\lambda) = L_*(\lambda)L(\lambda)$ *with* $L(\infty) = I$ *such that* $L(\lambda)$ *has no poles in the open upper half-plane and* $\det L(\lambda) \neq 0$ *for* $\mathrm{Im}\,\lambda > 0$.

In connection with Theorem 3.12 observe that for a rational matrix function $V(\lambda)$, if a point $\lambda_0 \in \mathbb{C}$ is neither a pole of $V(\lambda)$ nor a pole of $V(\lambda)^{-1}$, then $\det V(\lambda_0) \neq 0$.

The results of this section can also be extended to hermitian rational matrix functions $W(\lambda)$ such that $W(\lambda_0)$ is positive definite for some real λ_0 (cf. the remark at the end of Section 3.6).

3.8 Minimal Factorizations of Real Hermitian Rational Matrix Functions

Consider a minimal node $\theta = (A,B,C,I;\mathbb{C}^m,\mathbb{C}^n)$ in which A, B, C are *real* matrices of sizes $m \times m$, $m \times n$ and $n \times m$, respectively (ref. Appendix to this chapter). The transfer function $W(\lambda)$ for such a node is easily seen to be an $n \times n$ real, hermitian, rational matrix function, and so is the transfer function for the node $\theta^* = (A^*,C^*,B^*,I;\mathbb{C}^m,\mathbb{C}^n)$. Furthermore, by Theorem 3.1, the nodes θ and θ^* are similar and the similarity is achieved by a *real* invertible hermitian matrix H. This is simply because H is the unique solution of a system of linear equations with real coefficients: $HA = A^*H$, $HB = C^*$, $C = B^*H$ (cf. Proposition A.1 of the Appendix).

Using Theorem A.3 of the Appendix one can easily reproduce results of Sections 3.1-3.6 (except for Corollary 3.8) for the real case, as described above. The real analogue of Corollary 3.8, as well as real analogues of the results of Section 3.7, hold under the additional assumption that the real matrices A and $A - BC$ from a minimal real node $\theta = (A,B,C,I;\mathbb{C}^m,\mathbb{C}^n)$ with the transfer function $W(\lambda)$, have m real eigenvalues (counting mulciplicities). This condition ensures the existence of the subspaces needed. The following example shows that this condition is not automatic.

EXAMPLE 3.1 Let $W(\lambda) = 1 + \lambda^{-2}$ be a scalar nonnegative real rational function with value 1 at infinity. The function $W(\lambda)$ does not admit real factorizations of type $W(\lambda) = L_*(\lambda)L(\lambda)$ (but of course admits a non-real one, as it should be by Theorem 3.11: $1 + \lambda^{-2} = (1+i\lambda^{-1})(1-i\lambda^{-1})$). As a minimal real node for $W(\lambda)$ pick $A = \left(\begin{bmatrix} 0 & 1 \\ 0 & 0 \end{bmatrix}, \begin{bmatrix} 0 \\ 1 \end{bmatrix}, [1 \;\; 0]; ¢, ¢\right)$. Note that $A = \begin{bmatrix} 0 & 1 \\ 0 & 0 \end{bmatrix}$ has 2 real eigenvalues, but $A - BC = \begin{bmatrix} 0 & 1 \\ -1 & 0 \end{bmatrix}$ has no real eigenvalues at all. So the condition mentioned above does not hold in this example. □

APPENDIX TO CHAPTER 3

RATIONAL MATRIX FUNCTIONS

In the interest of a self-contained development of the theory of rational matrix-valued functions, this appendix is devoted to a review of the basic ideas and results for such functions, thus laying the foundations for the treatment of hermitian rational functions in the main body of the chapter.

A.1 Linear Systems, Their Transfer Functions and Nodes

Consider a linear system of differential equations

$$\begin{cases} \frac{dx(t)}{dt} = Ax(t) + Bu(t) \ ; \quad x(0) = 0 \\ y(t) \ = Cx(t) + Du(t) \end{cases} \tag{A.1}$$

where $x(t)$, $y(t)$, $u(t)$ are vector functions of the real parameter t, and A, B, C, D are constant matrices of sizes $m \times m$, $m \times n$, $p \times m$, $p \times n$ respectively (so the vectors $x(t)$, $y(t)$, $u(t)$ have m, n, p components respectively). One associates $x(t)$, $y(t)$ and $u(t)$ with the state, output and control of the system, respectively. Equations of type (A.1) are used to study electrical or mechanical systems which can be described by a finite number of independent parameters and whose physical characteristics are time-independent (see any book on linear systems theory for more detail). A standard method for the solution of (A.1) takes advantage of the Laplace transform. For a vector function $f(t)$ let $F(\lambda) = \int_0^\infty x(t)e^{-\lambda t}dt$, $\lambda \in \mathbb{R}$ be its Laplace transform (we assume, of course, that the integral $\int_0^\infty x(t)e^{-\lambda t}dt$ converges at least for λ large enough). Denoting by capital Latin letters the Laplace transforms of the functions denoted by the corresponding small Latin letter, we have from (A.1):

$$\lambda X(\lambda) = AX(\lambda) + BU(\lambda) ;$$
$$Y(\lambda) = CX(\lambda) + DU(\lambda) .$$

Consequently, $X(\lambda) = (\lambda I-A)^{-1}BU(\lambda)$, and

$$Y(\lambda) = [C(\lambda I-A)^{-1}B + D]U(\lambda) = W(\lambda)U(\lambda) ,$$

where

$$W(\lambda) = C(\lambda I-A)^{-1}B + D \tag{A.2}$$

is a rational matrix function with finite value $(= D)$ at infinity, called the *transfer function* of the system (A.1). Thus, in terms of Laplace transforms, the output of the system is obtained by applying the transfer function to the control.

From the point of view of system theory the *hermitian* rational matrix function (studied in Chapter 3) appear as transfer functions of *reciprocal* networks, i.e. such that the influence of the i-th element on the j-th element is the same as the influence of the j-th element of the i-th element (see [1] for more detail).

In the design of electrical or mechanical systems with prescribed response to given controls, the following problem is important: given a rational $p \times n$ matrix function $W(\lambda)$ with finite value at infinity, find matrices A, B, C, D such that $W(\lambda)$ is the transfer function of (A.1). In the language of linear systems theory this is described as finding a "state space realization". A solution of the state space realization problem always exists, but solutions of particular interest are the *minimal realizations*, i.e. such that the size of the matrix A is as small as possible (this means that the designed system is as economical as possible, in terms of the number of independent parameters which describe the state of the system).

It turns out (see, e.g., Theorem 2.1 in [4]) that any $n \times n$ rational matrix function $W(\lambda)$ with finite limit $\lim_{\lambda\to\infty} W(\lambda)$ admits a representation (A.2). It is convenient to summarize the four constant matrices in the representation and the spaces on which they act in the form $\theta = (A,B,C,D;\mathbb{C}^m,\mathbb{C}^n)$ and call this a *node* for $W(\lambda)$. The matrix A is referred to as the *main matrix* of θ. Conversely, given a node θ the rational matrix function given by (A.2) is called the *transfer function* of the node θ and written

$W_\theta(\lambda)$. Observe that the transfer function of the node $\theta = (A,B,C,D;\mathbb{C}^m,\mathbb{C}^n)$ has no poles outside the spectrum of A.

Let $W(\lambda)$ be a rational matrix function such that $W(\infty) = D$ is invertible. A node $\theta = (A,B,C,D;\mathbb{C}^m,\mathbb{C}^n)$ such that $W(\lambda)$ is its transfer function, is not determined uniquely. For instance, the nodes θ and

$$\left(\begin{bmatrix} A & 0 \\ 0 & 0 \end{bmatrix}, [\,C \quad 0\,], \begin{bmatrix} B \\ 0 \end{bmatrix}, D ; \mathbb{C}^{m+1}, \mathbb{C}^n\right)$$

have the same transfer function. We say that the node θ is *minimal* if its main matrix has the least possible size among all nodes with the transfer function $W_\theta(\lambda)$. Minimal nodes can be characterized as follows: $\theta = (A,B,C,D;\mathbb{C}^m,\mathbb{C}^n)$ is minimal if and only if

$$\operatorname{rank}[B,AB,\cdots,A^{m-1}B] = \operatorname{rank}\begin{bmatrix} C \\ CA \\ \cdot \\ \cdot \\ \cdot \\ CA^{m-1} \end{bmatrix} = m .$$

Moreover, two minimal nodes $\theta_i = (A_i,B_i,C_i,D_i;\mathbb{C}^m,\mathbb{C}^n)$, $i = 1,2$ with the same transfer function are *similar*, i.e.

$$A_2 = SA_1S^{-1} \ ; \quad B_2 = SB_1 \ ; \quad C_2 = C_1S^{-1} \ ; \quad D_2 = D_1$$

for some invertible $m \times m$ matrix S which is determined uniquely by θ_1 and θ_2. More exactly, the following proposition holds:

PROPOSITION A.1 *Let* $\theta_i = (A_i,B_i,C_i,D_i;\mathbb{C}^m,\mathbb{C}^n)$, $i = 1,2$, *be minimal nodes with the same transfer function. Then* $D_1 = D_2$ *and the system of equations*

$$A_2S = SA_1 \ , \quad B_2 = SB_1 \ , \quad C_2S = C_1 \tag{A.3}$$

have a unique solution S; *furthermore,* S *is an invertible matrix.*

PROOF. Put

$$\Omega_k = \begin{bmatrix} C_k \\ C_kA_k \\ \cdot \\ \cdot \\ \cdot \\ C_kA_k^{m-1} \end{bmatrix}, \quad k = 1,2 ;$$

$$\Delta_k = [B_k, A_kB_k, \cdots, A_k^{m-1}B_k] , \quad k = 1,2 .$$

Since θ_k is a minimal node, the columns of Ω_k are linearly independent, and therefore there exists a left inverse $\Omega_k^{(-1)}$ of Ω_k:

$$\Omega_1^{(-1)}\Omega_1 = \Omega_2^{(-1)}\Omega_2 = I .$$

Similarly, there exists a right inverse $\Delta_k^{(-1)}$ of Δ_k, $k = 1,2$. On the other hand, comparing the Laurent expansions in a neighbourhood of infinity of the transfer functions of θ_1 and θ_2, we obtain

$$C_1A_1^jB_1 = C_2A_2^jB_2 , \quad j = 0,1,2,\cdots . \tag{A.4}$$

It follows that $\Omega_1\Delta_1 = \Omega_2\Delta_2$. But then $\Omega_2^{(-1)}\Omega_1 = \Delta_2\Delta_1^{(-1)} = S$, say. The matrix S is invertible with the inverse $\Omega_1^{(-1)}\Omega_2 = \Delta_1\Delta_2^{(-1)}$. Indeed,

$$\Omega_2^{(-1)}\Omega_1\Delta_1\Delta_2^{(-1)} = \Omega_2^{(-1)}\Omega_2\Delta_2\Delta_2^{(-1)} = I .$$

Further, (A.4) implies $\Omega_2A_2\Delta_2 = \Omega_1A_1\Delta_1 = \Omega_1\Delta_1\Delta_1^{(-1)}A_1\Delta_1 = \Omega_2\Delta_2\Delta_1^{(-1)}A_1\Delta_1$. Premultiply by $\Omega_2^{(-1)}$ and postmultiply by $\Delta_2^{(-1)}$ to obtain $A_2S = SA_1$. Similarly, one checks $B_2 = SB_1$, $C_2S = C_1$; so (A.3) holds. Finally, the solution of (A.3) is unique (indeed, the difference between two solutions S' and S'' satisfies

$$(S'-S'')[B_1, A_1B_1, \cdots, A_1^{m-1}B_1] = 0 ,$$

and consequently $S' = S''$. □

Let $\theta = (A,B,C,D;¢^m,¢^n)$ be a node. Then $\theta^\times = (A-BD^{-1}C, BD^{-1}, -D^{-1}C, D^{-1};¢^m,¢^n)$ is again a node, which is called the *associated* node of θ. One checks easily that if θ is minimal, then so is $\theta^\times$, and

$$W_{\theta^\times}(\lambda) = (W_\theta(\lambda))^{-1} .$$

Further, let $\theta_i = (A_i,B_i,C_i,D_i;\mathbb{C}^{m_i},\mathbb{C}^n)$, $i = 1,2$ be nodes. Their product is defined as the node

$$\theta_1\theta_2 = \left(\begin{bmatrix} A_1 & B_1C_2 \\ 0 & A_2 \end{bmatrix}, \begin{bmatrix} B_1D_2 \\ B_2 \end{bmatrix}, [C_1,D_1C_2], D_1D_2 ; \mathbb{C}^{m_1+m_2}, \mathbb{C}^n\right).$$

It is easily seen that the multiplication of nodes is associative: $(\theta_1\theta_2)\theta_3 = \theta_1(\theta_2\theta_3)$. Also,

$$W_{\theta_1\theta_2}(\lambda) = W_{\theta_1}(\lambda)W_{\theta_2}(\lambda) ,$$

but, in general the product of two minimal nodes need not to be minimal. For instance, the product of two minimal nodes $(0,1,1,1;\mathbb{C},\mathbb{C})$ and $(-1,1,-1,1;\mathbb{C},\mathbb{C})$ is not minimal. On the other hand, if the product $\theta_1\theta_2$ is a minimal node, then so are θ_1 and θ_2.

The notion of a node allows us to study general rational matrix functions as well. Thus, let $W(\lambda)$ be an $n \times n$ rational matrix function and let $W(\lambda) = \sum_{j=-\infty}^{q} W_j\lambda^j$ be the Laurent series for $W(\lambda)$ at infinity. Since $W(\lambda)$ is rational, only a finite number of non-zero terms with positive degree (in λ) will occur in this Laurent series. Denoting by $V(\lambda)$ the matrix polynomial $-I + \sum_{j=0}^{q} W_j\lambda^j$, we obtain the following decomposition for $W(\lambda)$:

$$W(\lambda) = \widetilde{W}(\lambda) + V(\lambda) , \tag{A.5}$$

where $V(\lambda)$ is a matrix polynomial, and $\widetilde{W}(\lambda)$ is a rational matrix function with value I at infinity. The decomposition (A.5) is unique. Indeed, let

$$W(\lambda) = \widetilde{W}_1(\lambda) + V_1(\lambda)$$

be another representation of $W(\lambda)$, where $V_1(\lambda)$ is a polynomial, and $\widetilde{W}_1(\lambda)$ is a rational function with value I at infinity. Then

$$\widetilde{W}(\lambda) - \widetilde{W}_1(\lambda) = V(\lambda) - V_1(\lambda) ,$$

and both sides of this equality represent a matrix function which must be a polynomial and has finite value at infinity. But then this polynomial must be a constant, so $\widetilde{W}(\lambda) - \widetilde{W}_1(\lambda) = A$, a constant matrix. Since $\widetilde{W}(\infty) = \widetilde{W}_1(\infty)$

$= I$, we obtain $A = 0$; so $\widetilde{W}(\lambda) = \widetilde{W}_1(\lambda)$ and $V(\lambda) = V_1(\lambda)$.

Now let $\theta = (A,B,C,I;\mathbb{C}^m,\mathbb{C}^n)$ be a node with transfer function $\widetilde{W}(\lambda)$. Then

$$W(\lambda) = C(\lambda I-A)^{-1}B + U(\lambda) , \tag{A.6}$$

where $U(\lambda) = V(\lambda) + I$ is a matrix polynomial. Representation (A.6) is called a *realization* of the rational matrix function $W(\lambda)$. This realization is *minimal* if the node θ is minimal.

A.2 The Local Smith Form and Partial Multiplicities

Let $W(\lambda)$ be an $n \times n$ rational matrix function with $\det W(\lambda) \not\equiv 0$. In a neighbourhood of each point $\lambda_0 \in \mathbb{C}$ the function $W(\lambda)$ admits the representation, called the *local Smith form* of $W(\lambda)$ at λ_0:

$$W(\lambda) = E_1(\lambda)\operatorname{diag}[(\lambda-\lambda_0)^{\nu_1},\cdots,(\lambda-\lambda_0)^{\nu_n}]E_2(\lambda) , \tag{A.7}$$

where $E_1(\lambda)$ and $E_2(\lambda)$ are rational matrix functions which are defined and invertible at λ_0, and $\nu_1,\cdots,\nu_n$ are integers. Let us verify this.

Consider first the case when $W(\lambda)$ is a matrix polynomial with the *Smith form* (see, e.g., [19b], [14]):

$$W(\lambda) = Q_1(\lambda)\operatorname{diag}[d_1(\lambda),\cdots,d_n(\lambda)]Q_2(\lambda) ,$$

where $Q_1(\lambda)$ and $Q_2(\lambda)$ are matrix polynomials with constant non-zero determinant, and $d_i(\lambda)$ are scalar polynomials determined by the equalities $d_i(\lambda) = \dfrac{p_i(\lambda)}{p_{i-1}(\lambda)}$, $i = 1,\cdots,n$. Here $p_0(\lambda) \equiv 1$ and $p_i(\lambda)$, $i = 1,\cdots,n$ is the greatest common divisor of the (not identically zero) minors of $W(\lambda)$ of order i (because of the assumption $\det W(\lambda) \not\equiv 0$, there are minors of $W(\lambda)$ of any order i $(i = 1,2,\cdots,n)$ which are not identically zero.) Put $d_i(\lambda) = (\lambda-\lambda_0)^{\nu_i}f_i(\lambda)$, where $f_i(\lambda)$ is a scalar polynomial with $f_i(\lambda_0) \neq 0$, and ν_i is a nonnegative integer. Now with $E_1(\lambda) = Q_1(\lambda)$, $E_2(\lambda) = \operatorname{diag}[f_1(\lambda),\cdots,f_n(\lambda)]Q_2(\lambda)$ the equality (A.7) holds.

In the general case, write

$$W(\lambda) = p(\lambda)^{-1}\,\widetilde{W}(\lambda) , \tag{A.8}$$

where $\widetilde{W}(\lambda)$ and $p(\lambda)$ are matrix and scalar polynomials, respectively. As the existence of representation (A.7) was already proved for $\widetilde{W}(\lambda)$, then using equation (A.8) we prove (A.7) for $W(\lambda)$.

The integers $\nu_1,\cdots,\nu_n$ in (A.7) are uniquely determined by $W(\lambda)$ at λ_0 up to permutation, and do not depend on the particular choice of the local Smith form (A.7). To see this, assume $\nu_1 \leq \cdots \leq \nu_n$, and define the multiplicity of a scalar rational function $g(\lambda) \not\equiv 0$ at λ_0 as the integer ν such that the function $g(\lambda)(\lambda-\lambda_0)^{-\nu}$ is analytic and non-zero at λ_0. Then, using the Cauchy-Binet formulas for the minors of the product of matrices (see [14], Section I.2), we easily see that $\nu_1 + \cdots + \nu_i$ is the minimal multiplicity at λ_0 of the not identically zero minors of size $i \times i$ of $W(\lambda)$, $i = 1,\cdots,n$. Thus, the numbers $\nu_1 + \cdots + \nu_i$, $i = 1,\cdots,n$, and, consequently, $\nu_1,\cdots,\nu_n$ are uniquely determined by $W(\lambda)$.

The integers $\nu_1,\cdots,\nu_n$ from the local Smith form (A.7) of $W(\lambda)$ are called the *partial multiplicities* of $W(\lambda)$ at λ_0.

Note that $\lambda_0 \in \mathbb{C}$ is a pole of $W(\lambda)$ (i.e. a pole of at least one entry in $W(\lambda)$) if and only if $W(\lambda)$ has a negative partial multiplicity at λ_0. Indeed, the minimal partial multiplicity of $W(\lambda)$ at λ_0 coincides with the minimal multiplicity at λ_0 of the not identically zero entries of $W(\lambda)$. Also, $\lambda_0 \in \mathbb{C}$ is an *eigenvalue* of $W(\lambda)$, (by definition, this means that λ_0 is a pole of $W(\lambda)^{-1}$) if and only if $W(\lambda)$ has a positive partial multiplicity. In particular, for every $\lambda_0 \in \mathbb{C}$, except for the finite number of points, all partial multiplicities are zeros.

There is a close relation between the partial multiplicities of $W(\lambda)$ and the minimal realization of $W(\lambda)$. Namely, let $W(\lambda)$ be a rational $n \times n$ matrix function with $\det W(\lambda) \not\equiv 0$, and let

$$W(\lambda) = C(\lambda I-A)^{-1}B + U(\lambda)$$

be a minimal realization of $W(\lambda)$, as explained in the preceding section. Then $\lambda_0 \in \mathbb{C}$ is a pole of $W(\lambda)$ if and only if λ_0 is an eigenvalue of A. Moreover, for a fixed pole λ_0 of $W(\lambda)$ the number of negative partial multiplicities of $W(\lambda)$ at λ_0 coincides with the number of Jordan blocks with eigenvalue λ_0 in the Jordan normal form of A, and the absolute values of these partial multiplicities coincide with the sizes of these Jordan blocks. Similarly, let

$$W(\lambda)^{-1} = C_1(\lambda I - A_1)^{-1}B_1 + U_1(\lambda)$$

be a minimal realization of $W(\lambda)^{-1}$. Then $\lambda_0 \in ¢$ is an eigenvalue of A_1 if and only if λ_0 is an eigenvalue of $W(\lambda)$, and the positive partial multiplicities of $W(\lambda)$ at λ_0 coincide with the sizes of Jordan blocks of A_1 with eigenvalue λ_0. For more detail and proofs of these statements we refer the reader to [4], especially Theorem 3.3.

A.3 Minimal Factorizations

In this section we shall describe minimal factorizations of rational matrix functions.

Such factorizations appear naturally in linear systems theory. Consider two systems of type (A.1):

$$\left.\begin{aligned} &\begin{cases} \dfrac{dx_1}{dt} = A_1x_1(t) + B_1u_1(t) ; & x_1(0) = 0 ; \\ y_1(t) = C_1x_1(t) + D_1u_1(t) ; & \end{cases} \\ &\begin{cases} \dfrac{dx_2}{dt} = A_2x_2(t) + B_2u_2(t) ; & x_2(0) = 0 ; \\ y_2(t) = C_2x_2(t) + D_2u_2(t) . & \end{cases} \end{aligned}\right\} \tag{A.9}$$

Putting together the systems in a cascade form (so that the output y_2 of the second system becomes the control u_1 of the first system), we obtain a new system whose transfer function $W(\lambda)$ is a product of transfer functions $W_1(\lambda)$ and $W_2(\lambda)$ of the given systems (A.9): $W(\lambda) = W_1(\lambda)W_2(\lambda)$. So a cascade of two systems has a natural description in terms of transfer functions.

Conversely, a factorization $W(\lambda) = W_1(\lambda)W_2(\lambda)$ of a given transfer function, where $W_1(\lambda)$, $W_2(\lambda)$ are rational matrix functions with finite value at infinity, can be interpreted as a decomposition of a given linear system into a cascade of two systems.

Factorization of a rational matrix function $W(\lambda)$ (which is finite at infinity) in the form $W(\lambda) = W_1(\lambda)W_2(\lambda)$ allows us to obtain a state space realization for $W(\lambda)$ if state space realizations for $W_1(\lambda)$ and $W_2(\lambda)$ are known. This is done using the cascade of the two systems (see above). However, the realization for W is not always minimal even if we take minimal realizations for W_1 and W_2. A minimal realization for W is obtained if and only if the factorization $W(\lambda) = W_1(\lambda)W_2(\lambda)$ is *minimal*. This means, roughly speaking, that pole-zero cancellation does not occur.

Let us give the exact definition of minimal factorization. It will be assumed from now on that the rational matrix functions involved have value I at infinity. A factorization

$$W(\lambda) = W_1(\lambda)W_2(\lambda) , \tag{A.10}$$

where W, W_1, W_2 are rational $n \times n$ matrix functions, is called *minimal* if, whenever θ_1 and θ_2 are minimal nodes with transfer functions W_1 and W_2 respectively, the product $\theta_1\theta_2$ is a minimal node (with transfer function W). This definition will now be reformulated in terms of eigenvalues and eigenvectors of the rational matrix function W. We have defined in the preceding section an eigenvalue of $W(\lambda)$ as a pole of $W(\lambda)^{-1}$. It can also be described as follows. For $\lambda_0 \in \mathbb{C}$, let $W(\lambda) = \sum_{j=-q}^{\infty} (\lambda-\lambda_0)^j A_j$ be the Laurent series of $W(\lambda)$ in a neighbourhood of λ_0. Then λ_0 is an eigenvalue of $W(\lambda)$ if and only if there exist vectors $x_0, x_1, \cdots, x_q$ with $x_0 \neq 0$ such that

$$\begin{bmatrix} A_{-q} & 0 & \cdots & 0 \\ A_{-q+1} & A_{-q} & \cdots & 0 \\ \cdot & & & \cdot \\ \cdot & \cdot & \cdot & \cdot \\ \cdot & & & \cdot \\ A_0 & A_{-1} & \cdots & A_{-q} \end{bmatrix} \begin{bmatrix} x_0 \\ x_1 \\ \cdot \\ \cdot \\ \cdot \\ x_q \end{bmatrix} = 0 .$$

In this case x_0 is called an *eigenvector* corresponding to λ_0. Clearly, the number of eigenvalues of $W(\lambda)$ is finite. It is easily seen that the set of all eigenvectors of $W(\lambda)$ corresponding to the eigenvalue λ_0, together with the zero vector, is a linear subspace of $\mathbb{C}^n$; denote this subspace by $\operatorname{Ker}(W;\lambda_0)$. If $\lambda_0 \in \mathbb{C}$ is not an eigenvalue of W, put formally $\operatorname{Ker}(W;\lambda_0) = \{0\}$. If λ_0 is not a pole of $W(\lambda)$, then $\operatorname{Ker}(W;\lambda_0) = \operatorname{Ker} W(\lambda_0)$.

It turns out that the factorization (A.10) is *minimal* if and only if, for every $\lambda \in \mathbb{C}$, $\operatorname{Ker}(W_1;\lambda) \cap \operatorname{Ker}(W_2^{-1};\lambda) = \{0\}$ and $\operatorname{Ker}(W_1^{T-1};\lambda) \cap \operatorname{Ker}(W_2^T;\lambda) = \{0\}$. Here $W^T(\lambda) = (W(\lambda))^T$ and the superscript "T" stands for "transposed".

Observe that if $W = W_1W_2$ is a minimal factorization of W, then $W^{-1} = W_2^{-1}W_1^{-1}$ is a minimal factorization of W^{-1}, and $W^T = W_2^TW_1^T$ is a minimal factorization of W^T.

A factorization

$$W(\lambda) = W_1(\lambda) \cdots W_p(\lambda) \tag{A.11}$$

of $W(\lambda)$ into a product of p rational matrix functions $W_i(\lambda)$, $i = 1,2,\cdots,p$ with $W_i(\infty) = I$ is called *minimal* if, whenever $\theta_1,\cdots,\theta_p$ are minimal nodes of $W_1,\cdots,W_p$ respectively, their product $\theta_1 \cdots \theta_p$ is a minimal node. Observe that factorization (A.11) is minimal if and only if each one of the factorizations $W(\lambda) = K_{p-1}(\lambda)W_p(\lambda)$, $K_{p-1}(\lambda) = K_{p-2}(\lambda)W_{p-1}(\lambda),\cdots$, $K_2(\lambda) = W_1(\lambda)W_2(\lambda)$ is minimal (or, equivalently, if each one of the factorizations $W(\lambda) = W_1(\lambda)L_2(\lambda)$, $L_2(\lambda) = W_2(\lambda)L_3(\lambda),\cdots,L_{p-1}(\lambda) = W_{p-1}(\lambda)W_p(\lambda)$ is minimal).

We now give a description of minimal factorizations in the geometric language of subspaces. To this end we introduce the notion of supporting projector of a node. Let $\theta = (A,B,C,I;¢^m,¢^n)$ be a node, and let $\pi : ¢^m \to ¢^m$ be a projector (i.e. a linear transformation with the property $\pi^2 = \pi$). Then $¢^m$ admits the decomposition into a direct sum $¢^m = \text{Ker}\,\pi \dotplus \text{Im}\,\pi$. With respect to this decomposition one writes matrices A, B, C as follows (choosing fixed bases in $\text{Ker}\,\pi$ and $\text{Im}\,\pi$):

$$A = \begin{bmatrix} A_{11} & A_{12} \\ A_{21} & A_{22} \end{bmatrix}, \quad B = \begin{bmatrix} B_1 \\ B_2 \end{bmatrix}, \quad C = [\,C_1 \quad C_2\,] .$$

For instance $A_{11} : \text{Ker}\,\pi \to \text{Ker}\,\pi$. The node $pr_\pi(\theta) = (A_{22},B_2,C_2,I;\text{Im}\,\pi,¢^n)$ will be called the *projector node* of θ associated with π. If, in addition, π satisfies the properties

$$A(\text{Ker}\,\pi) \subset \text{Ker}\,\pi ;$$

$$(A-BC)(\text{Im}\,\pi) \subset \text{Im}\,\pi ,$$

then π is called a *supporting projector* of θ. As the following theorem shows, supporting projectors describe the minimal factorizations of the transfer function corresponding to θ.

THEOREM A.2 *Let* $\theta = (A,B,C,I;¢^m,¢^n)$ *be a minimal node with transfer function* $W(\lambda) = I + C(\lambda I-A)^{-1}B$.

(i) *If* π *is a supporting projector for* θ, $W_1(\lambda)$ *is the transfer function of* $pr_{I-\pi}(\theta)$ *and* $W_2(\lambda)$ *is the transfer function of* $pr_\pi(\theta)$, *then* $W(\lambda) = W_1(\lambda)W_2(\lambda)$ *is a minimal factorization of* $W(\lambda)$.

(ii) *If* $W(\lambda) = W_1(\lambda)W_2(\lambda)$ *is a minimal factorization of* $W(\lambda)$, *then there exists a unique supporting projector* π *for the node* θ *such that* $W_1(\lambda)$ *and* $W_2(\lambda)$ *are transfer functions of* $pr_{I-\pi}(\theta)$ *and* $pr_\pi(\theta)$, *respectively.*

For the proof see Section 4.3 of [4].

The matrix $A - BC$ which appears in the definition of the supporting projector of $\theta = (A,B,C,I;\mathbb{C}^m,\mathbb{C}^n)$, is called the matrix associated with the node θ, and is often denoted $A^\times$. It is the main matrix in the minimal node $\theta^\times = (A-BC,B,-C,I;\mathbb{C}^m,\mathbb{C}^n)$ with transfer function $W(\lambda)^{-1}$.

In the degenerate cases when $\pi = 0$ or $\pi = I$ (in this case π, of course, is a supporting projector) we obtain the following trivial minimal factorizations of $W(\lambda)$: $W(\lambda) = W(\lambda)\cdot I$ for $\pi = I$; $W(\lambda) = I\cdot W(\lambda)$ for $\pi = 0$.

Consider now a minimal factorization

$$W(\lambda) = W_1(\lambda)W_2(\lambda)W_3(\lambda) = W_1(\lambda)K(\lambda)$$

where $K(\lambda) = W_2(\lambda)W_3(\lambda)$. By Theorem A.2, there exists a supporting projector π_1 of the node $\theta = (A,B,C,I;\mathbb{C}^m,\mathbb{C}^n)$ with the transfer function $W(\lambda)$ corresponding to the minimal factorization $W = W_1K$. Also there exists a supporting projector $\pi_2 : \operatorname{Im}\pi_1 \to \operatorname{Im}\pi_1$ of $pr_{\pi_1}(\theta)$ corresponding to the minimal factorization $K = W_2W_3$. It turns out that the supporting projector of θ corresponding to the minimal factorization $W = LW_3$, where $L = W_1W_2$, is just $\pi_2\pi_1 : \mathbb{C}^m \to \mathbb{C}^m$ (here we consider π_2 as a mapping from $\operatorname{Im}\pi_1$ to $\operatorname{Im}\pi_1 \subset \mathbb{C}^m$). Indeed, since $\pi_2 : \operatorname{Im}\pi_1 \to \operatorname{Im}\pi_1$ is a projector, certainly $\pi_2\pi_1$ is a projector in $\mathbb{C}^m$. Choose $x \in \operatorname{Ker}(\pi_2\pi_1) = \operatorname{Ker}\pi_1 \dotplus \operatorname{Ker}\pi_2$. If $x \in \operatorname{Ker}\pi_1$, then $Ax \in \operatorname{Ker}\pi_1 \subset \operatorname{Ker}(\pi_2\pi_1)$. If $x \in \operatorname{Ker}\pi_2$, then $Ax = (I-\pi_1)Ax + \pi_1Ax$; now $(I-\pi_1)Ax \in \operatorname{Ker}\pi_1$ and $\pi_1Ax = \pi_1(A|_{\operatorname{Im}\pi_1})x \in \operatorname{Ker}\pi_2$ as $\operatorname{Ker}\pi_2$ is $(\pi_1A|_{\operatorname{Im}\pi_1})$-invariant. So $A(\operatorname{Ker}(\pi_2\pi_1)) \subset \operatorname{Ker}(\pi_2\pi_1)$. Choose now $x \in \operatorname{Im}(\pi_2\pi_1) = \operatorname{Im}\pi_2 \subset \operatorname{Im}\pi_1$. As $A^\times(\operatorname{Im}\pi_1) \subset \operatorname{Im}\pi_1$, we have $A^\times x = \pi_1A^\times|_{\operatorname{Im}\pi_1}x = (\pi_1A|_{\operatorname{Im}\pi_1})^\times x$. Then, since π_2 is a supporting projector for $pr_{\pi_1}(\theta)$, we have $(\pi_1A|_{\operatorname{Im}\pi_1})^\times(\operatorname{Im}\pi_2) \subset \operatorname{Im}\pi_2$, and consequently $A^\times x \in \operatorname{Im}(\pi_2\pi_1)$. So $\operatorname{Im}(\pi_2\pi_1)$ is $A^\times$-invariant, and $\pi_2\pi_1$ is indeed a supporting projector. Since $pr_{\pi_2\pi_1}(\theta) = pr_{\pi_2}(pr_{\pi_1}(\theta))$, the transfer function of $pr_{\pi_2\pi_1}(\theta)$ is just $W_3(\lambda)$.

So far, we have considered factorizations of rational matrix functions which attain the value I at infinity. The general rational matrix function with determinant not identically zero can be easily reduced to this case by applying a suitable Moebius transformation

$$\varphi(\lambda) = \frac{p\lambda+q}{r\lambda+s}, \quad p,q,r,s \in \mathbb{C}, \quad ps - qr \neq 0,$$

and by multipyling, if necessary, by a constant invertible matrix. Indeed, let $W(\lambda)$ be a rational $n \times n$ matrix function such that λ_0 is not a pole of $W(\lambda)$ and $W(\lambda_0)$ is invertible. Then the rational function $W(\lambda_0)^{-1}W(\varphi(\lambda))$, where $\varphi(\lambda)$ is any Moebius transformation such that $\varphi(\infty) = \lambda_0$ (for instance, $\varphi(\lambda) = \frac{\lambda_0\lambda+1}{\lambda}$), has value I at infinity.

A.4 Minimal Factorization of Real Rational Functions

We consider now the factorization of real rational functions.

THEOREM A.3 *Let* $W(\lambda)$ *be a real rational* $n \times n$ *matrix function with value* I *at infinity. Then there exists a minimal node* $(A_0,B_0,C_0,I;\mathbb{C}^m,\mathbb{C}^n)$ *with the transfer function* $W(\lambda)$ *such that the matrices* A_0, B_0, C_0 *are real.*

PROOF. Let

$$W(\lambda) = I + C(\lambda I-A)^{-1}B,$$

where $\theta = (A,B,C,I;\mathbb{C}^m,\mathbb{C}^n)$ is a minimal node with the transfer function $W(\lambda)$. Define the rational $n \times n$ matrix function $W_-(\lambda)$ by $W_-(\lambda) = \overline{W(\bar{\lambda})}$ (the upper bar denotes complex conjugation of every entry in a matrix). Since $W(\lambda)$ is real, we have $W_-(\lambda) = W(\lambda)$ for real λ which is not a pole of $W(\lambda)$, and therefore $W_-(\lambda) = W(\lambda)$ for all λ (in the domain of W). Further, $(\bar{A},\bar{B},\bar{C},I;\mathbb{C}^m,\mathbb{C}^n)$ is clearly a minimal node with the transfer function $W_-(\lambda) = W(\lambda)$. By Proposition A.1 there exists an invertible (in general, complex) matrix U such that

$$U^{-1}AU = \bar{A}, \quad A^{-1}B = \bar{B}, \quad CU = \bar{C}. \tag{A.12}$$

Put

$$Q = \begin{bmatrix} C \\ CA \\ \cdot \\ \cdot \\ \cdot \\ CA^{m-1} \end{bmatrix} .$$

Then $QU = \bar{Q}$. The matrix Q has linearly independent columns (as θ is minimal); choose $1 \leq i_1 < i_2 < \cdots < i_m \leq mn$ such that the submatrix S of Q formed by the rows with ordinal numbers $i_1, i_2, \cdots, i_m$ is invertible. Define $Q^{(-1)}$ to be the $m \times mn$ matrix all of whose columns are zeros except those with ordinal numbers $i_1, i_2, \cdots, i_m$, and these exceptional columns form the matrix S^{-1}. Then $Q^{(-1)}Q = I$, and consequently $U = Q^{(-1)}\bar{Q} = S^{-1}\bar{S}$. Using this in (A.12), we get

$$S^{-1}AS = \overline{SAS^{-1}} , \quad SB = \overline{SB} , \quad CS^{-1} = \overline{CS^{-1}} .$$

So the minimal node $(S^{-1}AS, SB, CS^{-1}, I; ¢^m, ¢^n)$ of $W(\lambda)$ is real. □

Finally, we describe real factorizations of real rational matrix functions. Let

$$W(\lambda) = W_1(\lambda)W_2(\lambda) , \tag{A.13}$$

where $W(\lambda)$, $W_1(\lambda)$, $W_2(\lambda)$ are rational matrix functions with value I at infinity. We say that the factorization (A.13) is *minimal real* if it is minimal and the matrix functions $W_1(\lambda)$ and $W_2(\lambda)$ are real. The description of all minimal factorizations is given by Theorem A.2. The minimal real factorizations are distinguished in terms of supporting projectors as follows.

THEOREM A.4 *Let* $\theta = (A,B,C,I;¢^m,¢^n)$ *be a minimal node with real matrices* A, B, C *and transfer function* $W(\lambda) = I + C(\lambda I - A)^{-1}B$. *Let* π *be a supporting projector of* θ *with corresponding minimal factorization* $W(\lambda) = W_1(\lambda)W_2(\lambda)$. *This factorization is real if and only if* π *is a real matrix.*

PROOF. One checks without difficulty that $\bar{\pi}$ is also a supporting projector of θ. The corresponding minimal factorization is $W = W_{1-}W_{2-}$, where $W_{i-}(\lambda) = \overline{(W_i(\bar{\lambda}))}$, $i = 1,2$. The desired result is now follows immediately from Theorem A.2. □

CHAPTER 4

THE ALGEBRAIC RICCATI EQUATION

This chapter is devoted to the study of the matrix equation

$$XDX + XA + A^*X - C = 0 \qquad (4.1)$$

where A, C, D are given $n \times n$ complex matrices with the properties that $C^* = C$, D is non-negative definite and

$$\text{rank}[D, AD, \cdots, A^{n-1}D] = n \ .$$

The $n \times n$ matrix X is to be found, and we note that the rank condition on A, D is trivially satisfied if D is positive definite. The conditions placed on A, C and D will apply throughout the chapter. An important problem of optimal control gives rise to a symmetric Riccati equation of the form (4.1) and, in this case, it is *hermitian* solution matrices X which are of physical interest. An expository account of this problem is given in the first section, and the reader who is already familiar with this may wish to begin with the systematic theory starting in the second section.

It will be shown that questions of existence and uniqueness of solutions of (4.1) and the description of the solution sets is reduced to the analysis of the $2n \times 2n$ matrix

$$M = i\begin{bmatrix} A & D \\ C & -A^* \end{bmatrix}$$

which is selfadjoint with respect to either of the matrices

$$H_1 = i\begin{bmatrix} 0 & -I \\ I & 0 \end{bmatrix}, \quad \text{or} \quad H_2 = \begin{bmatrix} -C & A^* \\ A & D \end{bmatrix},$$

(the question of invertibility of H_2 being addressed later). In fact, it will be shown that the above questions concerning solutions of (4.1) can all be reduced to properties of M-invariant subspaces.

4.1 An Optimal Control Problem

In this section we review a well-known optimal control problem whose solution depends on knowledge of certain hermitian solutions of (4.1).

Consider the time-invariant linear system

$$\frac{dx(t)}{dt} = Ax(t) + Bu(t) \; ; \quad x(0) = x_0 \, , \tag{4.2}$$

where A is an $n \times n$ real matrix and B is an $n \times r$ real matrix. The optimal control problem is to find an r-dimensional real vector function $u(t)$ which minimizes the "cost" functional

$$J(u) = \int_0^\infty (x^* C^* Cx + u^* Ru)dt \, ,$$

where R is an $r \times r$ positive definite real matrix, and C is a $p \times n$ real matrix.

If the system (4.2) is controllable (i.e. the rows of $[B, AB, \cdots, A^{n-1}B]$ are linearly independent), and observable (i.e. the columns of

$$\begin{bmatrix} C \\ CA \\ \cdot \\ \cdot \\ \cdot \\ CA^{n-1} \end{bmatrix}$$

are linearly independent), then it is known that the solution of the optimal control problem exists and is given by the formula

$$u_0(t) = -R^{-1}B^*\hat{X}x(t) \, , \tag{4.3}$$

where $\hat{X}$ is the unique $n \times n$ positive definite matrix which satisfies the equation

$$\hat{X}BR^{-1}B^*\hat{X} - \hat{X}A - A^*\hat{X} - C^*C = 0 \, , \tag{4.4}$$

and this equation is clearly of type (4.1). (It is easily seen that the condition $\text{rank}[D,AD,\cdots,A^{n-1}D] = n$ for equation (4.1) follows from the controllability of (4.2)). Moreover, the minimum cost $J(u_0)$ is given by the formula

$$J(u_0) = \frac{1}{2}\, x(t)^* \hat{X} x(t) ,$$

where $x(t) = \exp\{(A-BR^{-1}B^*\hat{X})t\}x_0$ is the solution of (4.2) under the optimal control. For a lucid exposition of this and related results see [7].

In fact, the controllability condition on (4.2) can be weakened. Consider a slightly more general optimal control problem with the cost functional

$$\int_0^\infty (y^*Qy + u^*Ru)dt ,$$

where R is an $r \times r$ positive definite real matrix, $Q = Q^*$ is a $p \times p$ real matrix, and $y(t) = Fx(t)$ is the output of system (4.2) with constant matrix F. Assume that:

(i) every root subspace of A corresponding to an eigenvalue λ_0 with $\mathit{Im}\,\lambda_0 \geq 0$, is contained in the subspace spanned by the columns of $[B,AB,\cdots,A^{n-1}B]$;

(ii) the kernel of

$$\begin{bmatrix} C \\ CA \\ \cdot \\ \cdot \\ \cdot \\ CA^{n-1} \end{bmatrix}$$

is contained in the sum of all root subspaces of A corresponding to eigenvalues λ_0 with $\mathit{Im}\,\lambda < 0$. Then an optimal control exists, it is given by formula (4.3), and $\hat{X}$ is now the unique *non-negative* definite solution of equation (4.4), where Q is replaced by C^*QC. For the proof of this statement we refer to [29].

4.2 General Solutions of the Riccati Equation

Consider equation (4.1), where A, C, D are $n \times n$ matrices and have the properties $C^* = C$, D is non-negative definite, and

$$\operatorname{rank}[D, AD, \cdots, A^{n-1}D] = n \ . \tag{4.5}$$

Let

$$M = i \begin{bmatrix} A & D \\ C & -A^* \end{bmatrix}, \quad H = \begin{bmatrix} -C & A^* \\ A & D \end{bmatrix}. \tag{4.6}$$

Clearly, H is hermitian. Without loss of generality we can (and will) assume that H is also invertible and the signature of H is zero. Indeed, the solutions of (4.1) remain the same if A is replaced by $A + \alpha i I$, where α is real. Taking α large enough, the corresponding matrix

$$H_\alpha = \begin{bmatrix} -C & A^* - \alpha i I \\ A + \alpha i I & D \end{bmatrix}$$

is invertible and has signature zero (one can consider $\alpha^{-1}H_\alpha$ as a small perturbation of the matrix

$$\begin{bmatrix} 0 & -iI \\ iI & 0 \end{bmatrix}$$

which certainly is invertible with signature zero).

Note that the replacement of A by $A + \alpha i I$ does not disturb the condition (4.5) because

$$\operatorname{rank}[D, AD, \cdots, A^{n-1}D] = \operatorname{rank}[D, (A+zI)D; \cdots, (A+zI)^{n-1}D]$$

for every complex number z. Also, the new matrix M would be

$$i \begin{bmatrix} A + \alpha i I & D \\ C & -A^* + \alpha i I \end{bmatrix} = i \begin{bmatrix} A & D \\ C & -A^* \end{bmatrix} - \alpha I \ ,$$

so the invariant subspaces of the matrix M do not change if we replace A by $A + \alpha i I$. As we shall see later, certain M-invariant subspaces play a crucial role in the description of the solutions of equation (4.1).

We begin our analysis with the observation that, with the above understandings, M is H-selfadjoint. This follows from the straighforward verification of the relation $HM = M^*H$. Now recall the concept of M-invariant maximal H-nonpositive subspaces as developed in Chapter I.3. The first result demonstrates a connection between certain subspaces of $\mathbb{C}^{2n}$ of this kind and certain solutions of (4.1). Note, in particular, that hermitian solutions of (4.1) (when they exist) automatically satisfy the condition (4.8).

Observe that, since an n-dimensional, M-invariant, maximal H-nonpositive subspace always exists, the theorem asserts the existence of a solution (possibly non-hermitian) of equation (4.1).

THEOREM 4.1 *Let* L *be an* n-*dimensional* M-*invariant maximal* H-*nonpositive subspace. Then*

$$L = \operatorname{Im}\begin{bmatrix} I \\ X \end{bmatrix},$$

where X *is a solution of the equation*

$$XDX + XA + A^*X - C = 0 \tag{4.7}$$

such that

$$(X^*-X)(A+DX) \tag{4.8}$$

is nonpositive definite. Conversely, if X *is a solution of* (4.7) *such that* $(X^*-X)(A+DX)$ *is nonpositive definite, then the subspace*

$$\operatorname{Im}\begin{bmatrix} I \\ X \end{bmatrix}$$

is n-*dimensional*, M-*invariant, and maximal* H-*nonpositive.*

PROOF. Let X be a solution of (4.7) such that $(X^*-X)(A+DX)$ is nonpositive definite. It is easily verified that (4.7) implies

$$i\begin{bmatrix} A & D \\ C & -A^* \end{bmatrix}\begin{bmatrix} I \\ X \end{bmatrix} = \begin{bmatrix} I \\ X \end{bmatrix} i(A+DX)\ , \tag{4.9}$$

so $\operatorname{Im}\begin{bmatrix} I \\ X \end{bmatrix}$ is M-invariant. Further

$$[\ I \quad X^*\ ]\begin{bmatrix} -C & A^* \\ A & D \end{bmatrix}\begin{bmatrix} I \\ X \end{bmatrix} = -C + X^*A + A^*X + X^*DX\ , \tag{4.10}$$

and since X is a solution of (4.7) this is equal to $(X^*-X)(A+DX)$. So

$$[\ I \quad X^*\ ]\begin{bmatrix} -C & A^* \\ A & D \end{bmatrix}\begin{bmatrix} I \\ X \end{bmatrix}$$

is nonpositive definite which means that $\operatorname{Im}\begin{bmatrix} I \\ X \end{bmatrix}$ is an H-nonpositive subspace. As $\dim \operatorname{Im}\begin{bmatrix} I \\ X \end{bmatrix} = n$ and the signature of H is zero, the subspace $\operatorname{Im}\begin{bmatrix} I \\ X \end{bmatrix}$ is maximal H-nonpositive (see Theorem I.1.3).

Conversely, assume L is an n-dimensional M-invariant maximal H-nonpositive subspace. Write

$$L = \operatorname{Im} \begin{bmatrix} X_1 \\ X_2 \end{bmatrix} \tag{4.11}$$

for some $n \times n$ matrices X_1 and X_2. We are going to prove that X_1 is invertible.

First, observe that M-invariance of L means that

$$\begin{bmatrix} A & D \\ C & -A^* \end{bmatrix} \begin{bmatrix} X_1 \\ X_2 \end{bmatrix} = \begin{bmatrix} X_1 \\ X_2 \end{bmatrix} T$$

for some $n \times n$ matrix T. In other words,

$$AX_1 + DX_2 = X_1 T ; \tag{4.12}$$

$$CX_1 - A^* X_2 = X_2 T . \tag{4.13}$$

Then H-nonpositivity of L means that the matrix

$$[\, X_1^* \quad X_2^* \,] \begin{bmatrix} -C & A^* \\ A & D \end{bmatrix} \begin{bmatrix} X_1 \\ X_2 \end{bmatrix}$$

$$= X_2^* D X_2 + X_1^* A^* X_2 + X_2^* A X_1 - X_1^* C X_1 \tag{4.14}$$

is non-positive definite.

Let $K = \operatorname{Ker} X_1$. Since (4.14) is nonpositive definite, we have for every $x \in K$:

$$0 \geq x^* X_2^* D X_2 x + x^* X_1 A^* X_2 x + x^* X_2 A X_1 x - x X_1^* C X_1 x = x^* X_2^* D X_2 x ,$$

and since D is non-negative definite, $X_2 x \in \operatorname{Ker} D$, i.e.

$$X_2 K \subset \operatorname{Ker} D . \tag{4.15}$$

Further, equation (4.12) implies that

$$TK \subset K . \tag{4.16}$$

Indeed, for $x \in K$ we have in view of (4.12) and (4.15):

$$X_1 T x = AX_1 x + DX_2 x = 0 .$$

Now equation (4.13) gives for every $x \in K$:

$$A^* X_2 x = -CX_1 x + A^* X_2 x = -X_2 Tx \in X_2 K$$

and so

$$A^* X_2 K \subset X_2 K \ . \tag{4.17}$$

We see from (4.12) that $A^* X_2 K \subset \text{Ker } D$, and we now claim, more generally, that

$$A^{*r} X_2 K \subset \text{Ker } D \ , \quad r = 0,1,2,\cdots \ . \tag{4.18}$$

We have already proved this inclusion for $r = 0$ and $r = 1$. Assuming, inductively, that (4.18) holds for $r - 1$, and using (4.17), it is found that

$$A^{*r}(X_2 K) = A^{*r-1}(A^* X_2 K) \subset A^{*r-1} X_2 K \subset \text{Ker } D \ ;$$

so (4.18) holds. Now for every $x \in K$:

$$\begin{bmatrix} D \\ DA^* \\ \cdot \\ \cdot \\ \cdot \\ DA^{*n-1} \end{bmatrix} (X_2 x) = 0 \ ,$$

or $(X_2 x)^* [D, AD, \cdots, A^{n-1}D] = 0$. But $\text{rank}[D, AD, \cdots, A^{n-1}D] = n$, so $X_2 x = 0$. But the only n-dimensional vector x for which $X_1 x = X_2 x = 0$, is the zero vector; otherwise $\dim L < n$, which contradicts our assumptions. So $K = \{0\}$ and X_1 is invertible. Then we can write

$$L = \text{Im} \begin{bmatrix} I \\ X \end{bmatrix} ,$$

where $X = X_2 X_1^{-1}$. Now the M-invariance of L means

$$\begin{bmatrix} A & D \\ C & -A^* \end{bmatrix} \begin{bmatrix} I \\ X \end{bmatrix} = \begin{bmatrix} I \\ X \end{bmatrix} T_0 \ ,$$

which gives $T_0 = A + DX$, and

$$C - A^* X = XT_0 = X(A+DX) \ ,$$

i.e. X is a solution of (4.7). Further, using (4.10) H-nonpositivity of L implies that

$$X^*DX + A^*X + X^*A - C = (X^*-X)(A+DX)$$

is a non-positive definite matrix. □

In particular, Theorem 4.1 shows that there is a one-to-one correspondence between the n-dimensional M-invariant maximal H-nonpositive subspaces L and solutions of (4.1) with the property that $(X^*-X)(A+DX)$ is non-positive definite. This correspondence is given by the formula

$$L = \operatorname{Im}\begin{bmatrix} I \\ X \end{bmatrix}.$$

Now take advantage of the constructive development of M-invariant H-nonpositive subspaces of Section I.3.12. Let C be a maximal set of non-real eigenvalues of M with the property that if $\lambda \in C$ then $\bar{\lambda} \notin C$. Then Theorem I.3.20 ensures the existence of an n-dimensional M-invariant H-nonpositive subspace L_C such that $\sigma(M|_{L_C}) \setminus \mathbb{R} = C$. By Theorem 4.1

$$L_C = \operatorname{Im}\begin{bmatrix} I \\ X \end{bmatrix},$$

where X is a solution of (4.1) such that $(X^*-X)(A+DX)$ is nonpositive definite. Further, (4.9) shows that $\sigma(M|_{L_C}) = \sigma(i(A+DX))$. So we obtain the following result:

THEOREM 4.2 *For every maximal set* C' *of non-real eigenvalues of* M *with the property that* $\lambda \in C'$ *implies* $\bar{\lambda} \notin C'$, *there exists a solution* X *of* (4.1) *such that* $(X^*-X)(A+DX)$ *is nonpositive definite and* $\sigma(i(A+DX)) \setminus \mathbb{R} = C'$. *In particular, there exist solutions* X_1 *and* X_2 *of* (4.1) *such that* $(X_i^*-X_i)(A+DX_i)$ *is nonpositive definite,* $i = 1,2$, *and*

$$\mathrm{Re}\,\sigma(A+DX_1) \leq 0 \ ; \quad \mathrm{Re}\,\sigma(A+DX_2) \geq 0 \ .$$

4.3 Existence of Hermitian Solutions of the Riccati Equation

We continue the analysis of equation (4.1) with the properties of A, C, D as before, i.e. $C^* = C$, D is nonnegative definite, and

$$\operatorname{rank}[D, AD, \cdots, A^{n-1}D] = n \ . \tag{4.19}$$

In this section we are interested in hermitian solutions X of (4.1) i.e. such that $X = X^*$. Although Theorem 4.1 guarantees the existence of some solution X for (4.1), a hermitian solution does not necessarily exist. For instance, in the scalar case the equation $x^2 + 1 = 0$ does not have hermitian solutions, i.e. solutions in real numbers. The following result gives a description of all hermitian solutions of (4.1), if they exist. As in the preceding section, assume that H is invertible with signature zero.

THEOREM 4.3 *The following statements are equivalent:*

(i) *there is a solution* $X = X^*$ *of* (4.1);

(ii) *the partial multiplicities of the real eigenvalues of* M *(if any) are all even;*

(iii) *there exists an* n*-dimensional* M*-invariant* H*-neutral subspace* L.

In connection with Theorem 4.3 note that the matrix M is invertible (so the real eigenvalues of M cannot include zero). This follows from the relation

$$H = FM , \tag{4.20}$$

where $F = i\begin{bmatrix} 0 & I \\ -I & 0 \end{bmatrix}$, and the invertibility of H (which has been assumed). One checks easily that M is also F-selfadjoint. Now equality (4.20) implies (see Theorem I.6.10) that an M-invariant subspace is H-neutral if and only if it is F-neutral. Hence one can replace H by F in the statement (iii) of Theorem 4.3. However, since M-invariant maximal H-nonpositive subspaces and M-invariant maximal F-nonpositive subspaces are not the same, it is not immediately clear how to reformulate the results of Section 4.1 in terms of M and F.

Note also that an n-dimensional H-neutral subspace is actually maximal H-neutral (in view of the assumption that $\operatorname{sig} H = 0$).

Proof of Theorem 4.3. Let $X = X^*$ be a solution of (4.1). Then the relations (4.9) and (4.10) imply that the subspace $L = \operatorname{Im}\begin{bmatrix} I \\ X \end{bmatrix}$ is M-invariant and H-neutral. Thus, it is proved that (i)$\Rightarrow$(iii).

Let L be n-dimensional M-invariant H-neutral subspace. By Theorem 4.1,

$$L = \operatorname{Im}\begin{bmatrix} I \\ X \end{bmatrix}$$

for some solution X of (4.1) such that $(X^*-X)(A+DX)$ is nonpositive definite. By H-neutrality of L, we have

$$0 = [I \quad X^*] \begin{bmatrix} -C & A^* \\ A & D \end{bmatrix} \begin{bmatrix} I \\ X \end{bmatrix} = (X^*-X)(A+DX) . \tag{4.21}$$

Now recall that M is invertible (cf. (4.20)). But equation (4.9) implies
that $i(A+DX)$ is similar to $M|_L$ and hence the matrix $A + DX$ is invertible
as well. So (4.21) now implies that $X^* = X$, i.e. X is hermitian and it
is proved that $(iii) \Rightarrow (i)$.

Using Theorem I.3.21 the implications $(ii) \Rightarrow (iii)$ is obtained immediately.

Finally, in order to prove that $(i) \Rightarrow (ii)$ we need two lemmas. The
first is Lemma 8.4 of [19a]; see also Lemma 3.4 of [19b].

LEMMA 4.4 Let $J_1, \cdots, J_k$ be the nilpotent Jordan blocks of sizes
$\alpha_1 \geq \cdots \geq \alpha_k$ respectively. Let

$$J_0 = \begin{bmatrix} J_1 & 0 & \cdot & \cdot & \cdot & 0 \\ 0 & J_2 & \cdot & \cdot & \cdot & 0 \\ \cdot & & & & & \cdot \\ \cdot & \cdot & \cdot & \cdot & \cdot & \cdot \\ \cdot & & & & & \cdot \\ 0 & 0 & \cdot & \cdot & \cdot & J_k \end{bmatrix}$$

and

$$\Phi = \begin{bmatrix} J_0 & \Phi_0 \\ 0 & J_0^T \end{bmatrix}$$

where $\Phi_0 = [\phi_{ij}]_{i,j=1}^{\alpha}$ is a matrix of size $\alpha \times \alpha$ $(\alpha = \alpha_1 + \cdots + \alpha_k)$.
Let

$$\beta_i = \alpha_1 + \cdots + \alpha_i , \quad i = 1, \cdots, k$$

and suppose that the $k \times k$ submatrix $\Psi = [\phi_{\beta_i \beta_j}]_{j,j=1}^k$ of Φ_0 is invertible and $\Psi = \Psi_1 \Psi_2$, where Ψ_1 is a lower triangular matrix and Ψ_2 is an
upper triangular matrix. Then the sizes of Jordan blocks in the Jordan form
of Φ are $2\alpha_1, \cdots, 2\alpha_k$.

LEMMA 4.5 Let

$$V = [v_{ij}]_{i,j=1}^m$$

be a nonnegative definite matrix such that $v_{mm} = 0$. *Then* $v_{im} = v_{mi} = 0$ *for* $i = 1,\cdots,m$.

PROOF. If $e_m = \langle 0,\cdots,0,1\rangle$ then $v_{mm} = 0$ implies that $0 = (Ve_m,e_m) = ||V^{\frac{1}{2}}e_m||^2$ where $V^{\frac{1}{2}}$ is the non-negative square root of V. Hence $V^{\frac{1}{2}}e_m = 0$ and $Ve_m = 0$, from which the result follows. □

We prove now that (i)⇒(ii). A positive (resp. nonnegative) definite matrix Q will be denoted $Q>0$ (resp. $Q\geq 0$). Let $X=X^*$ be a solution of (4.1); then

$$\begin{bmatrix} I & 0 \\ -X & I \end{bmatrix}\begin{bmatrix} A & D \\ C & -A^* \end{bmatrix}\begin{bmatrix} I & 0 \\ X & I \end{bmatrix} = \begin{bmatrix} A+DX & D \\ 0 & -(A^*+XD) \end{bmatrix}. \tag{4.22}$$

So we can consider the matrix

$$i\begin{bmatrix} A+DX & D \\ 0 & -(A^*+XD) \end{bmatrix}$$

in place of M. Let Z be a Jordan form of $A + DX$ and let S be the reducing matrix. $S(A+DX)S^{-1} = Z$. Then, since $X^* = X$, we can write $A^* + XD = (A+DX)^*$ and

$$\begin{bmatrix} A+DX & D \\ 0 & -(A^*+XD) \end{bmatrix} = \begin{bmatrix} S^{-1} & 0 \\ 0 & S^* \end{bmatrix}\begin{bmatrix} Z & D_0 \\ 0 & -Z^* \end{bmatrix}\begin{bmatrix} S & 0 \\ 0 & S^{-1*} \end{bmatrix},$$

where $D_0 = SDS^* \geq 0$, and it is sufficient to prove that all the imaginary eigenvalues of $\begin{bmatrix} Z & D_0 \\ 0 & -Z^* \end{bmatrix}$ (if any) have even multiplicities.

Let λ_0 be such a pure imaginary eigenvalue. Let $Z_1,\cdots,Z_k$ be the Jordan blocks of Z corresponding to λ_0, let their sizes be $\alpha_1,\cdots,\alpha_k$, and denote

$$\alpha = \alpha_1 + \cdots + \alpha_k .$$

Without loss of generality we can suppose that these blocks are in the north-west corner of Z. So we can write

$$\begin{bmatrix} Z & D_0 \\ 0 & -Z^* \end{bmatrix} = \begin{bmatrix} Z' & 0 & D_1 & D_2 \\ 0 & Z'' & D_2^* & D_3 \\ 0 & 0 & -Z'^* & 0 \\ 0 & 0 & 0 & -Z''^* \end{bmatrix}$$

where $Z' = Z_1 \oplus \cdots \oplus Z_k$, Z'' is the 'rest' of Z, and

$$D_0 = \begin{bmatrix} D_1 & D_2 \\ D_2^* & D_3 \end{bmatrix}$$

is the corresponding partition of D_0. Since $\sigma(Z') \cap \sigma(-Z''^*) = \emptyset$, and applying if necessary a similarity transformation with similarity matrix of the type

$$\begin{bmatrix} I & 0 & 0 & T \\ 0 & I & 0 & 0 \\ 0 & 0 & I & 0 \\ 0 & 0 & 0 & I \end{bmatrix}$$

we can suppose that $D_2 = 0$. Let $D_1 = (D_{1ij})_{i,j=1}^{k}$ be the partition of D_1 consistent with the partitioning of Z'. It is enough to prove that in the Jordan form of the matrix

$$\begin{bmatrix} Z_1-\lambda_0 I & 0 & \cdot & 0 & D_{111} & D_{112} & \cdot & D_{11k} \\ 0 & Z_2-\lambda_0 I & \cdot & 0 & D_{121} & D_{122} & \cdot & D_{12k} \\ \cdot & \cdot & \cdot & \cdot & \cdot & \cdot & \cdot & \cdot \\ 0 & 0 & \cdot & Z_k-\lambda_0 I & D_{1k1} & D_{1k2} & \cdot & D_{1kk} \\ 0 & 0 & \cdot & 0 & -Z_1^*-\lambda_0 I & 0 & \cdot & 0 \\ 0 & 0 & \cdot & 0 & 0 & -Z_2^*-\lambda_0 I & \cdot & 0 \\ \cdot & \cdot & \cdot & \cdot & \cdot & \cdot & \cdot & \cdot \\ 0 & 0 & \cdot & 0 & 0 & 0 & \cdot & -Z_k^*-\lambda_0 I \end{bmatrix} \tag{4.23}$$

the blocks with eigenvalue 0 have sizes $2\alpha_1, \cdots, 2\alpha_k$. Let f_{ij} $(i,j = 1,\cdots,k)$ be the entry in the south-east corner of D_{1ij}; consider the matrix $F = [f_{ij}]_{i,j=1}^{k}$ formed by all these entries. Since F is a principal submatrix of D_1, and hence of D_0, $F \geq 0$. Let us show that F is invertible. Suppose not; then there exists an invertible matrix $U = [u_{ij}]_{i,j=1}^{k}$ such that UFU^* has a zero in the south-east corner. Let $G = [g_{ij}]_{i,j=1}^{\alpha}$ be an $\alpha \times \alpha$ invertible matrix of the following structure

$$g_{\beta_i\beta_j} = u_{ij} \quad \text{where} \quad \beta_i = \alpha_1 + \cdots + \alpha_i \; ; \quad i,j = 1,\cdots,k$$

$$g_{qq} = 1 \quad \text{for} \quad q \notin \{\beta_1,\cdots,\beta_k\}$$

$$g_{pq} = 0 \quad \text{for} \quad p \neq q \;\text{ and }\; \{p,q\} \not\subset \{\beta_1,\cdots,\beta_k\} .$$

Then the matrix GD_1G^* has a zero in the south-east corner, and since $GD_1G^* \geq 0$, by Lemma 4.5 the last column and last row of GD_1G^* are also zeros. On the other hand, from the structure of G it is seen (bearing in mind that the β_1-th,$\cdots$,β_k-th rows of $Z' - \lambda_0 I$ are zeros) that the last row of $G(Z'-\lambda_0 I)G^{-1}$ is also zero. Now let

$$\tilde{G} = \begin{bmatrix} G & 0 \\ 0 & I \end{bmatrix}$$

be an $n \times n$ matrix where I is the $(n-\alpha) \times (n-\alpha)$ unit matrix. It is clear that the β_k-th row of $\tilde{G}D_0\tilde{G}^*$ is zero, as well as the β_k-th row of $\tilde{G}(Z-\lambda_0 I)\tilde{G}^{-1}$. So

$$\begin{aligned} n &> \operatorname{rank}[\tilde{G}D_0\tilde{G}^*, \tilde{G}(Z-\lambda_0 I)\tilde{G}^{-1}\cdot\tilde{G}D_0\tilde{G}^*, \cdots, [\tilde{G}(Z-\lambda_0 I)\tilde{G}^{-1}]^{n-1}\cdot\tilde{G}D_0\tilde{G}^*] \\ &= \operatorname{rank}[D_0, (Z-\lambda_0 I)D_0, \cdots, (Z-\lambda_0 I)^{n-1}D_0] \qquad (4.24) \\ &= \operatorname{rank}[D_0, ZD_0, \cdots, Z^{n-1}D_0] = \operatorname{rank}[D, (A+DX)D, \cdots, (A+DX)^{n-1}D] . \end{aligned}$$

One proves easily by induction on j, that

$$(A+DX)^j D = A^j DY_{j0} + A^{j-1}DY_{j1} + \cdots + DY_{jj} , \quad j = 0,1,\cdots,n-1 ,$$

for some matrices Y_{jk} (with $Y_{j0} = I$). Consequently,

$$[D, (A+DX)D, \cdots, (A+DX)^{n-1}D] =$$

$$= [D, AD, \cdots, A^{n-1}D] \begin{bmatrix} I & Y_{11} & \cdot & \cdot & \cdot & Y_{n-1,n-1} \\ & I & & & & \cdot \\ & & \cdot & & & \cdot \\ & & & \cdot & & \cdot \\ & & & & \cdot & Y_{n-1,1} \\ & 0 & & & & I \end{bmatrix} , \qquad (4.25)$$

and in view of (4.24)

$$\operatorname{rank}[D, AD, \cdots, A^{n-1}D] < n ,$$

a contradiction with our basic assumption (4.19).

So the matrix F is invertible, and since $F > 0$ it is easy to see that F can be represented as a product of upper and lower triangular matrices. Thus, Lemma 4.4 is applicable, and our assertion about the matrix (4.23) follows. □

4.4 Hermitian Solutions and Nonnegative Rational Functions

In this section the characterization of solutions of (4.1) will be approached from another direction. The hypotheses on (4.1) are just those discussed in Section 4.2 and include, it will be recalled, the facts that the hermitian matrix H of (4.6) is invertible and has signature zero.

In this section we shall prove that the existence of hermitian solutions of (4.1) can also be expressed in terms of nonnegativeness of a certain hermitian rational matrix function, as follows. Let D_0 be any $n \times n$ matrix such that $D = D_0 D_0^*$.

THEOREM 4.6 *The equation* (4.1) *has hermitian solutions if and only if the rational matrix function*

$$Z(\lambda) = I + D_0^*(\lambda I + iA^*)^{-1} C(\lambda I - iA)^{-1} D_0$$

is nonnegative on the real axis, i.e. $(Z(\lambda)x, x) \geq 0$ *for every* $x \in \mathbb{C}^n$ *and every real* λ *which is not a pole of* $Z(\lambda)$.

Note that the rational matrix function $Z(\lambda)$ is hermitian (namely, $Z(\lambda) = (Z(\lambda))^*$ for all real λ at which $Z(\lambda)$ is defined).

PROOF. Let X be a hermitian solution of (4.1). Then the equality

$$X(i\lambda I + A) + (-i\lambda I + A^*)X + XDX = C$$

holds for all complex λ. Premultiplying by $D_0^*(-i\lambda I + A^*)^{-1}$, postmultiplying by $(i\lambda I + A)^{-1} D_0$, and adding I to both parts, we have

$$\begin{aligned} I + D_0^*(-i\lambda I+A^*)^{-1}XD_0 + D_0^*X(i\lambda I+A)^{-1}D_0 + D_0^*(-i\lambda I+A^*)^{-1}XD_0D_0^*X(i\lambda I+A)^{-1}D_0 \\ = I + D_0^*(-i\lambda I+A^*)^{-1}C(i\lambda I+A)^{-1}D_0 \end{aligned} \tag{4.26}$$

for every λ which is not an eigenvalue of iA or of $-iA^*$. The right hand side of (4.26) is just $Z(\lambda)$, while the left hand side is equal to

$$(I + D_0^*(-i\lambda I+A^*)^{-1}XD_0)(I + D_0^*X(i\lambda I+A)^{-1}D_0) ,$$

which is nonnegative definite for real $\lambda \notin \sigma(iA) \cup \sigma(-iA^*)$ in view of the equality $X = X^*$. So $Z(\lambda)$ is nonnegative on the real axis.

Conversely, assume $Z(\lambda)$ is nonnegative on the real axis. Then, evidently, the function

$$W(\lambda) = I + (\lambda I - iA)^{-1} D_0 Z(\lambda) D_0^* (\lambda I + iA^*)^{-1}$$

is also nonnegative on the real axis. Rewrite $W(\lambda)$ in the form

$$\begin{aligned} W(\lambda) &= I + (\lambda I - iA)^{-1}(I + D(\lambda I + iA^*)^{-1} C(\lambda I - iA)^{-1})^{-1} D(\lambda I + iA^*)^{-1} \\ &= I + (\lambda I - iA + D(\lambda I + iA^*)^{-1} C)^{-1} D(\lambda I + iA^*)^{-1} \\ &= I - iV(\lambda) , \end{aligned}$$

where

$$V(\lambda) = -(\lambda I - iA + D(\lambda I + iA^*)^{-1} C)^{-1}(-iD)(\lambda I + iA^*)^{-1} .$$

Now observe that $V(\lambda)$ is just the $n \times n$ upper right quarter in the $2n \times 2n$ matrix

$$\begin{bmatrix} \lambda I - iA & -iD \\ -iC & \lambda I + iA \end{bmatrix}^{-1} = (\lambda I - M)^{-1} ,$$

and M is defined in (4.6). Indeed, this follows from the general fact that the inverse of an $(n+m) \times (n+m)$ matrix

$$S = \begin{bmatrix} S_{11} & S_{12} \\ S_{21} & S_{22} \end{bmatrix} ,$$

where the size of S_{11} (resp. S_{22}) is n (resp. m), is given by the formula

$$S^{-1} = \begin{bmatrix} T^{-1} & -T^{-1} S_{12} S_{22}^{-1} \\ -S_{22}^{-1} S_{21} T^{-1} & S_{22}^{-1} S_{21} T^{-1} S_{12} S_{22}^{-1} + S_{22}^{-1} \end{bmatrix} , \tag{4.27}$$

where $T = S_{11} - S_{12} S_{22}^{-1} S_{21}$, provided the matrices S_{22} and T are invertible. So

$$W(\lambda) = I - i[I \quad 0](\lambda I - M)^{-1} \begin{bmatrix} 0 \\ I \end{bmatrix} = I + R^* F (\lambda I - M)^{-1} R , \tag{4.28}$$

where $R = \begin{bmatrix} 0 \\ I \end{bmatrix}$ and $F = i \begin{bmatrix} 0 & I \\ -I & 0 \end{bmatrix}$. We shall prove below that (4.28) is a minimal realization of $W(\lambda)$. As $W(\lambda)$ is nonnegative, it follows from Theorem 3.11(iii) (see also Section A.2 in the appendix to Chapter 3) that all the Jordan blocks of M corresponding to real eigenvalues are of even size. By Theorem 4.3 this means that equation (4.1) has a hermitian solution.

It remains to prove that the realization (4.28) is minimal. Since

M is F-selfadjoint, i.e. $FM = M^*F$, we have only to check that

$$\operatorname{rank}[R,MR,\cdots,M^mR] = 2n$$

for m large enough (cf. Section A.1 in the appendix to Chapter 3). As we know (Theorem 4.1) equation (4.1) always has a (not necessarily hermitian) solution X. Then

$$\begin{bmatrix} I & 0 \\ -X & I \end{bmatrix} M \begin{bmatrix} I & 0 \\ X & I \end{bmatrix} = i \begin{bmatrix} A+DX & D \\ 0 & -XD-A^* \end{bmatrix}$$

and clearly it will suffice to prove that

$$\operatorname{rank}[R,M_0R,\cdots,M_0^mR] = 2n \ . \tag{4.29}$$

for m large enough, where

$$M_0 = \begin{bmatrix} A+DX & D \\ 0 & -XD-A^* \end{bmatrix} .$$

A simple induction argument shows that

$$M_0^kR = \begin{bmatrix} \sum_{j=0}^{k-1} (A+DX)^j D(-XD-A^*)^{k-1-j} \\ * \end{bmatrix}, \quad k = 1,2,\cdots . \tag{4.30}$$

Denoting $Y_1 = A + DX$, $Y_2 = -XD - A^*$, we have

$$\left[D, Y_1D+DY_2, \cdots, \sum_{j=0}^{k-1} Y_1^j D Y_2^{k-1-j}\right] \begin{bmatrix} I & -Y_2 & \cdot & \cdot & \cdot & 0 \\ 0 & I & \cdot & & & \cdot \\ \cdot & \cdot & & \cdot & & \cdot \\ \cdot & \cdot & & & \cdot & \cdot \\ \cdot & \cdot & & & & -Y_2 \\ 0 & 0 & \cdot & \cdot & \cdot & I \end{bmatrix}$$

$$= [D, Y_1D, \cdots, Y_1^{k-1}D] \ ;$$

so

$$\operatorname{rank}\left[D, Y_1D+DY_2, \cdots, \sum_{j=0}^{k-1} Y_1^j D Y_2^{k-1-j}\right] = \operatorname{rank}[D, Y_1D, \cdots, Y_1^{k-1}D] \tag{4.31}$$

for $k = 1,2,\cdots$. But in view of the condition $\operatorname{rank}[D,AD,\cdots,A^{n-1}D] = n$ we have also $\operatorname{rank}[D,Y_1D,\cdots,Y_1^{n-1}D] = n$ (see (4.25)). So (4.29) follows from (4.30) and (4.31). □

As a byproduct of the proof of Theorem 4.6 we obtain the following corollary which will be used later.

COROLLARY 4.7 *Assume that* (4.1) *has a hermitian solution. Then the sign characteristic of* (M,F), *where* $F = i\begin{bmatrix} 0 & I \\ -I & 0 \end{bmatrix}$, *consists of* +1'*s only.*

Indeed, the proof of Theorem 4.6 shows that the rational matrix function $W(\lambda) = I + R^*F(\lambda I-M)^{-1}R$, where $R = \begin{bmatrix} 0 \\ I \end{bmatrix}$, is nonnegative definite on the real axis, and the node $(M,R^*F,R;\mathbb{C}^m,\mathbb{C}^n)$ is minimal (here m is the size of M). By Theorem 3.11(iii) the sign characteristic of $W(\lambda)$ consists only of +1's which implies, by definition of the sign characteristic of a selfadjoint rational matrix function, that the sign characteristic of (M,F) also consists of only +1's.

4.5 Description of Hermitian Solutions

In the two preceding sections necessary and sufficient conditions have been studied for the existence of hermitian solutions of (4.1). Returning to the concepts used in Theorem 4.1, and assuming hermitian solutions exist, they are now to be described in terms of certain M-invariant subspaces (where M is defined in (4.6)). The first result was actually obtained in the course of proof of Theorem 4.3.

THEOREM 4.8 *Every* n-*dimensional* M-*invariant* H-*neutral subspace* L *has the form*

$$L = \operatorname{Im}\begin{bmatrix} I \\ X \end{bmatrix} \tag{4.32}$$

where $X = X^*$ *is a solution of* (4.1), *and in this case* $M|_L$ *is similar to* i(A+DX).

Conversely, for every hermitian solution X *of* (4.1) *the subspace* L *given by* (4.32) *is* n-*dimensional,* M-*invariant, and* H-*neutral.*

Observe that this theorem establishes a one-to-one correspondence between hermitian solutions of (4.1) and n-dimensional, M-invariant, H-neutral subspaces L, given explicitly by (4.32). As remarked after the proof of Theorem 4.3, the matrix H can be replaced here by $F = i\begin{bmatrix} 0 & I \\ -I & 0 \end{bmatrix}$ since L is H-neutral if and only if it is F-neutral.

Combining Corollary 4.7 and Theorem I.3.22, we obtain the following description of the selfadjoint solutions of (4.1) in terms of all M-invariant subspaces N such that the restriction $M|_N$ has its spectrum in the open upper half-plane.

THEOREM 4.9 *Assume equation* (4.1) *has a hermitian solution. Let* N_+ *be the spectral subspace of* M *corresponding to the eigenvalues in the open upper half-plane. Then for every* M-*invariant subspace* $N \subset N_+$ *there exists a unique hermitian solution* X *of* (4.1) *such that*

$$\operatorname{Im}\begin{bmatrix} I \\ X \end{bmatrix} \cap N_+ = N .$$

Conversely, if X *is a hermitian solution of* (4.1) *then* $\operatorname{Im}\begin{bmatrix} I \\ X \end{bmatrix} \cap N_+$ *is* M-*invariant.*

Note that Theorem 4.9 also holds when M is not invertible (replace A by $A+i\alpha I$, where α is real). Also, the subspace N_+ of this theorem can be replaced by the spectral subspace of M corresponding to any maximal set C of eigenvalues with the property that $\lambda_0 \in C$ implies $\bar{\lambda}_0 \notin C$. Indeed, the following particular case of Theorem 4.9 justifies a separate statement. It is obtained by taking N_+ in Theorem 4.9 to be the spectral subspace of M corresponding to C, and $N = N_+$; the assertion about partial multiplicities of $A + DX$ follows from Theorem I.3.21.

THEOREM 4.10 *Assume* (4.1) *has a hermitian solution. Then for every set* C *of non-real eigenvalues of* M *which is maximal with respect to the property that if* $\lambda_0 \in C$ *then* $\bar{\lambda}_0 \notin C$, *there exists a unique hermitian solution* X *of* (4.1) *such that* $\sigma(i(A+DX)) \setminus \mathbb{R} = C$. *Furthermore the partial multiplicities for every real eigenvalue* λ_0 (*if any*) *of* $\lambda I - i(A+DX)$ *are equal to* $m_1, \cdots, m_k$, *where* $2m_1, \cdots, 2m_k$ *are the partial multiplicities of* $\lambda I - M$ *corresponding to* λ_0 (*which are all even by Theorem* 4.3). *In particular, the matrices* $i(A+DX)|_{L(X)}$, *where* $L(X)$ *is the spectral subspace of* $i(A+DX)$ *corresponding to the real eigenvalues, and* $X = X^*$ *is any solution of the Riccati equation* (4.1), *are all similar.*

Theorem 4.9 allows us to determine the number of hermitian solutions of (4.1), by equating it to the number of invariant subspaces of the restriction $M|_{N_+}$. Let J_+ be the Jordan normal form of $M|_{N_+}$; clearly, J_+ and $M|_{N_+}$ have the same number of invariant subspaces. Now a Jordan block of size k has exactly $k + 1$ invariant subspaces. Hence, if for each eigenvalue of J_+ there is exactly one Jordan block with this eigenvalue, then J_+ has exactly $\prod_{i=1}^{\alpha}(k_i+1)$ invariant subspaces, where $k_i = \dim \operatorname{Ker}(\lambda_i I - J_+)$, $i = 1,\cdots,\alpha$, and $\lambda_1,\cdots,\lambda_\alpha$ are all distinct eigenvalues of J_+.

If $\dim \operatorname{Ker}(\lambda I - J_+) \geq 2$ for some λ, then J_+ has a continuum of invariant subspaces. To see this, let x and y be linearly independent eigenvectors of J_+ corresponding to the eigenvalue λ; then the 1-dimensional subspaces $\operatorname{Span}\{x+cy\}$, $c \in \mathbb{C}$ are all different and J_+-invariant. We have proved the following corollary:

COROLLARY 4.11 *Assume (4.1) has at least one hermitian solution. If* $\dim \operatorname{Ker}(\lambda I - M) \leq 1$ *for all* λ *in the open upper half-plane, then the number of hermitian solutions of (4.1) is exactly* $\prod_{i=1}^{\alpha}(k_i+1)$, *where* $k_1,\cdots,k_\alpha$ *are the multiplicities of all distinct eigenvalues* $\lambda_1,\cdots,\lambda_\alpha$ *of* M *lying in the open upper half-plane, respectively. If* $\dim \operatorname{Ker}(\lambda I - M) \geq 2$ *for some non-real* λ, *then (4.1) has a continuum of hermitian solutions.*

Observe that equation (4.1) has a *unique* hermitian solution if and only if all eigenvalues of M are real and all their partial multiplicities are even.

4.6 Extremal Hermitian Solutions

As we have seen in Section 4.1 the maximal hermitian solution of (4.1) is used in the solution of the optimal control problem. In this section we shall prove the existence of the maximal and the minimal solution.

There is a natural order relation in the set of all hermitian matrices. Namely, $X_1 \leq X_2$ for hermitian matrices X_1 and X_2 means that $X_2 - X_1$ is nonnegative definite. A hermitian solution X_+ (resp. X_-) of (4.1) is called *maximal* (resp. *minimal*) if $X_- \leq X \leq X_+$ for every hermitian solution X of (4.1). Obviously, if a maximal (resp. minimal) solution exists, it is unique. The following theorem establishes the existence of extremal hermitian solutions and characterizes them in spectral terms.

Everywhere in this section it will be assumed that (4.1) has at least one hermitian solution.

THEOREM 4.12 *There exists a maximal hermitian solution,* X_+, *and a minimal hermitian solution,* X_-, *of (4.1). The solution* X_+ *(resp.* X_-*) is the unique hermitian solution of (4.1) for which* $\sigma(A+DX_+)$ *lies in the closed right (resp. left) half-plane, and is obtained by taking* $N = N_+$ *(resp.* $N = \{0\}$*) in Theorem 4.9.*

For convenience, we state and prove a lemma which will be used in the proof of Theorem 4.12.

LEMMA 4.13 *Let* Q, R *be* $n \times n$ *complex matrices such that* R *is nonnegative definite and*

$$\operatorname{rank}[R, QR, \cdots, Q^{n-1}R] = n . \tag{4.33}$$

Then the matrix

$$\Omega(t) = -\int_0^t e^{-\tau Q} R e^{-\tau Q^*} d\tau$$

is negative (resp. positive) definite for all $t > 0$ *(resp.* $t < 0$*) and the matrix* $\hat{Q} = Q + R\Omega(t)^{-1}$, $t > 0$ *is stable (i.e.* $\sigma(\hat{Q})$ *lies in the open left half-plane)*.

PROOF. Evidently, $\Omega(0) = 0$ and $\Omega(t_1) \leq \Omega(t_2)$ for $t_1 \geq t_2$. In particular, $\Omega(t)$ is nonpositive (resp. nonnegative) definite for $t > 0$ (resp. $t < 0$). To prove that $\Omega(t) > 0$ (resp. $\Omega(t) < 0$) for $t < 0$ (resp. $t > 0$) it is enough to check that $(\Omega(t)x, x) = 0$ only if $t = 0$ or $x = 0$. Assume the contrary, i.e. $(\Omega(t)x, x) = 0$ for some $t \neq 0$ (say, $t > 0$) and $x \neq 0$. Since

$$(\Omega(t)x, x) = -\int_0^t ||R^{\frac{1}{2}} e^{-\tau Q^*} x||^2 d\tau ,$$

it follows that $R^{\frac{1}{2}} e^{-\tau Q^*} x = 0$ for all $\tau \in (0,t)$. Differentiating with respect to τ we deduce that

$$\begin{bmatrix} R^{\frac{1}{2}} \\ R^{\frac{1}{2}} Q^* \\ \cdot \\ \cdot \\ \cdot \\ R^{\frac{1}{2}} Q^{*n-1} \end{bmatrix} e^{-\tau Q^*} x = 0 .$$

Consequently,

$$\operatorname{rank}[R, QR, \cdots, Q^{n-1}R] \leq \operatorname{rank}[R^{\frac{1}{2}}, QR^{\frac{1}{2}}, \cdots, Q^{n-1}R^{\frac{1}{2}}] < n ,$$

which contradicts (4.33).

To prove the last statement observe that for $t > 0$ we have

$$Q\Omega(t) + \Omega(t)Q^* = \int_0^t [\frac{d}{d\tau} e^{-\tau Q}Re^{-\tau Q^*}]d\tau$$
$$= e^{-tQ}Re^{-tQ^*} - R ,$$

and hence

$$\hat{Q}\Omega(t) + \Omega(t)\hat{Q}^* = e^{-tQ}Re^{-tQ^*} + R .$$

Let λ_0 be an eigenvalue of $\hat{Q}^*$ and x a corresponding eigenvector. Then (for $t > 0$)

$$(\lambda_0+\bar{\lambda}_0)x^*\Omega(t)x = x^*(\hat{Q}\Omega(t) + \Omega(t)\hat{Q}^*)x$$
$$= x^*(e^{-tQ}Re^{-tQ^*} + R)x \geq 0 . \qquad (4.34)$$

But $\Omega(t)$ is negative definite. So (4.34) is possible only if $\lambda_0 + \bar{\lambda}_0 \leq 0$. Further, if $\lambda_0 + \bar{\lambda}_0 = 0$, then $Rx = 0$, and the definition of $\hat{Q}^*$ implies $Q^*x = \lambda_0 x$. Hence

$$\Omega(t)x = -\int_0^t e^{-\tau Q}Rxe^{-\lambda_0\tau} d\tau = 0 , \quad t > 0$$

a contradiction with the negative definite property of $\Omega(t)$. So $\lambda_0 + \bar{\lambda}_0 < 0$ for every eigenvalue λ_0 of $\hat{Q}^*$, and hence for every eigenvalue of $\hat{Q}$. □

Proof of Theorem 4.12 Let X be a hermitian solution of (4.1). Denote $\tilde{A} = A + DX$ and

$$U(t) = -\int_0^t e^{-\tau\tilde{A}}De^{-\tau\tilde{A}^*} d\tau , \quad t \in \mathbb{R} .$$

It is easy to see that $\text{rank}[D,\tilde{A}D,\cdots,\tilde{A}^{n-1}D] = n$ (cf. the contradiction to formula (4.24) in the proof of Theorem 4.3). By Lemma 4.13, $U(t)$ is positive (resp. negative) definite for $t < 0$ (resp. $t > 0$), so the limit $\lim_{t\to-\infty} U(t)^{-1}$ exists and is nonnegative definite.

For any positive real T and $t < T$ define

$$X_T(t) = X + U^{-1}(t-T) .$$

Since $X_T(t) = [XU(t-T)+I]U(t-T)^{-1}$, it follows that $X_T(t)$ is invertible for $t \in [T-\delta,T)$, where $\delta > 0$ is small enough, and

$$\lim_{t\to T} X_T(t)^{-1} = 0 . \qquad (4.35)$$

By a direct computation, using the facts that $X = X^*$ is a solution of the algebraic Riccati equation (4.1) and that $U(t)$ satisfies the differential equation

$$U(t)' = -\tilde{A}U - U\tilde{A}^* - D ,$$

one checks easily that $X_T(t)$ satisfies the differential Riccati equation

$$X_T(t)' + C - X_T(t)A - A^*X_T(t) - X_T(t)DX_T(t) = 0 , \quad t < T . \tag{4.36}$$

Equations (4.35) and (4.36) determine $X_T(t)$ uniquely as a quantity independent of X. Indeed, $V_T(t) \overset{\text{def}}{=} X_T(t)^{-1}$ satisfies the differential equation

$$-V_T(t)' + V_T(t)CV_T(t) - AV_T(t) - V_T(t)A^* - D = 0 ,$$

for $T - \delta < t < T$, with the initial condition $\lim_{t \to T} V(t) = 0$, and therefore $V_T(t)$, and consequently $X_T(t)$, is uniquely determined for $t \in (T-\delta,T)$. But then $X_T(t)$ is uniquely determined for all $t < T$.

As $T \to \infty$, the matrix function $X_T(t)$ converges uniformly on compact intervals to the constant matrix

$$X_+ = X + \lim_{t \to -\infty} U(t)^{-1} . \tag{4.37}$$

Of course, X_+ is independent of X. Now (4.37) implies that $X_+ = X_+^*$ and $X \leq X_+$. Passing to the limit $T \to \infty$ in (4.36) we find that X_+ is a solution of (4.1). Since $X \leq X_+$ for any hermitian solution X of (4.1), X_+ is maximal.

Furthermore,

$$A + DX_+ = \lim_{T \to \infty}(A+DX_T(0)) , \tag{4.38}$$

and

$$A + DX_T(0) = \tilde{A} + D(U(-T))^{-1} = \tilde{A} + D\left[\int_0^{-T} e^{-\tau\tilde{A}}De^{-\tau\tilde{A}^*}d\tau\right]^{-1} =$$

$$= -\left[(-\tilde{A}) + D\left(\int_0^T e^{-\tau(-\tilde{A})}De^{-\tau(-\tilde{A}^*)}d\tau\right)^{-1}\right] .$$

Since $\operatorname{rank}[D,-\tilde{A}D,\cdots,(-1)^{n-1}\tilde{A}^{n-1}D] = n$, the second part of Lemma 4.13 ensures that $\sigma(A+DX_T(0))$ lies in the open right half-plane. Taking into account (4.38) we find that $Re\,\lambda_0 \geq 0$ for every $\lambda_0 \in \sigma(A+DX_+)$. Uniqueness of a hermitian solution X of (4.1) with the additional property that $Re\,\sigma(A+DX) \geq 0$ follows from Theorem 4.10.

Applying these results, concerning maximal solutions of (4.1), to the equation

$$XDX - XA - A^*X - C = 0 \tag{4.39}$$

and noting that X is a solution of (4.39) if and only if $-X$ is a solution of (4.1), we obtain the corresponding results for minimal solutions. □

4.7 Real Symmetric Solutions of the Algebraic Riccati Equation with Real Coefficients

In this section we shall consider the algebraic Riccati equation (4.1) with the additional property that A, D, C are real matrices.

As before, define

$$M = i\begin{bmatrix} A & D \\ C & -A^* \end{bmatrix}, \quad H = \begin{bmatrix} -C & A^* \\ A & D \end{bmatrix}.$$

We know already that M is H-selfadjoint and, in particular, the spectrum of M is symmetric relative to the real axis. But since A, D, C are real, the spectrum of M is also symmetric with respect to the imaginary axis. So if $\lambda_0 \in \sigma(M)$, then $\bar{\lambda}_0$, $-\lambda_0$ and $-\bar{\lambda}_0$ are also in $\sigma(M)$, and the partial multiplicities of $\lambda I - M$ corresponding to λ_0, $\bar{\lambda}_0$, $-\lambda_0$ and $-\bar{\lambda}_0$, are the same.

A description of real symmetric solutions of equation (4.1) with real coefficients is given by the following theorem:

THEOREM 4.14 *Assume that equation* (4.1) *with real matrices* D, A, C *has a hermitian (not necessarily real) solution. Let* $\tilde{N}_+$ *be the spectral subspace of* M *corresponding to the eigenvalues in the quadrant* $\{\lambda \in \mathbb{C} \mid Im\,\lambda > 0,\ Re\,\lambda \geq 0\}$. *Then for every M-invariant subspace* $\tilde{L} \subset \tilde{N}_+$ *there exist a unique real symmetric solution* X *of* (4.1) *such that*

$$\mathrm{Range}\begin{bmatrix} I \\ X \end{bmatrix} \cap \tilde{N}_+ = \tilde{L}\ . \tag{4.40}$$

Conversely, if X *is a real symmetric solution of* (4.1), *then* $\mathrm{Range}\begin{bmatrix} I \\ X \end{bmatrix} \cap \tilde{N}_+$ *is M-invariant.*

PROOF. Denote by N_{++} (resp. N_+) the spectral subspace of M

corresponding to the eigenvalues in the open quadrant $\{\lambda \in ¢ \mid Im\ \lambda > 0, Re\ \lambda > 0\}$ (resp. in the open upper half-plane). So $N_{++} \subset \tilde{N}_+ \subset N_+$.

Given an M-invariant subspace $\tilde{L} \subset \tilde{N}_+$, let $L_{++} = \tilde{L} \cap N_{++}$, and let L be the sum of two subspaces: $\tilde{L}$ and $\bar{L}_{++} = \{<x_1,\cdots,x_{2n}> \in ¢^{2n} \mid <\bar{x}_1,\cdots,\bar{x}_{2n}> \in L_{++}\}$. It is easily seen that if the 2n-dimensional vectors $f_0,\cdots,f_k$ form a Jordan chain of M corresponding to an eigenvalue λ_0, then the vectors $\bar{f}_0,-\bar{f}_1,\bar{f}_2,\cdots,\pm\bar{f}_k$ (where $\bar{f}_i$ is obtained from f_i by taking the complex conjugate of each coordinate) form a Jordan chain of M corresponding to $-\bar{\lambda}_0$. Hence the subspace $\bar{L}_{++}$ is M-invariant and $\bar{L}_{++} \cap \tilde{N}_+ = \{0\}$; moreover, the subspace L enjoys the property that

$$<x_1,\cdots,x_{2n}> \in L \quad \text{implies} \quad <\bar{x}_1,\cdots,\bar{x}_{2n}> \in L . \tag{4.41}$$

In particular, L is M-invariant and $L \cap \tilde{N}_+ = \tilde{L}$. By Theorem 4.9 there exists a unique hermitian solution X of (4.1) such that $\mathrm{Im}\begin{bmatrix} I \\ X \end{bmatrix} \cap N_+ = L$. We claim that X is real. Indeed, $\bar{X}$ is also a hermitian solution of (4.1), and both L and N_+ enjoy the property (4.41), hence $\mathrm{Im}\begin{bmatrix} I \\ \bar{X} \end{bmatrix} \cap N_+ = L$. By the uniqueness of X, the equality $X = \bar{X}$ follows. So we have found a real symmetric solution X of (4.1) such that (4.40) holds.

Assume now that

$$\mathrm{Im}\begin{bmatrix} I \\ X \end{bmatrix} \cap \tilde{N}_+ = \mathrm{Im}\begin{bmatrix} I \\ Y \end{bmatrix} \cap \tilde{N}_+ \tag{4.42}$$

for two real symmetric solutions X and Y of (4.1). Theorem 4.9 ensures that the subspace (4.42) is M-invariant. From (4.42) it follows (by taking complex conjugates) that

$$\mathrm{Im}\begin{bmatrix} I \\ X \end{bmatrix} \cap \tilde{N}_- = \mathrm{Im}\begin{bmatrix} I \\ Y \end{bmatrix} \cap \tilde{N}_- , \tag{4.43}$$

where $\tilde{N}_- = \{<x_1,\cdots,x_{2n}> \in ¢^{2n} \mid <\bar{x}_1,\cdots,\bar{x}_{2n}> \in \tilde{N}_+\}$. But $\tilde{N}_-$ is the spectral subspace of M corresponding to the eigenvalues in the quadrant $\{\lambda \in ¢ \mid Re\ \lambda \lesssim 0, Im\ \lambda > 0\}$ (cf. the property of Jordan chains of M mentioned above). Combining (4.42) and (4.43) we get

$$\mathrm{Im}\begin{bmatrix} I \\ X \end{bmatrix} \cap N_+ = \mathrm{Im}\begin{bmatrix} I \\ Y \end{bmatrix} \cap N_+ ,$$

and the uniqueness part of Theorem 4.9 ensures that X = Y. □

In Theorem 4.14 one can take instead of $\widetilde{N}_+$ the spectral subspace of M corresponding to any set C_+ of non-real eigenvalues of M which is maximal with respect to the following two properties:

(a) if $\lambda_0 \in C_+$ and is pure imaginary, then $\bar{\lambda}_0 \notin C_+$,

(b) if $\lambda_0 \in C_+$ and is not pure imaginary, then $\bar{\lambda}_0$, $-\lambda_0$ and $-\bar{\lambda}_0$ do not belong to C_+.

Now let us see how these results relate to the corresponding results for hermitian solutions. Recall that a c-set C of non-real eigenvalues of M is defined to contain no conjugate pairs of complex numbers and is maximal with respect to this property. It has been shown in Theorem 4.10 that, given the existence of at least one hermitian solution, every c-set C determines a unique hermitian solution X_C of (4.1) such that

$$\sigma(i(A+DX_C)) \setminus \mathbb{R} = C\ . \tag{4.44}$$

The next theorem gives necessary and sufficient conditions for X_C to be real.

THEOREM 4.15 *Let equation* (4.1) *have real coefficients and at least one hermitian solution (not necessarily real). Then a c-set* C *of eigenvalues of* M *has the property that* $\lambda_0 \in C$ *implies* $-\bar{\lambda}_0 \in C$ *if and only if the solution* X_C *of* (4.1) *satisfying* (4.44) *is real. In particular, the maximal and minimal solutions of* (4.1) *are real.*

The proof of Theorem 4.15 follows the same line of argument as that of Theorem 4.14 and is therefore omitted. The statement concerning maximal and minimal solutions follows from Theorem 4.12.

It is easily deduced from Theorem 4.15 that if equation (4.1) with real coefficients has a continuum of hermitian solutions, then it has a continuum of real symmetric solutions. The number of real symmetric solutions (provided they exist) is finite if and only if $\dim \operatorname{Ker}(M-\lambda I) \leq 1$ for every non-real λ, i.e. if and only if the number of hermitian solutions is finite. In that case the number of real symmetric solutions of equation (4.1) with real coefficients is $\prod_{i=1}^{t} (k_i+1)$, where $k_i = \dim \operatorname{Ker}(M-\lambda_i I)$, and $\lambda_1, \cdots, \lambda_t$ are all the distinct eigenvalues of M in the quadrant $\{\lambda \mid \operatorname{Im} \lambda > 0, \operatorname{Re} \lambda \geq 0\}$.

NOTES TO PART II

The presentation in Sections 1.1, 1.2, 1.4, 1.5 is based on [26,15] (see also Sections 6.1 - 6.5 in [17b]). The presentation of Sections 1.6, 1.7, 1.8 is based on [32, 33]. More general forms of the results of Chapter 1 for infinite dimensional Hamiltonian equations are presented in [26b], together with an extensive bibliography.

Chapter 2 (except for Section 2.4) is based on [19a]. More detailed exposition of this material and a guide to references is contained in Part III of [19b]. Section 2.4 is based on [19c].

Sections 3.2, 3.3, 3.4 are based on [19d], and Sections 3.1, 3.5, 3.6, 3.7 are based on [38]. For another approach to the sign characteristic of a real rational selfadjoint matrix function see [13a]. Proofs and other details of the results described in Appendix to Chapter 3 are found in Chapters 2 - 4 of [4].

Details about the optimal control problem from Section 4.1 can be found in many books, see e.g. [7]. The results of Sections 4.2 and 4.3 as well as Theorem 4.8 are taken from [31]. Theorem 4.6 is obtained in [47]. Here we give a new proof. Section 4.6 is based on [10]. Theorem 4.9 is proved in [40]. In the real case results similar to this theorem were obtained originally in [41].

PART III

PERTURBATIONS AND STABILITY

The content of Part III includes a natural continuation of both preceding parts. One objective is further development of the theory of Part I to include results concerning perturbations and stability of H-selfadjoint and H-unitary matrices. The other is the application of this theory to the problems developed in Part II. These include differential and difference equations with bounded solutions which remain bounded under perturbations, stability of factorizations of matrix-valued functions and the dependence of solutions of the symmetric algebraic Riccati equation on its coefficients.

Throughout Part III, admissible perturbations of a pair of matrices (A, H) where A is H-selfadjoint transform the pair to a neighbouring pair (B, G) where B is G-selfadjoint. General perturbations of this kind are studied as well as perturbations with an additional property: addition to A of an H-

definite matrix (in Chapter 2), and perturbations which preserve Jordan structure (in Chapter 5). The behaviour of eigenvalues, eigenvectors and invariant subspaces under these perturbations are discussed, and special sections are devoted to analytic perturbation theory.

For the reader who is interested only in the development of perturbation theory it is possible to focus on Chapters 1 and 2 and the first sections of Chapters 4 and 5 without reference to Part II.

In this part it is necessary to introduce a metric into the linear space of subspaces of a finite-dimensional space. An appendix to Part III includes some essential properties of such a metric space and an account of the concepts of continuous and analytic families of subspaces.

CHAPTER 1

GENERAL PERTURBATIONS. STABILITY OF DIAGONABLE MATRICES

If A is an H-selfadjoint matrix, a "general perturbation" of the pair (A,H) results in a pair (B,G) in which B is G-selfadjoint and is close to the unperturbed pair (A,H) in an appropriate sense. A similar convention applies to the perturbations of H-unitary matrices that we consider.

Identification of a quantity which is invariant under such perturbations is one of the main results of the chapter. This general theorem admits the characterization of all diagonable H-selfadjoint matrices with real spectrum which retain these properties after a general perturbation. Also a description of those cases in which analytic perturbations of H-selfadjoint matrices retain spectral properties which are familiar from the classical hermitian case is obtained. Analogous results for perturbations of H-unitary matrices are also discussed.

1.1 General Perturbations of H-Selfadjoint Matrices

Recall that the signature, $\operatorname{sig} H$, of a hermitian matrix H is defined as the difference between the number of positive eigenvalues of H and the number of negative eigenvalues of H (in both cases counting multiplicities); zero eigenvalues of H, if any, do not count.

For a given $n \times n$ matrix A and $\lambda \in \mathbb{C}$, let $E_A(\lambda) = \{x \in \mathbb{C}^n \mid (A-\lambda I)^n x = 0\}$. So $E_A(\lambda) \neq \{0\}$ if and only if λ is an eigenvalue of A, and in this case $E_A(\lambda)$ is the root subspace of A corresponding to λ. The orthogonal projection onto $E_A(\lambda)$ is denoted by $P_A(\lambda)$.

Observe that if $\lambda \in \sigma(A)$ is real and A is H-selfadjoint, then $P_A(\lambda)HP_A(\lambda)$ determines the quadratic form on $E_A(\lambda)$ associated with a restriction of H. The main theorem shows that an invariant of general perturbations of the pair (A,H) is determined by the signatures of quadratic forms of this kind.

THEOREM 1.1 *Let* A *be* H*-selfadjoint and* $\Omega \subset \mathbb{R}$ *be any open set such that the boundary of* Ω *does not intersect* $\sigma(A)$. *Then for some sufficiently small neighbourhoods* $\mathcal{U}_A$ *of* A *and* $\mathcal{U}_H$ *of* H *the equality*

$$\sum_{\lambda\in\Omega} \operatorname{sig} P_A(\lambda)HP_A(\lambda) = \sum_{\mu\in\Omega} \operatorname{sig} P_B(\mu)GP_B(\mu) \tag{1.1}$$

holds for every $B \in \mathcal{U}_A$ *which is* G*-selfadjoint for an invertible selfadjoint* $G \in \mathcal{U}_H$.

Moreover, for such a B, *the number* $\nu_\Omega(B)$ *of eigenvalues in* Ω *(counting multiplicities), satisfies the inequality*

$$\nu_\Omega(B) \geqslant \sum_{\lambda\in\Omega} |\operatorname{sig} P_A(\lambda)HP_A(\lambda)| , \tag{1.2}$$

and in every neighbourhood $\mathcal{U} \in \mathcal{U}_A$ *of* A *there exists an* H*-selfadjoint matrix* B *for which equality holds in* (1.2).

PROOF. We first prove (1.1). Evidently, it is sufficient to consider the case $\Omega = \{\mu \in \mathbb{R} \mid \mu_2 < \mu < \mu_1\}$, where $\mu_i \notin \sigma(A)$, $i = 1,2$.

Let us compute the signature of the hermitian matrix $\mu H - HA$ where $\mu \in \mathbb{R} \setminus \sigma(A)$. Passing to the canonical form $(J,P_{\varepsilon,J})$ of (A,H) (Theorem I.3.3) one sees easily that

$$\operatorname{sig}(\mu H-HA) = \sum_{\lambda\in\sigma(A)\cap\mathbb{R}} \operatorname{sgn}(\mu-\lambda) \sum_{i=1}^{k(\lambda)} \frac{\varepsilon_i(\lambda)}{2} [1-(-1)^{m_i(\lambda)}] , \tag{1.3}$$

where $m_1(\lambda),\cdots,m_{k(\lambda)}(\lambda)$ are the sizes of Jordan blocks in J with eigenvalue λ, and $\varepsilon_1(\lambda),\cdots,\varepsilon_{k(\lambda)}(\lambda)$ are the corresponding signs in the sign characteristic of (A,H). As usual, $\operatorname{sgn}(\mu-\lambda) = 1$ if $\mu - \lambda > 0$ and $\operatorname{sgn}(\mu-\lambda) = -1$ if $\mu - \lambda < 0$. From (1.3) we find that

$$\operatorname{sig}(\mu_1 H-HA) - \operatorname{sig}(\mu_2 H-HA) = \sum_{\substack{\mu_1<\lambda<\mu_2\\ \lambda\in\sigma(A)}} \sum_{i=1}^{k(\lambda)} \varepsilon_i(\lambda)[1-(-1)^{m_i(\lambda)}] , \tag{1.4}$$

which is equal to $\sum_{\mu_1<\lambda<\mu_2} \operatorname{sig}(P_A(\lambda)HP_A(\lambda))$. Since the signature of an invertible hermitian matrix is stable under small perturbations, there exist neighbourhoods $\mathcal{U}_A$ of A and $\mathcal{U}_H$ of H such that, for every G-selfadjoint

$B \in \mathcal{U}_A$ with $G \in \mathcal{U}_H$, we have:

$$\mathrm{sig}(\mu_i H - HA) = \mathrm{sig}(\mu_i G - GB) .$$

For such B and G (1.1) follows from (1.4).

Inequality (1.2) is a direct consequence of (1.1). Indeed, the canonical form of B shows that

$$\nu_\Omega(B) \geq \sum_{\mu\in\Omega} |\mathrm{sig}(P_B(\mu) G P_B(\mu))| .$$

To prove the last part of the theorem let $(J, P_{\varepsilon,J})$ be the canonical form of (A,H). Consider the part J_0 of the Jordan matrix J which corresponds to a fixed real eigenvalue λ_0 of A and let P_{ε,J_0} be the corresponding part of $P_{\varepsilon,J}$. Clearly, it is sufficient to find a P_{ε,J_0}-selfadjoint matrix K in every neighbourhood of J_0 with the property that the number of real eigenvalues of K is exactly $|\mathrm{sig}\ P_0 H P_0|$, where $P_0 = P_A(\lambda_0)$. For simplicity of notation assume $\lambda_0 = 0$.

Consider first the construction of K in three particular cases:

(i) J_0 is a Jordan block of even size α;

(ii) J_0 is a Jordan block of odd size α;

(iii) $J_0 = J_1 \oplus J_2$ consists of two Jordan blocks J_1 and J_2 of odd sizes α_1 and α_2 respectively, and with opposite signs in the sign characteristic.

Denote by $J(\pm i, \beta)$ the Jordan block of size β with eigenvalue $\pm i$, and let ξ be a small positive number. In case (i) put $K = J_0 + \xi\ \mathrm{diag}[J(i,\frac{\alpha}{2}), J(-i,\frac{\alpha}{2})]$. It is easy to check that K is P_{ε,J_0}-selfadjoint with all eigenvalues non-real.

In case (ii) put $K = J_0 + \xi\ \mathrm{diag}[J(i,\frac{\alpha-1}{2}), 0, J(-i,\frac{\alpha-1}{2})]$; then K is P_{ε,J_0}-selfadjoint with exactly $\alpha - 1$ non-real eigenvalues.

In case (iii) put

$$K = J_0 + \xi\ \mathrm{diag}\left[J(i,\tfrac{\alpha_1-1}{2}),\ \begin{bmatrix} 1 & 0 & 0 & 1 \\ 0 & J(-i,\frac{\alpha_1-1}{2}) & 0 & 0 \\ 0 & 0 & J(i,\frac{\alpha_2-1}{2}) & 0 \\ -1 & 0 & 0 & 1 \end{bmatrix},\ J(-i,\tfrac{\alpha_2-1}{2})\right],$$

then K is P_{ε,J_0}-selfadjoint, and

$$\det(\lambda I-K) = (\lambda^2+\xi^2)^{\frac{1}{2}(\alpha_1+\alpha_2)-1}[(\lambda-\xi)^2 + \xi^2] ,$$

so that all eigenvalues of K are non-real.

In the general case we apply the construction of case (i) to each Jordan block of J_0 of even size, the construction of case (iii) to each pair of Jordan blocks of J_0 of odd size and different signs, and if Jordan blocks of odd size are left, we apply the construction of case (ii) to each of them. It is easily seen that in this way we produce a P_{ε,J_0}-selfadjoint matrix K in every neighbourhood of J_0 such that $\nu_{\mathbb{R}}(K) = |\mathrm{sig}\ P_0HP_0|$. This completes the proof. □

The following special case of the main theorem will be useful subsequently. For a real eigenvalue λ_0 there is an associated set of signs $\varepsilon(\lambda_0) \subset \varepsilon$, the sign characteristics of (A,H). The statement of the corollary concerns the number $k(\lambda_0)$ which is defined to be the minimum of the number of positive signs in $\varepsilon(\lambda_0)$ and the number of negative signs in $\varepsilon(\lambda_0)$.

COROLLARY 1.2 *Let* A *be* H-*selfadjoint and assume that all elementary divisors of* A *corresponding to* λ_0 *are linear. Then for every* $\delta > 0$ *there exist neighbourhoods* $\mathcal{U}_A$ *of* A *and* $\mathcal{U}_H$ *of* H *such that, for every pair* $(B,G) \in \mathcal{U}_A \times \mathcal{U}_H$ *with* B G-*selfadjoint, the number* $s_c(B)$ *of non-real eigenvalues of* B *in the disc* $\{\lambda \mid |\lambda-\lambda_0| < \delta\}$ *does not exceed* $2k(\lambda_0)$, *counting multiplicities.*

Moreover, in every neighbourhood $\mathcal{U}$ *of* A *contained in* $\mathcal{U}_A$ *there exists an* H-*selfadjoint matrix* B *such that* $s_c(B) = 2k(\lambda_0)$.

1.2 Stably Diagonable H-Selfadjoint Matrices

Let A be an H-selfadjoint matrix, and let Ω be an open subset of the real line. We say that A is *Ω-diagonable* if for every $\lambda_0 \in \Omega \cap \sigma(A)$ the multiplicity of λ_0 as a zero of $\det(I\lambda-A)$ coincides with $\dim \mathrm{Ker}(\lambda_0 I-A)$. In other words, the restriction of A to a spectral subspace corresponding to the eigenvalues of A in Ω, is similar to a diagonal matrix.

Next, we need to consider matrices for which all neighbouring matrices (with similar symmetries) are also Ω-diagonable. More formally, we call matrix A *stably Ω-diagonable* if there exist neighbourhoods $\mathcal{U}_A$ of A and $\mathcal{U}_H$ of H such that, whenever B is G-selfadjoint and $(B,G) \in \mathcal{U}_A \times \mathcal{U}_H$ it follows that B has the same number of eigenvalues as A in Ω (counting multiplicities) and is Ω-diagonable. Note that, in particular, A must be Ω-diagonable.

We will also use the corresponding notion in which the matrix H is kept fixed. Thus, an H-selfadjoint matrix A is called *H-stably Ω-diagonable* if there exists a neighbourhood $\mathcal{U}_A$ of A such that every H-selfadjoint matrix B in $\mathcal{U}_A$ is Ω-diagonable. In the next theorem we assume that Ω is an open subset of the real line such that its boundary does not contain eigenvalues of A.

THEOREM 1.3 *Let* A *be H-selfadjoint. Then the following statements are equivalent:*

(i) A *is stably Ω-diagonable;*

(ii) A *is H-stably Ω-diagonable;*

(iii) *the quadratic form* (Hx,x) *is either positive definite or negative definite on the subspace* $\mathrm{Ker}(\lambda_0 I-A)$, *for every* $\lambda_0 \in \sigma(A) \cap \Omega$.

We shall call the real eigenvalue λ_0 of an H-selfadjoint matrix A *definite* if the quadratic form (Hx,x) is either positive definite or negative definite on the root subspace of A corresponding to λ_0. Thus, statement (iii) above says that each eigenvalue of A in Ω is definite.

The canonical form (Theorem I.3.3) shows that the real eigenvalue λ_0 is definite if and only if the elementary divisors of $\lambda I - A$ corresponding to λ_0 are all linear, and the signs in the H-sign characteristic of A corresponding to λ_0 are either all equal to +1, or all equal to -1.

The proof of Theorem 1.3 will indicate some additional properties of an H-selfadjoint stably Ω-diagonable matrix A. First, there exist neighbourhoods $\mathcal{U}_A$ of A, $\mathcal{U}_H$ of H such that if B is G-selfadjoint and $(B,G) \in \mathcal{U}_A \times \mathcal{U}_H$, then B is *stably* Ω-diagonable (and not only Ω-diagonable as the definition of "stably Ω-diagonable" requires). Second, if a real eigenvalue λ_0 of A "splits" under the perturbation and produces eigenvalues $\mu_1,\cdots,\mu_r$ of B (all of them real), then the sign of μ_j in the G-sign characteristic of B $(j = 1,\cdots,r)$ is just the sign of λ_0 in the H-sign characteristic of A.

Note also that in every neighbourhood of an H-selfadjoint matrix which is Ω-diagonable but not stably Ω-diagonable, there exists an H-selfadjoint B with non-real eigenvalues.

<u>Proof of Theorem 1.3</u>. (iii)$\Rightarrow$(i) We are given that each eigenvalue of A in Ω is definite so that, in particular, A is Ω-diagonable. Let

$\lambda_1 < \cdots < \lambda_r$ be the eigenvalues of A in Ω and let

$$\delta = \min\left\{\frac{1}{3}(\lambda_2-\lambda_1),\cdots,\frac{1}{3}(\lambda_r-\lambda_{r-1}),\tilde{\lambda}_1,\cdots,\tilde{\lambda}_r\right\}$$

where $\tilde{\lambda}_i$ is the distance from λ_i to the boundary of Ω.

If B is a perturbation of A, we write P_i (Q_μ) for the orthogonal projector onto the root subspace of A (of B) associated with λ_i (with $\mu \in \mathbb{R}$), and $\nu_i(B)$ for the number of real eigenvalues (counting multiplicities) of B whose distance from λ_i is less than δ.

Using Theorem 1.1 neighbourhoods $\mathcal{U}_A$ of A and $\mathcal{U}_H$ of H can be found so that, if $(B,G) \in \mathcal{U}_A \times \mathcal{U}_H$ and B is G-selfadjoint then, for $i = 1,2,\cdots,r$,

$$\nu_i(B) \geq |\Sigma \operatorname{sig} Q_\mu G Q_\mu| = |\operatorname{sig} P_i H P_i| , \tag{1.5}$$

and the summation is over all *real* μ whose distance from λ_i is less than δ. Since λ_i is a definite eigenvalue of A, the last term in (1.5) is just the dimension of the root subspace $E_A(\lambda_i)$ of λ_i. But for B close to A we obviously have $\nu_i(B) \leq \dim E_A(\lambda_i)$ and so, taking $\mathcal{U}_A$ smaller, if necessary, the inequality in (1.5) is in fact an equality which means that B is Ω-diagonable and all its eigenvalues are definite. Further, the relation $\nu_i(B) = \dim E_A(\lambda_i)$ shows that B has the same number of eigenvalues as A in Ω (counting multiplicities). So A is stably Ω-diagonable.

(i)⇒(ii) is evident.

(ii)⇒(iii) Assume that A is Ω-diagonable, but its eigenvalue $\lambda_0 \in \Omega$ is not definite. By Theorem 1.1, in every neighbourhood of A there exists an H-selfadjoint matrix B such that the number of real eigenvalues of B in a neighbourhood of λ_0 is less than $\dim \operatorname{Ker}(\lambda_0 I - A)$. This means that B has non-real eigenvalues, a contradiction to (ii). □

The case $\Omega = \mathbb{R}$ in Theorem 1.3 will be of particular interest for us. An H-selfadjoint matrix A is called *diagonable with real eigenvalues* (in short, r-*diagonable*) if A is similar to a diagonal matrix with real eigenvalues, i.e. A is Ω-diagonable with $\Omega = \mathbb{R}$ and all eigenvalues of A are real. The definition of *stably* r-*diagonable* matrices, and of H-*stably* r-*diagonable* matrices are now evident. The following result is proved in the same way as Theorem 1.3.

THEOREM 1.4 *Let* A *be* H-*selfadjoint. Then the following state-*

ments are equivalent:

(i) A *is stably* r-*diagonable;*

(ii) A *is* H-*stably* r-*diagonable;*

(iii) *all eigenvalues of* A *are real and definite.*

The remarks concerning additional properties of stably Ω-diagonable matrices (stated before the proof of Theorem 1.3) apply for the stably r-diagonable matrices as well.

1.3 General Perturbations and Stably Diagonable H-Unitary Matrices

We describe here results for H-unitary matrices which are analogous to those obtained in Sections 1.1 and 1.2 for H-selfadjoint matrices.

We continue to use the notation $P_A(\lambda)$ for the orthogonal projector on the root subspace for A associated with λ. The unit circle is denoted $\mathbb{T}$.

THEOREM 1.5 *Let* U *be* H-*unitary, and let* $\Omega \subset \mathbb{T}$ *be an open set (relative to* $\mathbb{T}$*) whose boundary does not intersect* $\sigma(U)$. *Then for some neighbourhoods* $\mathcal{U}_U$ *of* U *and* $\mathcal{U}_H$ *of* H *we have*

$$\sum_{\lambda\in\Omega} \operatorname{sig} P_\lambda(U)HP_\lambda(U) = \sum_{\mu\in\Omega} \operatorname{sig} P_\mu(V)GP_\mu(V)$$

for every $V \in \mathcal{U}_U$ *which is* G-*unitary for an invertible hermitian* $G \in \mathcal{U}_H$. *Moreover, the number* $\nu_\Omega(V)$ *of eigenvalues of such a matrix* V *in* Ω *(counting multiplicities) satisfies the inequality*

$$\nu_\Omega(V) \geq \sum_{\lambda\in\Omega} |\operatorname{sig} P_\lambda(U)HP_\lambda(U)| , \tag{1.6}$$

and in every neighbourhood $\mathcal{U} \subset \mathcal{U}_U$ *of* U *there exists an* H-*unitary* V *for which equality holds in* (1.6).

This result can be obtained from Theorem 1.1 by using the Cayley transform.

Let U be an H-unitary matrix and let Ω be a relatively open subset of the unit circle. The matrix U is called *Ω-diagonable* if for every $\lambda_0 \in \Omega \cap \sigma(A)$ the multiplicity of λ_0 as a zero of $\det(\lambda I-A)$ is just $\dim \operatorname{Ker}(\lambda_0 I-A)$. The matrix U is *H-stably Ω-diagonable* if every H-unitary matrix V sufficiently close to U has the same number of eigenvalues in Ω as U, and is Ω-diagonable. The matrix U is *stably Ω-diagonable* if this property holds for every G-unitary V such that G (resp. V) is sufficiently close to H (resp. U). In the following theorem we assume that the boundary of Ω does not intersect $\sigma(U)$.

THEOREM 1.6 *Let* U *be an* H-*unitary matrix. The following statements are equivalent:*

(i) U *is stably* Ω-*diagonable;*

(ii) U *is* H-*stably* Ω-*diagonable;*

(iii) *the quadratic form* (Hx,x) *is either positive definite or negative definite on the subspace* $\mathrm{Ker}(\lambda_0 I-U)$, *for every* $\lambda_0 \in \sigma(U) \cap \Omega$.

It is easy to see that Theorems 1.3 and 1.6 can be obtained one from another by using the Cayley transform and its inverse. A direct proof of Theorem 1.6 can also be obtained from the general perturbation Theorem 1.5.

An important particular case of Theorem 1.6 arises when Ω is the whole unit circle $\mathbb{T}$. An H-unitary matrix U is called *diagonable with unimodular eigenvalues* (in short, u-*diagonable*) if U is similar to a diagonal matrix with unimodular entries on the diagonal, i.e. U is Ω-diagonable with $\Omega = \mathbb{T}$, and all eigenvalues of U lie on the unit circle $\mathbb{T}$. The meaning of the notions of *stably* u-*diagonable* matrices and H-*stably* u-*diagonable* matrices is clear.

THEOREM 1.7 *Let* U *be an* H-*unitary matrix. The following statements are equivalent:*

(i) U *is* u-*diagonable;*

(ii) U *is* H-*stably* u-*diagonable;*

(iii) *all eigenvalues of* U *lie on the unit circle, and the quadratic form* (Hx,x) *is either positive definite or negative definite on* $\mathrm{Ker}(\lambda_0 I-U)$, *for every* $\lambda_0 \in \sigma(U)$.

1.4 Analytic Perturbations and Eigenvalues

In Section 1.2 we have studied H-selfadjoint matrices A which are stably r-diagonable. Observe that in the classical case, when H is positive definite, every H-selfadjoint matrix is stably r-diagonable. So stably r-diagonable matrices can be viewed as H-selfadjoint matrices which behave like hermitian ones with respect to small perturbations.

Hermitian matrices are noted also for their special properties with respect to analytic perturbations. Namely, if $A(\tau) = \sum_{j=0}^{\infty} \tau^j A_j$ is an analytic matrix function of a real parameter τ with hermitian coefficients A_j, then the eigenvalues of $A(\tau)$ are analytic functions of τ (see, e.g. Theorem II.3.3 and Section 2.6 in [25]). Note that in general (i.e. without the assumption of the hermitian property) the eigenvalues of $A(\tau)$ need not be analytic; one can claim only their continuity.

This point of view of analytic perturbations leads naturally to the following definition. Let A_0 be H_0-selfadjoint and let λ_0 be a real eigenvalue of A_0. We say that λ_0 is *analytically extendable* by analytic perturbations if for any pair of matrix functions $A(\tau)$, $H(\tau)$ which are analytic in the real variable τ on a neighbourhood $\mathcal{U}$ of 0 and such that $A(\tau)$ is $H(\tau)$-selfadjoint for all $\tau \in \mathcal{U}$ and $A(0) = A_0$, $H(\tau) = H_0$, the eigenvalues of $A(\tau)$ which tend to λ_0 as $\tau \to 0$ can be chosen analytic functions on $\mathcal{U}$.

When H_0 is positive definite, every eigenvalue λ_0 of an H_0-selfadjoint matrix A_0 is analytically extendable. Indeed, if H_0 is not perturbed (i.e. $H(\tau) \equiv H_0$) the result mentioned above ([25], Section 2.6) applies. The general analytic perturbation $A(\tau)$, $H(\tau)$ can be easily reduced to this case by considering an analytic matrix function $S(\tau)$ such that $H(\tau) = S(\tau)^*S(\tau)$ and replacing $A(\tau)$ by $S(\tau)A(\tau)S(\tau)^{-1}$.

It will be clear from Theorem 1.8 below (in view of Theorem 1.3) that stably r-diagonable matrices, and only they, have all eigenvalues analytically extendable. So, with respect to analytic perturbations as well, the stably r-diagonable matrices behave like hermitian ones.

THEOREM 1.8 *Let* A_0 *be* H_0*-selfadjoint. Then a real eigenvalue* λ_0 *of* A_0 *is analytically extendable if and only if the quadratic form* (H_0x,x) *is either positive definite or negative definite on the subspace* $\mathrm{Ker}(\lambda_0 I - A_0)$.

PROOF. Assume that the quadratic form (H_0x,x) is either positive or negative definite on $\mathrm{Ker}(\lambda_0 I - A)$. Let Δ be an open disc with center λ_0 such that $\bar{\Delta} \cap \sigma(A) = \{\lambda_0\}$. By Theorem 1.3 there exists an $\varepsilon > 0$ such that, for every G-selfadjoint matrix B such that $||A-B|| + ||H-G|| < \varepsilon$, all the eigenvalues of B in Δ are real. Now let $A(\tau)$, $H(\tau)$ be $n \times n$ matrix functions with the properties described in the definition of the analytic extendability of eigenvalues. Then the eigenvalues $\lambda_1(\tau),\cdots,\lambda_\nu(\tau)$ of $A(\tau)$ which tend to λ_0 as $\tau \to 0$ are real for τ sufficiently close to 0 (namely, those τ for which $\lambda_i(\tau) \in \Delta$ and $||A(\tau)-A|| + ||H(\tau)-H|| < \varepsilon$). This implies that, for $j = 1,2,\cdots,n$, $\lambda_j(\tau)$ is analytic in τ on a neighbourhood of zero. Indeed, $\lambda_j(\tau)$ is a zero of the polynomial $\det(\lambda I-A(\tau))$ with coefficients analytic in τ and, as such, admits expansion in a series of fractional powers of τ. More exactly, there exist positive integers $\alpha_1,\cdots,\alpha_m$ such that $\nu = \alpha_1 + \cdots + \alpha_m$ and (maybe after a reordering of $\lambda_j(\tau)$)

$$\lambda_p(\tau) = \lambda_0 + \sum_{k=1}^{\infty} c_k^{(q)}(x_p)^k , \quad \alpha_1+\cdots+\alpha_{q-1} < p \leq \alpha_1+\cdots+\alpha_q \tag{1.7}$$

where $x_p = x_p(\tau) = |t|^{\frac{1}{\alpha_q}}\left\{\cos\left[\frac{1}{\alpha_q}(\arg\tau + 2\pi i(p - \sum_{i=1}^{q-1}\alpha_i))\right] + \right.$ $\left. + i\sin\left[\frac{1}{\alpha_q}(\arg\tau + 2\pi i(p - \sum_{i=1}^{q-1}\alpha_i))\right]\right\}$, and $c_k^{(q)}$ are complex numbers (so x_p is an α_q-th root of τ); by definition, $\alpha_0 = 0$. For details see [5], for example. Consider

$$\lambda_{\alpha_1}(\tau) = \lambda_0 + \sum_{k=1}^{\infty} c_k^{(1)}(x_{\alpha_1})^k$$

and let k_1 be the smallest index such that $c_{k_1}^{(1)} \neq 0$ (if all $c_k^{(1)} = 0$, then $\lambda_1(t) \equiv \lambda_0$ is obviously analytic in t for $p = 1,\cdots,\alpha_1$). Then

$$c_{k_1}^{(1)} = \lim_{\tau\to+0} \frac{\lambda_1(\tau)-\lambda_0}{(x_{\alpha_1})^{k_1}} . \tag{1.8}$$

We find that, because $\lambda_1(\tau) - \lambda_0$ is real, so is $c_{k_1}^{(1)}$. A similar argument shows that all non-zero $c_k^{(1)}$ are real. As the imaginary part of $\lambda_{\alpha_1}(\tau)$ is zero, we obtain

$$\sum_{k=1}^{\infty} c_k^{(1)}|\tau|^{\frac{k}{\alpha_1}}\sin(\frac{k}{\alpha_1}\arg\tau) = 0$$

for all $|\tau| < \delta$. In particular, for $\tau < 0$ this implies that $c_k^{(1)}\sin(\frac{k\pi}{\alpha_1}) = 0$ for $k = 1,2,\cdots$, and means that k is an integer multiple of α_1 if $c_k^{(1)} \neq 0$. So all $\lambda_p(\tau)$, $1 \leq p \leq \alpha_1$, are analytic on a neighbourhood of $\tau = 0$. The same argument shows that all $\lambda_1(\tau),\cdots,\lambda_\nu(\tau)$ are analytic in τ on a neighbourhood of zero. Hence λ_0 is analytically extendable.

Assume now that the form (H_0x,x), is neither positive definite nor negative definite on $\mathrm{Ker}(\lambda_0 I-A_0)$. We shall construct an analytic matrix function $A(\tau)$, $-\infty < \tau < \infty$, which is H_0-selfadjoint and satisfies $A(0) = A_0$, but has a non-analytic eigenvalue $\lambda_0(\tau)$ which is equal to λ_0 in the limit as $\tau \to 0$.

Without loss of generality we can assume that (A_0,H_0) is in the canonical form. From the condition on the quadratic form (H_0x,x) it follows that either A_0 has a Jordan block J_0 of size $m \geq 2$ corresponding to the eigenvalue λ_0, or A_0 has two Jordan blocks $J_0 \oplus J_0$, each of size 1 and with opposite signs in the sign characteristic of (A_0,H_0). Consider the first case:

$$A_0 = J_0 \oplus J_1 \; ; \quad H_0 = \pm P_0 \oplus P_1 \; ,$$

where P_0 is the $m \times m$ sip matrix, and (J_1,P_1) is the rest of (A_0,H_0). Then

$$A(\tau) = \begin{bmatrix} \lambda_0 & 1 & & & & \\ 0 & \lambda_0 & \cdot & & 0 & \\ \cdot & & \cdot & \cdot & & \\ \cdot & & & \cdot & \cdot & \\ \cdot & & & & \cdot & \\ 0 & \cdot & \cdot & \cdot & \cdot & 1 \\ \tau & 0 & & & 0 & \lambda_0 \end{bmatrix} \oplus J_1$$

and $\lambda_0(\tau) = \lambda_0 + \tau^{\frac{1}{m}}$ will do. In the second case let

$$A_0 = \mathrm{diag}[\lambda_0,\lambda_0,J_1] \; ; \quad H_0 = \mathrm{diag}[1,-1,P_1] \; .$$

Then

$$A(\tau) = \begin{bmatrix} \lambda_0+2\tau+\tau^2 & -\tau \\ \tau & \lambda_0 \end{bmatrix} \oplus J_1$$

and $\lambda_0(\tau) = \lambda_0 + \frac{1}{2}(2\tau+\tau^2+\tau(4\tau+\tau^2)^{\frac{1}{2}})$ is not analytic on a neighbourhood of $\tau = 0$. □

In connection with Theorem 1.8 several remarks can be made.

A) The proof of the theorem actually shows the following:

Let λ_0 *be a real eigenvalue of an* H_0*-selfadjoint* $n \times n$ *matrix* A_0. *If the quadratic form* (H_0x,x) *is not definite on the subspace* $\mathrm{Ker}(\lambda_0 I-A_0)$, *then there exists a quadratic polynomial* $A_0 + \tau A_1 + \tau^2 A_2$ *with*

H_0-selfadjoint coefficients which has a continuous non-analytic eigenvalue tending to λ_0 as $\tau \to 0$.

Actually one can also take A_1 and A_2 such that rank $A_1 = 2$, rank $A_2 = 1$ and $\mathrm{Im}\, A_2 \subset \mathrm{Im}\, A_1$.

B) *The statement A) is not true in general if we replace the quadratic polynomial by a linear polynomial $A_0 + \tau A_1$.*

Indeed, if $A_0 = 0$, then the eigenvalues of τA_1 are just τ times the eigenvalues of A_1 and therefore analytic in τ (and this is true for any $n \times n$ matrix A_1, not necessarily H_0-selfadjoint).

C) *The proof of Theorem 1.8 also show that if the quadratic form (H_0x,x) is degenerate on $\mathrm{Ker}(\lambda_0 I-A)$ then there exists an H_0-selfadjoint matrix A_1 of rank one such that $A_0 + \tau A_1$ has a non-analytic eigenvalue tending to λ_0 as $\tau \to 0$.*

D) *Let A_0, H_0 and λ_0 be as in remark A). Suppose the quadratic form (H_0x,x) is not definite on $\mathrm{Ker}(\lambda_0 I-A_0)$, and let m_+ (resp. m_-) be the number of positive (resp. negative) squares in the canonical representation of the quadratic form (H_0x,x) on the root subspace $E_{A_0}(\lambda_0) = \mathrm{Ker}(\lambda_0 I-A_0)^n$. Then a continuous eigenvalue $\lambda_0(\tau)$ of an $H(\tau)$-selfadjoint analytic matrix $A(\tau)$ (where $H(0) = H_0$, $A(0) = A_0$ and τ belongs to some real neighbourhood of zero) has a fractional power expansion*

$$\lambda_0(\tau) = \lambda_0 + \sum_{j=1}^{\infty} c_j\left(\tau^{\frac{1}{p}}\right)^j , \quad c_j \in \mathbb{C} \tag{1.9}$$

with

$$p \leqslant 2\min(m_+,m_-) + 1 . \tag{1.10}$$

Observe that the quadratic form (H_0x,x) is nondegenerate on $E_{A_0}(\lambda_0)$. (This follows, for instance, from the canonical form of (A_0,H_0)).

As it is well-known, (see [4], for example), $\lambda_0(\tau)$ has an expansion into fractional power series (1.9). It remains to prove the estimation (1.10) for p.

Assume $\lambda_0(\tau) \not\equiv \lambda_0$, and let φ be the number of non-real functions $\lambda_0(\tau)$ given by the formula (1.9) (in this formula p different values for $\tau^{\frac{1}{p}}$ are allowed). Using the arguments employed in the proof of Theorem 1.8, it is not difficult to see that $\varphi \geqslant p - 1$ if p is odd and $\varphi = p$ if p

is even. On the other hand, by Theorem 1.1, the number of non-real eigenvalues of $A(\tau)$ (for τ sufficiently close to zero) does not exceed $2\min(m_+,m_-)$. Hence the formula (1.10) follows. If (H_0x,x) is either positive or negative definite on $\mathrm{Ker}(\lambda_0 I-A)$, formula (1.10) gives $p = 1$, i.e. λ_0 is analytically extendable. This is the "if" part of Theorem 1.8.

1.5 Analytic Perturbations and Eigenvectors

Now let λ_0 be a real eigenvalue of an H_0-selfadjoint matrix A_0. We say that the eigenvectors of A_0 corresponding to λ_0 are *analytically extendable* by analytic perturbations if the following holds. Let $A(\tau)$, $H(\tau)$ be a pair of matrix functions which are analytic in τ on a real neighbourhood U of 0, are such that $A(\tau)$ is $H(\tau)$-selfadjoint for all $\tau \in U$, and satisfy $A(0) = A_0$, $H(0) = H_0$. Also let Γ be a circle with the center λ_0 and radius so small that λ_0 is the only eigenvalue of A_0 inside or on Γ. Then for each real τ sufficiently close to zero there exists an $H(\tau)$-orthonormal basis $x_1(\tau),\cdots,x_k(\tau)$ of eigenvectors of $A(\tau)$ in the subspace $K(\tau)$ defined by $\Sigma\mathrm{Ker}(\lambda I-A(\tau))$, where the sum is taken over all eigenvalues λ of $A(\tau)$ inside Γ, and the vector functions $x_1(\tau),\cdots,x_k(\tau)$ are analytic in τ. Recall that $H(\tau)$-orthonormality of $x_1(\tau),\cdots,x_k(\tau)$ means that $(H(\tau)x_i(\tau),x_j(\tau))$ is equal to 0 if $i \neq j$, and to ± 1 if $i = j$. As we shall see shortly, $K(\tau)$ is in fact the sum of the root subspaces of $A(\tau)$ corresponding to the eigenvalues of $A(\tau)$ inside Γ.

If the eigenvectors of A_0 corresponding to λ_0 are analytically extendable, then the eigenvalue λ_0 is necessarily analytically extendable. Indeed, assuming the contrary, Theorem 1.8 shows that the form (H_0x,x) is neither positive definite nor negative definite on $\mathrm{Ker}(\lambda_0 I-A_0)$. Then, arguing as in the second part of the proof of Theorem 1.8, we find that the eigenvectors of A_0 corresponding to λ_0 are not analytically extendable. In particular, the analytic extendability of the eigenvectors of A_0 at λ_0 implies that the number k coincides with the multiplicity of λ_0 as a zero of $\det(\lambda I-A)$. Consequently, all elementary divisors of $\lambda I - A(\tau)$ corresponding to the eigenvalues of $A(\tau)$ inside Γ are linear (in particular, all elementary divisors of $\lambda I - A_0$ corresponding to λ_0 are linear).

When H_0 is positive (or negative) definite, it is a well-known fact that the eigenvectors of A_0 are always analytically extendable. If H_0 is not definite, then in general the analytic extendability of eigenvectors fails. In Theorem 1.9 below we give necessary and sufficient conditions for analytic extendability of eigenvectors. It will be seen from this theorem that the eigenvectors of A_0 corresponding to each eigenvalue are analytically extendable if and only if A_0 is stably r-diagonable. So again, stably r-diagonable matrices behave like hermitian ones.

It turns out that analytic extendability of λ_0 and of the eigenvectors corresponding to λ_0 are in fact equivalent.

THEOREM 1.9 *Let* A_0 *be* H_0*-selfadjoint. Then the eigenvectors of* A_0 *corresponding to* $\lambda_0 \in \sigma(A_0) \cap \mathbb{R}$ *are analytically extendable if and only if the quadratic form* (H_0x,x) *is either positive definite or negative definite on the subspace* $\mathrm{Ker}(\lambda_0 I - A_0)$.

PROOF. In view of Theorem 1.8 and the remark preceding the theorem, we have only to show that if λ_0 is analytically extendable (or, equivalently, if the quadratic form (H_0x,x) is definite on $\mathrm{Ker}(\lambda_0 I-A_0)$), then the eigenvectors corresponding to λ_0 are analytically extendable as well.

Let $A(\tau)$, $H(\tau)$ be a pair of matrix functions as in the definition of analytically extendable eigenvectors defined for $|\tau| < \delta$ (δ is a positive number), and let Γ be a small circle with center λ_0. By Theorem 1.3 (see also remarks after the statement of Theorem 1.3) there is a $\delta_1 \in (0,\delta]$ such that for $|\tau| < \delta_1$ all eigenvalues of $A(\tau)$ inside Γ are real, all elementary divisors corresponding to these eigenvalues are linear and the quadratic form $(H(\tau)x,x)$ is definite on $\mathrm{Ker}(\lambda(\tau)I-A(\tau))$ for every eigenvalue $\lambda(\tau)$ of $A(\tau)$ inside Γ. In particular, $(H(\tau)x,x) \neq 0$ for every eigenvector x corresponding to $\lambda(\tau)$ $(|\tau| < \delta_1)$.

Let $\lambda(\tau)$ be an eigenvalue of $A(\tau)$ which is analytic for $|\tau| < \delta$ and such that $\lambda(0) = \lambda_0$. Choose a non-zero analytic vector function $x(\tau) \in \mathrm{Ker}(\lambda(\tau)I-A(\tau))$, $|\tau| < \delta$ (such an $x(\tau)$ exists in view of Theorem A.12 of the Appendix to Part III). As we have seen in the preceding paragraph, $(H(\tau)x(\tau),x(\tau)) \neq 0$ for $|\tau| < \delta_1$. Put $x_1(\tau) = |(H(\tau)x(\tau),x(\tau))|^{-\frac{1}{2}}x(\tau)$; then $x_1(\tau)$ is an analytic eigenvector of $A(\tau)$ with $(H(\tau)x_1(\tau),x_1(\tau)) = \pm 1$.

Consider now the $H(\tau)$-orthogonal companion $M(\tau)$ of $\mathrm{Span}\{x_1(\tau)\}$ $(|\tau| < \delta_1)$. Since $\mathrm{Span}\{x_1(\tau)\}$ is $H(\tau)$-nondegenerate, $M(\tau)$ is in fact a direct complement to $\mathrm{Span}\{x_1(\tau)\}$ in $\mathbb{C}^n$ (n is the size of A_0). Moreover, the family of subspaces $M(\tau)$ is analytic for $|\tau| < \delta_1$ (indeed, by Proposition A.11 of the Appendix the family of subspaces $\mathrm{Span}\{x_1(\tau)\}$ is analytic for $|\tau| < \delta_1$; then the same is easily seen to be true for

$$M(\tau) = H(\tau)^{-1}(\mathrm{Span}\{x_1(\tau)\}^{\perp})).$$

By Theorem A.12 in the Appendix there exists an analytic basis in $M(\tau)$, and applying the Gram-Schmidt orthogonalization, we obtain an analytic *orthonormal* basis $y_1(\tau),\cdots,y_{n-1}(\tau)$ in $M(\tau)$ $(|\tau| < \delta_1)$. Consider linear transformations $A(\tau)|_{M(\tau)} : M(\tau) \to M(\tau)$ and $P_{M(\tau)}H(\tau)|_{M(\tau)} : M(\tau) \to M(\tau)$, where $P_{M(\tau)}$ is the orthogonal projector on $M(\tau)$ (note that since $A(\tau)$ is $H(\tau)$-selfadjoint, the subspace $M(\tau)$ is $A(\tau)$-invariant). Writing these linear transformations in the basis $y_1(\tau),\cdots,y_{n-1}(\tau)$ we obtain $(n-1) \times (n-1)$ matrices $A_1(\tau)$ and $H_1(\tau)$ $(|\tau| < \delta_1)$ such that $H_1(\tau)$ is hermitian and invertible, $A_1(\tau)$ is $H_1(\tau)$-selfadjoint, and $A_1(\tau)$ and $H_1(\tau)$ are analytic in τ. (The analyticity of $A_1(\tau)$ follows from the analyticity of the *unique* solution $\{\alpha_{ij}\}_{i,j=1}^{n-1}$ of the system of linear equations

$$A(\tau)y_i(\tau) = \sum_{j=1}^{n-1} \alpha_{ij}y_j(\tau) \ , \qquad i = 1,\cdots,n-1$$

with analytic coefficients.) Apply the argument employed in the first part of the proof to produce an analytic eigenvector $x_2(\tau)$ of $A(\tau)|_{M(\tau)}$ for $|\tau| < \delta_2 \leq \delta_1$ such that $(H(\tau)x_2(\tau),x_2(\tau)) = \pm 1$, and so on. Eventually we obtain the analytic $H(\tau)$-orthonormal basis $x_1(\tau),\cdots,x_k(\tau)$ of eigenvectors of $A(\tau)$ in the subspace $\mathrm{Im}[\frac{1}{2\pi i}\int_\Gamma(\lambda I-A(\tau))^{-1}d\lambda]$. Hence the eigenvectors of A_0 corresponding to λ_0 are analytically extendable. □

The proof of Theorem 1.9 also shows the validity of the following statement (and this also follows from Theorem 1.1). Let the eigenvectors of an H_0-selfadjoint matrix A_0 corresponding to $\lambda_0 \in \sigma(A\ ) \cap \mathbb{R}$ be analytically extendable, let $A(\tau)$, $H(\tau)$ be as in the definition of analytically extendable eigenvectors, and let $x_1(\tau),\cdots,x_k(\tau)$ be an analytic $H(\tau)$-orthogonal basis of eigenvectors of $A(\tau)$ in $\mathrm{Im}[\frac{1}{2\pi i}\int_\Gamma(\lambda I-A(\tau))^{-1}d\lambda]$ (such a basis exists by the analytical extendability). Then $(H(\tau)x_i(\tau),x_i(\tau))$ is $+1$ (resp. -1) if the form (H_0x,x) is positive (resp. negative) definite on $\mathrm{Ker}(\lambda_0 I-A_0)$.

1.6 The Real Case

Consider the important case of an H-selfadjoint matrix A where both A and H are real and pairs (B,G) obtained from perturbations of (A,H) are also confined to real matrices. It is not difficult to see that the results of Sections 1.1 and 1.2 have precise analogues in this context. The proofs are also the same with two exceptions.

The first concerns the construction of matrices with non-real eigenvalues developed in three cases in the final part of the proof of Theorem 1.1, and needed to establish the case of equality in the relation (1.2). For case (i) the role played by $\mathrm{diag}[J(i,\frac{\alpha}{2}),J(-i,\frac{\alpha}{2})]$ is now played by a block of the real Jordan form:

$$\left[\begin{array}{cc|cc|cc|c|cc}
0 & 1 & 1 & 0 & 0 & 0 & & & \\
-1 & 0 & 0 & 1 & 0 & 0 & & & \\
\hline
 & & 0 & 1 & 1 & 0 & & & \\
 & & -1 & 0 & 0 & 1 & & & \\
\hline
 & & & & & & \ddots & & \\
\hline
 & & & & & & & 0 & 1 \\
 & & & & & & & -1 & 0
\end{array}\right]$$

with similar modifications for cases (ii) and (iii).

Using the real analogue of Theorem 1.1 the proofs of "real" versions of Theorems 1.3 and 1.4 are essentially the same.

Many results of Section 1.3 for the real case are obvious corollaries of the complex case. An exception is the feasibility of the equality sign in (1.6); a question which will not be pursued here.

1.7 The Real Skew-Symmetric Case

It frequently happens in applications that the matrix defining an indefinite scalar product has the form $H = iE$ where E is real and skew-symmetric, i.e. $E^* = -E$. The case $H = i\begin{bmatrix} 0 & I \\ -I & 0 \end{bmatrix}$ is of particular interest, for example. Also, matrices A which are H-selfadjoint are likely to be real as well. In the remainder of the chapter some of the topics already discussed are re-examined in this context.

Thus, E will now denote an $n \times n$ real, invertible, skew-symmetric matrix and we observe that these conditions on E imply that n is even. A real matrix X will be called E-*orthogonal* if $X^TEX = E$. In other words, a real matrix X is E-orthogonal if and only if X is H-unitary with the hermitian invertible matrix $H = iE$.

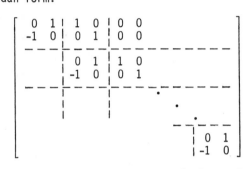

An E-orthogonal matrix X is called *u-diagonable* if X is similar (complex matrices are allowed in carrying out this similarity) to a diagonal matrix with unimodular eigenvalues. An E-orthogonal matrix X is called *stably u-diagonable* if every F-orthogonal matrix Y is diagonable provided the real invertible skew-symmetric matrix F is sufficiently close to E, and the matrix Y is sufficiently close to X; in particular, the matrix X itself should be u-diagonable. Finally, we introduce the definition of E-stably u-diagonable matrix: an E-orthogonal matrix X is *E-stably u-diagonable* if every E-orthogonal matrix Y sufficiently close to X is u-diagonable.

The main result of this section is the description of stably u-diagonable matrices, as given in the next theorem. Note that the equivalence of (i), (ii) and (iii) in this theorem is to be expected by analogy with Theorem 1.7. The statement (iv) is peculiar to the real skew-symmetric matrices E of this section.

THEOREM 1.10 *The following statements are equivalent for an E-orthogonal matrix* X:

(i) X *is E-stably u-diagonable;*

(ii) X *is stably u-diagonable;*

(iii) X *is u-diagonable and the quadratic form* (iEx,x), $x \in \operatorname{Ker}(X-\lambda_0 I)$ *is either positive definite or negative definite, for every* $\lambda_0 \in \sigma(X)$;

(iv) X *admits the following representation:*

$$X = G \begin{bmatrix} \cos\Theta & \sin\Theta \\ -\sin\Theta & \cos\Theta \end{bmatrix} G^{-1}, \tag{1.11}$$

where G *is E-orthogonal, and* $\Theta = \operatorname{diag}[\theta_1,\cdots,\theta_{\frac{n}{2}}]$, *where* θ_i *are real,* $|\theta_i| < \pi$ *and satisfy the property that* $\theta_i + \theta_j \neq 0$, $1 \le i,j \le \frac{n}{2}$.

From the proof it will be seen that $\theta_1,\cdots,\theta_{\frac{n}{2}}$ are the arguments of the eigenvalues λ_0 of X (counting multiplicities) for which the form (iEx,x) defined on $\operatorname{Ker}(\lambda_0 I-X)$ is positive definite.

The rest of the chapter is devoted to the proof of Theorem 1.10, including proofs of some auxiliary results.

The implication (ii)⇒(i) is trivial, while the implication (iii)⇒(ii) follows as a particular case of Theorem 1.7. We need some preliminary results to prove (i)⇒(iii) and (iii)⇔(iv), and these will be developed in the next section.

1.8 Auxiliary Results for the Real Skew-Symmetric Case

We start with the development of a canonical form for real skew-adjoint matrices under (real) orthogonal similarity transformations.

LEMMA 1.11 *Let* E *be a real skew-symmetric matrix. Then there exists an orthogonal matrix* U *such that* $U^{-1}EU$ *has the form*

$$\mathrm{diag}\left(\begin{bmatrix} 0 & \tau_1 \\ -\tau_1 & 0 \end{bmatrix}, \cdots, \begin{bmatrix} 0 & \tau_k \\ -\tau_k & 0 \end{bmatrix}, 0, \cdots, 0\right), \tag{1.12}$$

where $\tau_1,\cdots,\tau_k$ *are positive numbers. The form* (1.12) *is uniquely determined by* E, *up to permutation of blocks (in fact, the numbers* $\pm i\sqrt{\tau_j}$, $j = 1,\cdots,k$, *are the non-zero eigenvalues of* E).

PROOF. It follows from the real Jordan form of E, for instance, that there exists an E-invariant subspace $M \subset \mathbb{R}^n$ of dimension either 1 or 2. Moreover, $E|_M = \alpha$ and $\alpha \in \mathbb{R}$ if $\dim M = 1$, and $E|_M = \begin{bmatrix} \sigma & \tau \\ -\tau & \sigma \end{bmatrix}$ and $\sigma,\tau \in \mathbb{R}$, $\tau > 0$, if $\dim M = 2$ (here the linear transformation $E|_M$ is considered as a matrix in some basis in M; and we shall assume that the vectors in this basis have length one). The number α (if $\dim M = 1$), or the number $\sigma + i\tau$ (if $\dim M = 2$), is an eigenvalue of E. Now E is real skew-symmetric; so necessarily $\alpha = 0$ if $\dim M = 1$, or $\sigma = 0$ if $\dim M = 2$. In the latter case, the vectors x, y in the basis chosen for M are orthogonal. Indeed,

$$(x,y) = \frac{1}{\tau}(Ey,y) = \frac{1}{\tau}(y,E^Ty) = -\frac{1}{\tau}(y,Ey) = -\frac{1}{\tau}(Ey,y) = -(x,y) ,$$

so $(x,y) = 0$. The orthogonal complement $M^\perp$ of M in $\mathbb{R}^n$ is also E-invariant: for if $x \in M^\perp$ and $y \in M$, then

$$(Ex,y) = -(x,Ey) = 0 ,$$

since M is E-invariant. Choosing an orthonormal basis in $M^\perp$, we have (in this basis): either

$$E = \begin{bmatrix} 0 & 0 \\ 0 & E' \end{bmatrix}, \quad \text{or} \quad E = \begin{bmatrix} 0 & \tau & 0 \\ -\tau & 0 & 0 \\ 0 & 0 & E' \end{bmatrix},$$

where E' is real skew-symmetric. Continuing this process we eventually obtain an orthonormal basis in $\mathbb{R}^n$ in which E has the form (1.12), and this is equivalent to the statement of Lemma 1.11. □

From Lemma 1.11 one easily obtains the following fact.

COROLLARY 1.12 *Let* E *be an invertible real skew-symmetric* $n \times n$ *matrix. Then there exists an invertible real matrix* S *such that*

$$S^T E S = \begin{bmatrix} 0 & -I \\ I & 0 \end{bmatrix}$$

(here I *stands for the* $\frac{n}{2} \times \frac{n}{2}$ *unit matrix).*

The next simple lemma will also be useful.

LEMMA 1.13 *A diagonable* $n \times n$ *matrix* X *has all pure imaginary entries if and only if, for every eigenvalue* λ_0 *of* X *with an associated eigenvectors* $x \in \mathbb{C}^n$, $-\bar{\lambda}_0$ *is also an eigenvalue of* X *and has an associated eigenvector* $\bar{x}$.

PROOF. If X has pure imaginary entries then $\bar{X} = -X$ and, obviously, $Xx = \lambda_0 x$ implies $X\bar{x} = (-\bar{\lambda}_0)\bar{x}$.

Conversely, let $x_1, \cdots, x_n$ be a basis of eigenvectors of X for $\mathbb{C}^n$ and, for any $y \in \mathbb{C}^n$ write $y = \sum_{j=1}^{n} \eta_j x_j$. If $Xx_j = \lambda_j x_j$ for $j = 1,2,\cdots,n$ then $X\bar{x}_j = -\bar{\lambda}_j \bar{x}_j$ and so $\lambda_j x_j = -\bar{X} x_j$. Hence

$$Xy = \Sigma \eta_j x_j = \Sigma \eta_j \lambda_j x_j = -\Sigma \eta_j \bar{X} x_j = -\bar{X} y ,$$

i.e. $X = -\bar{X}$ as required. □

LEMMA 1.14 *Let* B *be an* $n \times n$ *invertible real matrix such that* iB *is* iE*-selfadjoint and is similar to a real diagonal matrix. Assume that for some* $\lambda_0 \in \sigma(iB)$, (iEx,x) *is not a definite form on* $\operatorname{Ker}(iB-\lambda_0 I)$. *Then for every* $\varepsilon > 0$ *there is a real matrix* $\widetilde{B}_\varepsilon$ *such that* $||\widetilde{B}_\varepsilon - B|| < \varepsilon$, $i\widetilde{B}_\varepsilon$ *is* iE*-selfadjoint and has non-real eigenvalues.*

PROOF. Observe that the eigenvalues of iB are non-zero and occur in real pairs $\pm\lambda$. We may assume $\lambda_0 \in \sigma(iB)$ and $\lambda_0 < 0$. Choose a basis $v_1, \cdots, v_m$ in $\operatorname{Ker}(iB-\lambda_0 I)$ such that $(iEv_j, v_k) = 0$ if $j \neq k$ and $(iEv_1, v_1) = -(iEv_2, v_2) = 1$. Let $M_0 = \operatorname{Span}\{v_1, v_2\}$, and let $P_{M_0} : \mathbb{C}^n \to M_0$ be the orthogonal projector on M_0. By Theorem 1.3 for every $\varepsilon > 0$ there exists a 2×2, $iP_{M_0} E|_{M_0}$-selfadjoint matrix A_ε (considered as a linear transformation on M_0) such that $||A_\varepsilon - (iB)|_{M_0}|| < \varepsilon$ and the two eigenvalues of A_ε are non-real. Let $\varphi_\varepsilon, \psi_\varepsilon \in M_0$ be eigenvectors of A_ε corresponding to the eigenvalues μ_ε ($\notin \mathbb{R}$), $\bar{\mu}_\varepsilon$ of A_ε, respectively.

Now define an $n \times n$ matrix X_ε as follows: for every eigenvector x of iB corresponding to a negative eigenvalue different from λ_0, define $X_\varepsilon x = iBx$, $X_\varepsilon \bar{x} = iB\bar{x}$; further, define $X_\varepsilon v_k = iBv_k$, $X_\varepsilon \bar{v}_k = iB\bar{v}_k$ for $k = 3,4,\cdots,m$; finally, put $X_\varepsilon \varphi_\varepsilon = A_\varepsilon \varphi_\varepsilon$ $(= \mu_\varepsilon \varphi_\varepsilon)$, $X_\varepsilon \psi_\varepsilon = A_\varepsilon \psi_\varepsilon$ $(= \bar{\mu}_\varepsilon \psi_\varepsilon)$, $X_\varepsilon \bar{\varphi}_\varepsilon = -\bar{\mu}_\varepsilon \bar{\varphi}_\varepsilon$, $X_\varepsilon \bar{\psi}_\varepsilon = -\mu_\varepsilon \bar{\psi}_\varepsilon$. Using Lemma 1.13 it is easily seen that X_ε is correctly defined, and $X_\varepsilon = i\widetilde{B}_\varepsilon$ satisfies all the requirements of Lemma 1.14. □

1.9 Proof of Theorem 1.10

First note the following property of E-stably u-diagonable matrices.

LEMMA 1.15 *Let* X *be a real* E-*stably* u-*diagonable matrix. Then the numbers* ± 1 *are not eigenvalues of* X.

PROOF. Assume the contrary, and let $M \neq \{0\}$ be the root subspace of X corresponding to the eigenvalue λ_0, where $\lambda_0 = 1$ or $\lambda_0 = -1$. Since X is real and diagonable, there is a basis $v_1,\cdots,v_m$ in M consisting of vectors with real coordinates. Let M' be the spectral subspace of X corresponding to the eigenvalues different from λ_0. It is easily seen that $x \in M'$ implies $\bar{x} \in M'$. Therefore, there exists a basis $v_{m+1},\cdots,v_n$ in M' consisting of vectors with real coordinates.

Let S be the real $n \times n$ matrix with columns $v_1,\cdots,v_n$. Evidently, it is sufficient to consider $\widetilde{X} = S^{-1}XS$ and $\widetilde{E} = S^TES$ in place of X and E respectively. By construction, $\widetilde{X} = \mathrm{diag}[\lambda_0 I_m, Y]$ for some matrix Y. Write $\widetilde{E} = \begin{bmatrix} E_1 & E_2 \\ E_3 & E_4 \end{bmatrix}$, where E_1 is of size $m \times m$. Then E_1 is real skew-symmetric and invertible (the invertibility of E_1 follows from the non-degeneracy of the quadratic form (iE_1x,x) on M). Replacing $\widetilde{E}$ by $(U^{-1} \oplus I)\widetilde{E}(U \oplus I)$ for a suitable orthogonal matrix U, and $\widetilde{X}$ by $(U^{-1} \oplus I)\widetilde{X}(U \oplus I)$, we can assume that, if $k = \frac{1}{2}m$,

$$E_1 = \mathrm{diag}\left(\begin{bmatrix} 0 & \sigma_1 \\ -\sigma_1 & 0 \end{bmatrix}, \cdots, \begin{bmatrix} 0 & \sigma_k \\ -\sigma_k & 0 \end{bmatrix}\right)$$

for some non-zero real numbers $\sigma_1,\cdots,\sigma_k$ (see Lemma 1.11). In particular, the number m is even. Now we shall obtain a contradiction to our assumption that X is E-stably u-diagonable by showing that the 2×2 matrix $\pm\begin{bmatrix} 1 & 0 \\ 0 & 1 \end{bmatrix}$ is not $\begin{bmatrix} 0 & \sigma \\ -\sigma & 0 \end{bmatrix}$-stably u-diagonable ($\sigma \in \mathbb{R} \setminus \{0\}$). Indeed, for small positive ε the matrix

$$I(\varepsilon) = \frac{1}{\sqrt{1-2\varepsilon^2}} \begin{bmatrix} 1-\varepsilon & \varepsilon \\ \varepsilon & 1+\varepsilon \end{bmatrix}$$

is close to I_2 and is $\begin{bmatrix} 0 & \sigma \\ -\sigma & 0 \end{bmatrix}$-orthogonal. However, the eigenvalues of $I(\varepsilon)$ are

$$\frac{1+\sqrt{2}\varepsilon}{\sqrt{1-2\varepsilon^2}}, \quad \frac{1-\sqrt{2}\varepsilon}{\sqrt{1-2\varepsilon^2}}$$

and are not unimodular. A similar argument applies to show that $-I_2$ is not $\begin{bmatrix} 0 & \sigma \\ -\sigma & 0 \end{bmatrix}$-stably u-diagonable. □

As we noted at the end of Section 1.7, only (i)⇒(iii) and (iii)⇔(iv) in Theorem 1.10 have to be proved.

<u>Proof of (i)⇒(iii) in Theorem 1.10</u>. Let X be an E-stably u-diagonable matrix, and assume that (iii) does not hold, i.e. there exists $\lambda_0 \in \sigma(X)$ for which the quadratic form (iEx,x) is not definite on $\mathrm{Ker}(X-\lambda_0 I)$. By Lemma 1.15 the numbers ± 1 are not eigenvalues of X. Let $A = i(X+I)(X-I)^{-1}$; then A is invertible and iE-selfadjoint, and the entries of A are pure imaginary. Since X is u-diagonable, A is similar to a real diagonal matrix. Also, the quadratic form (iEy,y), for $y \in \mathrm{Ker}(A-\mu I)$, where $\mu = i(\lambda_0+1)(\lambda_0-1)^{-1}$, is not definite. Now by Lemma 1.14 for every $\varepsilon > 0$ there exists an iE-selfadjoint pure imaginary matrix A_ε such that $||A_\varepsilon - A|| < \varepsilon$ and A_ε has non-real eigenvalues. Then the matrix $X_\varepsilon = (A_\varepsilon - i)^{-1}(A_\varepsilon + i)$ is real E-orthogonal matrix with non-unimodular eigenvalues, and $X_\varepsilon \to X$ as $\varepsilon \to 0$; a contradiction with the assumption on X.

<u>Proof of (iii)⇔(iv) in Theorem 1.10</u>. In view of Corollary 1.12, we can (and will) assume that

$$E = \begin{bmatrix} 0 & I \\ -I & 0 \end{bmatrix}.$$

Suppose X satisfies (iii). Let $\lambda_0 \in \sigma(X)$ and recall that λ_0 is unimodular and non-real (Lemma 1.15). Since (iEx,x) is definite on $\mathrm{Ker}(\lambda_0 I-X)$, there exists a basis $v_1,\cdots,v_k$ in $\mathrm{Ker}(\lambda_0 I-X)$ such that $(iEv_j,v_k) = \delta_{jk}\varepsilon(\lambda_0)$, where $\varepsilon(\lambda_0)$ is either 1 or -1. Then the vectors $\bar{v}_1,\cdots,\bar{v}_k$ form a basis in $\mathrm{Ker}(\bar{\lambda}_0 I-X)$ and $(iE\bar{v}_j,\bar{v}_k) = -\delta_{jk}\varepsilon(\lambda_0)$. Construct a basis in a similar way for each subspace $\mathrm{Ker}(\lambda_0 I-X)$, $\lambda_0 \in \sigma(X)$, on which the form

(iEx,x) is positive definite (so that the form (iEx,x) is negative definite on $\mathrm{Ker}(\bar{\lambda}_0 I - X)$). In this way we obtain a basis in $\mathbb{C}^n$ of the form $v_1,\cdots,v_m,v_{-1},\cdots,v_{-m}$, where $v_j = \bar{v}_{-j}$; $(iEv_j,v_k) = 0$ for $j \neq k$; and $(iEv_j,v_j) = 1$ for j positive, $(iEv_j,v_j) = -1$ for j negative.

Now let $e^{i\theta_j}$ $(1 \leq j \leq m;\ 0 < |\theta_j| < \pi)$ be the eigenvalue of X for which v_j is an eigenvector; the condition (iii) ensures that $\theta_j + \theta_k \neq 0$ $(1 \leq j,k \leq m)$. Also $e^{-i\theta_j}$ is the eigenvalue of X for which v_{-j} is an eigenvector. We have $Xv_k = e^{i\theta_k}v_k$; separating real and imaginary parts in this equality, we obtain

$$\left.\begin{aligned} Xg_k &= \cos\theta_k g_k - \sin\theta_k g_{k+m} \\ Xg_{k+m} &= \sin\theta_k g_k + \cos\theta_k g_{k+m} \end{aligned}\right\} \quad (1 \leq k \leq m) \tag{1.13}$$

where $g_k = \frac{1}{\sqrt{2}}(v_k+v_{-k})$; $g_{k+m} = \frac{1}{i\sqrt{2}}(v_k-v_{-k})$ are real. Let G be the real invertible matrix with columns $g_1,g_2,\cdots,g_{2m}$ (recalling that $n = 2m$). The equalities (1.13) give

$$XG = G\begin{bmatrix} \cos\theta & \sin\theta \\ -\sin\theta & \cos\theta \end{bmatrix},$$

where $\theta = \mathrm{diag}[\theta_1,\cdots,\theta_m]$. To prove (iv) it remains to show that G is E-orthogonal. But this follows immediately from the equalities

$$(iEg_j,g_k) = 0\ ; \quad |j-k| \neq m\ , \quad 1 \leq j,k \leq 2m\ ;$$

$$(iEg_k,g_{k+m}) = i\ , \quad (iEg_{k+m},g_k) = -i\ ; \quad k = 1,\cdots,m$$

taking into account that $E = \begin{bmatrix} 0 & I \\ -I & 0 \end{bmatrix}$.

Reversing the direction of this argument we find that (iv) implies (iii).

The proof of Theorem 1.10 is complete.

CHAPTER 2

APPLICATION TO DIFFERENTIAL AND DIFFERENCE EQUATIONS

The results concerning general perturbations developed in the preceding chapter are to be applied here to analysis of the solutions of systems of differential and difference equations with constant coefficients and certain symmetry properties. The main results concern description of equations for which all solutions are bounded and for which this property is retained after a small perturbation.

2.1 Differential Equations of First Order

Consider the differential equation

$$H \frac{dx}{dt} = iKx \;; \quad x = x(t) \in ¢^n , \quad t \in \mathbb{R} \tag{2.1}$$

where K and H are $n \times n$ hermitian matrices and H is invertible. This equation is the constant coefficient case of Hamiltonian equations with periodic coefficients considered in Part II. It has also been shown that, using the Floquet theorem, solutions of such constant coefficient systems can play an important part in the solution of systems with periodic coefficients.

It is clear that the matrix $A = H^{-1}K$ is H-selfadjoint, and the general solution of (2.1) is given by

$$x(t) = e^{iAt}x_0 , \quad x_0 = x(0) .$$

It is easily seen (by considering the Jordan normal form of A) that all solutions of (2.1) are bounded on the real line if and only if A is r-diagonable. This observation, together with Theorem 1.4, leads to the following result.

THEOREM 2.1 *The following statements are equivalent:*

(i) *the solutions of* (2.1) *are stably bounded, i.e. remain bounded under small hermitian perturbation of* H *and* K*;*

(ii) *the solutions of* (2.1) *are H-stably bounded, i.e. remain bounded under small hermitian perturbations of* K *(while* H *is kept fixed);*

(iii) $\sigma(H^{-1}K) \in \mathbb{R}$ *and quadratic form* (Hx,x) *is definite on the subspace* $\text{Ker}(\lambda_i I - H^{-1}K)$ *for every* $\lambda_i \in \sigma(H^{-1}K)$.

2.2 Differential Equations of Higher Order

Consider the system of differential equations with constant coefficients:

$$\sum_{j=0}^{\ell} i^j A_j \frac{d^j x}{dt^j} = 0 ; \quad t \in \mathbb{R} , \quad (i = \sqrt{-1}) \tag{2.2}$$

where A_j, $j = 0,\cdots,\ell$ are $n \times n$ hermitian matrices, and $x = x(t)$ is the unknown $\mathbb{C}^n$-valued function. We shall assume that the leading coefficient A_ℓ is invertible. The equation (2.1) studied in Section 2.1 is the particular case of (2.2) with $\ell = 1$.

Our main objective will be to characterize the cases when (2.2) has stably bounded solutions. Obviously, the properties of solutions of (2.2) are closely related to the properties of the matrix polynomial

$$L(\lambda) = \sum_{j=0}^{\ell} \lambda^j A_j , \tag{2.3}$$

and note that this matrix polynomial is hermitian: $A_j = A_j^*$ for $j = 0,1,$ $j = 0,1,\cdots,\ell$. A criterion for the stable boundedness of the solutions of (2.3) is given in terms of the polynomial $L(\lambda)$ in the next theorem.

THEOREM 2.2 *Equation* (2.2) *has stably bounded solutions (i.e. the solutions of* (2.2) *are bounded and remain bounded under any sufficiently small hermitian perturbations of the coefficients* $A_0,\cdots,A_\ell$*) if and only if* $\sigma(L)$ *is real and for every* $\lambda_0 \in \sigma(L)$*, the quadratic form* $(L'(\lambda_0)x,x)$ *is definite on the subspace* $\text{Ker}\, L(\lambda_0)$.

Here $L'(\lambda)$ denotes the derivative of $L(\lambda)$ with respect to λ. Theorem 2.2 will follow as an easy corollary of certain stability properties of the hermitian matrix polynomial $L(\lambda)$, which are independently interesting.

In Chapter II.2 some of the main ideas of the theory of matrix polynomials and hermitian matrix polynomials have been introduced, and their connections with the solutions of corresponding differential equations with constant coefficients have been developed.

Recall that the polynomial (2.3) is said to have *simple structure* if all its eigenvalues are real and all elementary divisors of $L(\lambda)$ are linear (the latter property means that $\dim \operatorname{Ker} L(\lambda_0)$ coincides with the multiplicity of λ_0 as a zero of $\det L(\lambda)$ for all $\lambda_0 \in \sigma(L)$). If these properties also hold for all polynomials obtained from (2.3) by sufficiently small hermitian perturbations of its coefficients $A_0,\cdots,A_\ell$, then the polynomial (2.3) is said to have *stable simple structure*. For example, every polynomial $L(\lambda)$ of the form (2.3) with $n\ell$ *distinct* real eigenvalues, has stable simple structure.

As an illustration, consider the "gyroscopic" systems arising in mechanics. In this case equation (2.3) applies with $\ell = 2$, A_2 positive definite, $A_1^* = A_1$ and A_0 negative definite. Then $L(\lambda)$ has stable simple structure (see Section 7.7 in [30a] and Ex. 7(a), p. 91 in [12] for the details).

The following theorem describes polynomials with stable simple structure in terms of definiteness of certain quadratic forms; an analogue of the description of Theorem 1.4.

THEOREM 2.3 *The polynomial* $L(\lambda)$ *given by* (2.3) *has stable simple structure if and only if* $\sigma(L) \subset \mathbb{R}$ *and for every* $\lambda_0 \in \sigma(L)$ *the quadratic form* $(L'(\lambda_0)x,x)$ *is definite on the subspace* $\operatorname{Ker} L(\lambda_0)$.

Note that Theorem 1.4 (more exactly, equivalence (i)⟺(iii) there) is a particular case of Theorem 2.3 which may be obtained by putting $L(\lambda) = \lambda H - HA$. Note also that the definiteness of the quadratic form $(L'(\lambda_0)x,x)$ on $\operatorname{Ker} L(\lambda_0)$ implies that all elementary divisors of L corresponding to λ_0 are linear (see Proposition II.2.2).

Theorem 2.2 follows from Theorem 2.3 immediately. Indeed, the solutions of (2.2) are bounded if and only if $L(\lambda)$ has simple structure (Theorem II.2.3).

<u>Proof of Theorem 2.3</u>. Together with the polynomial $L(\lambda)$ consider the block matrices

$$C_L = \begin{bmatrix} 0 & I & 0 & \cdots & 0 \\ 0 & 0 & I & \cdots & 0 \\ \vdots & & & \ddots & \vdots \\ 0 & 0 & 0 & \cdots & I \\ -A_\ell^{-1}A_0 & -A_\ell^{-1}A_1 & \cdots & & -A_\ell^{-1}A_{\ell-1} \end{bmatrix}; \quad B_L = \begin{bmatrix} A_1 & \cdots & A_{\ell-1} & A_\ell \\ \vdots & & \cdot^{\cdot^{\cdot}} & \vdots \\ A_{\ell-1} & A_\ell & 0 \cdots & 0 \\ A_\ell & 0 & \cdots & 0 \end{bmatrix} \tag{2.4}$$

Then C_L is B_L-selfadjoint; also the eigenvalues of $L(\lambda)$ coincide with those of C_L, and the elementary divisors of λ_0 as an eigenvalue of $L(\lambda)$, and as an eigenvalue of $\lambda I - C_L$, are the same (see Chapter II.2).

Assume that the quadratic form $(L'(\lambda_0)x,x)$ is definite on the subspace $\operatorname{Ker} L(\lambda_0)$ for every $\lambda_0 \in \sigma(L)$ so that, in particular, L has simple structure. Then C_L is r-diagonable and, moreover, the quadratic form $(B_L y,y)$ is definite on the subspace $\operatorname{Ker}(\lambda_0 I - C_L)$, for every $\lambda_0 \in \sigma(L)$ (see Proposition II.2.2). Hence, by Theorem 1.4, $L(\lambda)$ has stable simple structure.

Assume now that $L(\lambda)$ has simple structure, but for some $\lambda_0 \in \sigma(L)$, the quadratic form $(L'(\lambda_0)x,x)$ is not definite on $\operatorname{Ker} L(\lambda_0)$. We shall prove that by small hermitian perturbations of the coefficients of L one can make some eigenvalues of the perturbed polynomial non-real (and this will show that L does not have stable simple structure). Let

$$\mathbb{C}^n = \operatorname{Ker} L(\lambda_0) \oplus (\operatorname{Ker} L(\lambda_0))^{\perp} ;$$

and with respect to this decomposition write:

$$L(\lambda) = \begin{bmatrix} D_1(\lambda-\lambda_0) + E_1(\lambda)(\lambda-\lambda_0)^2 & D_2^*(\lambda-\lambda_0) + (E_2(\bar{\lambda}))^*(\lambda-\lambda_0)^2 \\ D_2(\lambda-\lambda_0) + E_2(\lambda)(\lambda-\lambda_0)^2 & E_3(\lambda) \end{bmatrix} .$$

Here $D_1 = D_1^*$; $E_1(\lambda)$, $E_2(\lambda)$, $E_3(\lambda)$ are matrix polynomials. Since all elementary divisors of L corresponding to λ_0 are linear, it follows that D_1 is invertible, and since $(L'(\lambda_0)x,x)$ is not definite on $\operatorname{Ker}(L(\lambda_0))$, D_1 has both negative and positive eigenvalues.

Replacing $L(\lambda)$ by $M(\lambda) = WL(\lambda)W^*$, where $W = \begin{bmatrix} I & 0 \\ -D_2D_1^{-1} & I \end{bmatrix}$, we can assume that $D_2 = 0$. Let x, y be orthonormal eigenvectors of D_1 corresponding to a positive eigenvalue α and a negative eigenvalue β, respectively. With respect to the decomposition $\mathbb{C}^n = \operatorname{Span}\{x\} + \operatorname{Span}\{y\} + (\operatorname{Span}\{x,y\})^{\perp}$ write:

$$M(\lambda) = \begin{bmatrix} \alpha & 0 & 0 \\ 0 & \beta & 0 \\ 0 & 0 & 0 \end{bmatrix}(\lambda-\lambda_0) + F(\lambda)(\lambda-\lambda_0)^2 + \begin{bmatrix} 0 & 0 & 0 \\ 0 & 0 & 0 \\ 0 & 0 & G(\lambda) \end{bmatrix}$$

where $F(\lambda)$, $G(\lambda)$ are selfadjoint matrix polynomials and the lower right entry of $F(\lambda)$ is zero. Now let ζ be a small positive number and put

$$M_\zeta(\lambda) = M(\lambda) + \begin{bmatrix} 0 & \alpha\beta & 0 \\ \alpha\beta & 0 & 0 \\ 0 & 0 & 0 \end{bmatrix}\zeta - F(\lambda)\alpha\beta\zeta^2 .$$

It is easily verified that $\lambda_0 + i\zeta(\alpha|\beta|)^{\frac{1}{2}}$ is a zero of $(\lambda-\lambda_0)^2 - \alpha\beta\zeta^2$ and a non-real eigenvalue of $M_\zeta(\lambda)$. □

Let $L(\lambda)$ be a hermitian matrix polynomial (2.3) which has simple structure but not stable simple structure. As the proof of Theorem 2.3 shows, there exists a perturbation of $L(\lambda)$ of type $L(\lambda,\zeta) = L(\lambda) + \zeta M_1 + \zeta^2 M_2(\lambda)$, where $\zeta > 0$ is small, with the following properties (we assume $n > 1$):

(i) M_1 is a hermitian matrix of rank 2;

(ii) $M_2(\lambda)$ is a hermitian matrix polynomial such that $\operatorname{Im} M_2(\lambda) \subset Q$ for all $\lambda \in \mathbb{C}$, where Q is a fixed (i.e. independent of λ) $\min(n,4)$-dimensional subspace in $\mathbb{C}^n$ containing $\operatorname{Im} M_1$;

(iii) $L(\lambda,\zeta)$ has a non-real eigenvalue $\lambda_0(\zeta)$ of the form $\lambda_0(\zeta) = \lambda_0 + i\zeta\gamma$, for all sufficiently small positive ζ, and where γ is a fixed real number.

2.3 The Strongly Hyperbolic Case

In this section we shall describe a class of differential equations (2.2) with stably bounded solutions.

An $n \times n$ matrix polynomial $L(\lambda) = \sum_{j=0}^{\ell} \lambda^j A_j$ will be called *strongly hyperbolic* if the leading coefficient A_ℓ is positive definite and for every $x \in \mathbb{C}^n \setminus \{0\}$ the scalar equation

$$(L(\lambda)x,x) = \sum_{j=0}^{\ell} \lambda^j (A_j x,x) = 0 \tag{2.5}$$

has ℓ distinct real zeros. All coefficients of a strongly hyperbolic polynomial $L(\lambda) = \sum_{j=0}^{\ell} \lambda^j A_j$ are hermitian (indeed, all the numbers $(A_j x,x)$, $x \in \mathbb{C}^n \setminus \{0\}$, $j = 1,\cdots,\ell$ are real, which implies the hermitian property of each A_j; see, e.g., [30b]).

THEOREM 2.4 *The differential equation*

$$\sum_{j=0}^{\ell} i^j A_j \frac{d^j x}{dt^j} = 0 , \quad t \in \mathbb{R}$$

has stably bounded solutions provided the matrix polynomial $L(\lambda) = \sum_{j=0}^{\ell} \lambda^j A_j$ *is strongly hyperbolic.*

The proof will be based on properties of strongly hyperbolic polynomials which will first be developed.

Let $L(\lambda)$ be a strongly hyperbolic $n \times n$ matrix polynomial of degree ℓ, and let

$$\lambda_1(x) > \cdots > \lambda_\ell(x)$$

be the zeros of $(L(\lambda)x,x) = 0$, $x \in ¢^n \setminus \{0\}$. As all zeros of the scalar polynomial $(L(\lambda)x,x) = 0$, $x \neq 0$ are real and distinct, and its leading coefficient is positive, we obtain

$$(-1)^{j-1}(L'(\lambda_j(x))x,x) > 0 , \quad j = 1,\cdots,\ell ; \quad x \neq 0 . \tag{2.6}$$

For $j = 1,2,\cdots,\ell$ define

$$\Delta_j = \{\lambda_j(x) \mid x \in ¢^n \setminus \{0\}\} = \{\lambda_j(x) \mid x \in ¢^n , ||x|| = 1\} .$$

The set Δ_j is called the j-th *spectral zone* of $L(\lambda)$. The spectral zones are compact and connected (because Δ_j is the image of the unit sphere under the continuous map λ_j, and the unit sphere is compact and connected); so in fact $\Delta_j = [\alpha_j,\beta_j]$ for some real numbers $\alpha_j \leqslant \beta_j$. Further, the spectrum $\sigma(L)$ of L is real and

$$\sigma(L) \subset \sum_{j=1}^{\ell} \Delta_j . \tag{2.7}$$

Indeed, for $\lambda \in ¢ \setminus \bigcup_{j=1}^{\ell} \Delta_j$ and x on the unit sphere we have

$$||L(\lambda)x|| \geq |(L(\lambda)x,x)| = (A_n x,x) \prod_{j=1}^{\ell} |\lambda-\lambda_j(x)| > 0 ,$$

so $L(\lambda)$ is invertible which means that $\lambda \notin \sigma(L)$.

The following property of the spectral zones is deeper.

LEMMA 2.5 *Different spectral zones of a strongly hyperbolic matrix polynomial do not intersect.*

PROOF. We have to prove that $\Delta_j \cap \Delta_k = \emptyset$ for $j \neq k$. It will suffice to show that $\Delta_j \cap \Delta_{j+1} = \emptyset$ for $j = 1,\cdots,\ell-1$. Assume the contrary, so $\Delta_j \cap \Delta_{j+1} \neq \emptyset$ for some j. So there exist vectors x and y of norm 1 and a real number $\alpha = \lambda_j(x) = \lambda_{j+1}(y)$ such that $(L(\alpha)x,x) = (L(\alpha)y,y) = 0$. By (2.6) we have $(L'(\alpha)x,x)(L'(\alpha)y,y) < 0$. Define the $n \times n$ matrix $C = L'(\alpha) + iL(\alpha)$. Then the non-zero numbers (Cx,x) and (Cy,y) are real and have opposite signs. Now we use the Toeplitz-Hausdorff theorem according to which the set $\{(Xf,f) \mid f \in \mathbb{C}^n, ||f|| = 1\}$ is convex for every $n \times n$ matrix X (see [16]). So there exists a vector z of norm 1 such that $(Cz,z) = 0$. This means that $(L(\alpha)z,z) = (L'(\alpha)z,z) = 0$; so α is at least a double zero of $(L(\lambda)z,z) = 0$, which contradicts the strongly hyperbolic property of $L(\lambda)$. □

Proof of Theorem 2.4. By Theorem 2.2, taking into account (2.7), we have only to check that for every $\lambda_0 \in \sigma(L)$ the quadratic form $(L'(\lambda_0)x,x)$ is definite on $\operatorname{Ker} L(\lambda_0)$. Evidently, $\lambda_0 \in \Delta_j$ for some j. Moreover, by Lemma 2.5 we have $\lambda_0 = \lambda_j(x)$ for every $x \in \operatorname{Ker} L(\lambda_0) \setminus \{0\}$. Now (2.6) shows that the quadratic form $(L'(\lambda_0)x,x)$ is positive definite on $\operatorname{Ker} L(\lambda_0)$ if j is odd and negative definite if j is even. □

2.4 Difference Equations

Consider the difference equation

$$A_0x_i + A_1x_{i+1} + \cdots + A_\ell x_{i+\ell} = 0, \quad i = 0,1,\cdots, \tag{2.8}$$

where $\{x_i\}_{i=0}^{\infty}$ is a sequence of n-dimensional vectors to be found, and $A_0,\cdots,A_\ell$ are given complex $n \times n$ matrices such that A_ℓ is invertible and $A_i^* = A_{\ell-i}$, $i = 0,\cdots,\ell$. The integer ℓ is supposed to be even. Such difference equations were considered in Section II.2.4. As we observed there, the solutions of (2.8) are intimately related to properties of the associated matrix polynomial $L(\lambda) = \sum_{j=0}^{\ell} \lambda^j A_j$.

We are interested in the case when the solutions of (2.8) are *stably bounded*, i.e. all solutions of (2.8) are bounded, and all solutions of every system

$$\tilde{A}_0y_i + \tilde{A}_1y_{i+1} + \cdots + \tilde{A}_\ell y_{i+\ell} = 0, \quad i = 0,1,\cdots,$$

with $\tilde{A}_j^* = \tilde{A}_{\ell-j}$ and $||\tilde{A}_j - A_j||$ small enough (for $j = 0,1,\cdots,\ell$) are also bounded.

THEOREM 2.6 *The solutions of* (2.8) *are stably bounded if and only if the spectrum of the associated polynomial* $L(\lambda)$ *lies on the unit circle, and the quadratic form*

$$(x, i\lambda_0(\lambda^{-\frac{\ell}{2}} L(\lambda))^{(1)}(\lambda_0)x), \quad x \in \operatorname{Ker} L(\lambda_0) \tag{2.9}$$

is either positive definite or negative definite for every $\lambda_0 \in \sigma(L)$.

PROOF. We shall use the following matrices introduced in Section II.2.4:

$$C_L = \begin{bmatrix} 0 & I & & \cdots & 0 \\ 0 & 0 & I & \cdots & 0 \\ \cdot & & & \ddots & \cdot \\ \cdot & & & & \cdot \\ 0 & 0 & 0 & \cdots & I \\ -A_\ell^{-1}A_0 & -A_\ell^{-1}A_1 & \cdots & & -A_\ell^{-1}A_{\ell-1} \end{bmatrix}$$

(this is just the companion matrix for the associated matrix polynomial $L(\lambda)$);

$$\hat{B}_L = i \left[\begin{array}{cccc|cccc} & & & & A_\ell & & & \\ & & & & A_{\ell-1} & \ddots & & 0 \\ & & 0 & & \vdots & \ddots & \ddots & \\ & & & & A_{k+1} & \cdots & A_{\ell-1} & A_\ell \\ \hline -A_0 & -A_1 & \cdots & -A_{k-1} & & & & \\ & -A_0 & \ddots & \vdots & & & & \\ & & \ddots & \vdots & & & 0 & \\ & 0 & & -A_0 & & & & \end{array}\right]; \quad k = \frac{\ell}{2}.$$

Assume that the spectrum of $L(\lambda)$ is unimodular, and the quadratic form (2.9) is definite for every $\lambda_0 \in \sigma(C_L)$. This means (see (II.2.20)) that for every $\lambda_0 \in \sigma(C_L)$ the quadratic form $(x,\hat{B}_L x)$, $x \in \mathrm{Ker}(\lambda_0 I - C_L)$ is either positive definite or negative definite. By Theorem 1.6 there exists an $\varepsilon > 0$ such that every H-unitary matrix U with $||U-C_L|| + ||H-\hat{B}_L|| < \varepsilon$ is similar to a diagonal matrix with unimodular spectrum. In particular, this is true for every matrix

$$\begin{bmatrix} 0 & I & \cdot & \cdot & \cdot & 0 \\ \cdot & \cdot & \cdot & & & \cdot \\ \cdot & \cdot & & \cdot & & \cdot \\ \cdot & \cdot & & & \cdot & \cdot \\ 0 & 0 & \cdot & \cdot & \cdot & I \\ -A_\ell'^{-1}A_0' & & \cdot & \cdot & \cdot & -A_\ell'^{-1}A_{\ell-1}' \end{bmatrix}$$

with $A_j'^{*} = A_{\ell-j}'$ for $j = 0,\cdots,\ell$ and $\sum_{j=0}^{\ell} ||A_j'-A_j||$ sufficiently small.

By Theorem II.2.3 the solutions of (2.8) are stably bounded.

Assume now that the solutions of (2.8) are stably bounded. Then using Theorem II.2.3 again, the polynomial $L(\lambda)$ has only unimodular eigenvalues with all elementary divisors linear, and every polynomial $\sum_{j=0}^{\ell} \lambda^j A_j'$ with $\sum_{j=0}^{\ell} ||A_j'-A_j||$ sufficiently small and $A_j' = A_{\ell-j}'^{*}$, $j = 0,\cdots,\ell$ has this property. Choose $w \in \mathbb{C}$, $|w| = 1$ such that $L(w)$ is invertible, and put

$$R(\lambda) = \sum_{j=0}^{\ell} (1+i\lambda)^j (1-i\lambda)^{\ell-j} B_j ,$$

where B_j are the coefficients of the polynomial $L_w(\lambda) = (-w)^{-\frac{\ell}{2}} L(-w\lambda) = \sum_{j=0}^{\ell} \lambda^j B_j$. Note that $L_w(\lambda)$ is selfadjoint with respect to the unit circle. Then $R(\lambda)$ is a hermitian (relative to the real line) matrix polynomial of degree ℓ with invertible leading coefficient. Clearly, all eigenvalues of $R(\lambda)$ are real with all elementary divisors linear. Moreover, all eigenvalues of hermitian matrix polynomials of degree ℓ which are close to $R(\lambda)$ also have this property. (This follows from the corresponding property for $L(\lambda)$, because

$$L(\mu) = \left[\frac{1-w^{-1}\mu}{2}\right]^{\ell} (-w)^k \, R\left(\frac{i(1+w^{-1}\mu)}{1-w^{-1}\mu}\right) , \quad k = \frac{\ell}{2} .)$$

By Theorem 2.3 the form $(x,R'(\lambda_0)x)$ is either positive definite or negative definite on $\operatorname{Ker} R(\lambda_0)$ for every $\lambda_0 \in \sigma(R)$. A computation shows that, putting $\hat{L}(\lambda) = \lambda^{-k}L(\lambda)$, we have

$$R'(\lambda) = 2\lambda k(1+\lambda^2)^{k-1} \hat{L}\left(\frac{-w(1+i\lambda)}{1-i\lambda}\right) + (-w)(1+\lambda^2)^k \frac{2i}{(1-i\lambda)^2} \hat{L}^{(1)}\left(\frac{-w(1+i\lambda)}{1-i\lambda}\right) .$$

Let $\lambda_0 \in \sigma(R) \cap \mathbb{R}$; then $\mu_0 \overset{\text{def}}{=} \frac{-w(1+i\lambda)}{1-i\lambda}$ is a unimodular eigenvalue of $L(\lambda)$ and $\operatorname{Ker} R(\lambda_0) = \operatorname{Ker} L(\mu_0) = \operatorname{Ker} \hat{L}(\mu_0)$. Hence, for every $x \in \operatorname{Ker} L(\mu_0)$ we have

$$i\mu_0 \hat{L}^{(1)}(\mu_0)x = \frac{1}{2}(1+\lambda_0^2)^{-k+1} R'(\lambda_0)x ,$$

so the form $(x, i\mu_0\hat{L}^{(1)}(\mu_0)x)$, $x \in \operatorname{Ker} \hat{L}(\mu_0)$ is positive definite or negative definite together with the form $(x,R'(\lambda_0)x)$, $x \in \operatorname{Ker} R(\lambda_0)$. □

2.5 Hamiltonian and Selfadjoint Equations

Consider the system of differential equations (as in Chapter II.1)

$$E \frac{dx}{dt} = iH(t)x , \tag{2.10}$$

where $E = E^*$ is an invertible complex matrix, and $H(t) = H(t)^*$ is a piecewise continuous complex matrix function with period ω. We have seen (Theorem II.1.15) that the solutions of (2.10) are bounded if and only if its monodromy matrix is similar to a diagonable matrix with unimodular eigenvalues. We say that the equation (2.10) has *stably bounded solutions* if any system

$$E \frac{dx}{dt} = i\widetilde{H}(t)x$$

with piecewise continuous hermitian complex matrix function $H(t)$ with period ω such that $\max_{0 \leq t \leq \omega} ||H(t)-\widetilde{H}(t)||$ is small enough also has only bounded solutions. In particular, the system (2.10) itself has only bounded solutions. Combining the dependence of the monodromy matrix of (2.10) on $H(t)$ developed in Section II.1.2 with Theorem 1.7, we obtain the following result.

THEOREM 2.7 *The equation* (2.10) *has stably bounded solution if and only if its monodromy matrix* U *has only unimodular eigenvalues, and the quadratic form* (Ex,x) *is either positive definite or negative definite on* $\operatorname{Ker}(\lambda_0 I-U)$, *for every* $\lambda_0 \in \sigma(U)$.

The real analogue of this result holds also (see Section II.1.4).

For selfadjoint systems of first order differential equations with periodic coefficients of the form

$$Q(t) \frac{dx}{dt} - (iH(t) - \frac{1}{2} \frac{dQ(t)}{dt})x = 0 , \tag{2.11}$$

where $Q(t) = Q(t)^*$ is invertible and piecewise continuously differentiable, $H(t) = H(t)^*$ is piecewise continuous, and both $Q(t)$ and $H(t)$ have a common period ω, a result similar to Theorem 2.7 holds (see Section II.1.6). Namely, the system (2.11) has stably bounded solutions, i.e. the solutions of any system

$$\widetilde{Q}(t) \frac{dx}{dt} - (i\widetilde{H}(t) - \frac{1}{2} \frac{d\widetilde{Q}(t)}{dt} x = 0$$

of type (2.11) with

$$\max_{0\leq t\leq\omega} ||\widetilde{Q}(t) - Q(t)|| + \max_{0\leq t\leq\omega} ||\frac{d\widetilde{Q}(t)}{dt} - \frac{dQ}{dt}|| + \max_{0\leq t\leq\omega} ||\widetilde{H}(t) - H(t)||$$

small enough, has bounded solutions, if and only if the monodromy matrix U of (2.11) has its eigenvalues on the unit circle, and the quadratic form $(Q(0)x,x)$ is either positive definite or negative definite on $\mathrm{Ker}(\lambda_0 I-U)$ for each $\lambda_0 \in \sigma(U)$. This characterization of equations (2.11) with stably bounded solutions holds in the real case as well.

CHAPTER 3

POSITIVE PERTURBATIONS

This chapter is devoted to the study of perturbations of H-selfadjoint matrices by the addition of a matrix which is definite with respect to H; a natural generalization of the problem of definite perturbations of hermitian matrices in the classical sense. It is shown that under such a perturbation, real eigenvalues of the unperturbed matrix remain real and the directions of their "motions" on the real axis can be described. Applications are made to Hamiltonian systems of differential equations with constant coefficients.

3.1 Positive Perturbations of H-Selfadjoint Matrices

Let H be an $n \times n$ invertible hermitian matrix, and let A be H-selfadjoint. An eigenvalue λ_0 of A will be called *simple* if $\dim \operatorname{Ker}(\lambda_0 I-A)$ coincides with the multiplicity of λ_0 as a zero of $\det(\lambda I-A)$, or, in other words, if the elementary divisors of $\lambda I - A$ corresponding to λ_0 are all linear. If λ_0 is a real simple eigenvalue of A; then the quadratic form (Hx,x) is non-degenerate on $\operatorname{Ker}(\lambda_0 I-A)$, i.e. zero is the only vector $x_0 \in \operatorname{Ker}(\lambda_0 I-A)$ with the property that $(Hx_0,y) = 0$ for all $y \in \operatorname{Ker}(\lambda_0 I-A)$. This follows easily from the canonical form for pairs (A,H) in which A is H-selfadjoint.

Let $r_+(\lambda_0)$ (resp. $r_-(\lambda_0)$) be the number of positive (resp. negative) squares in the canonical representation of the quadratic form (Hx,x) on $\operatorname{Ker}(\lambda_0 I-A)$, where λ_0 is simple. So, in particular,

$$r_+(\lambda_0) + r_-(\lambda_0) = \dim \operatorname{Ker}(\lambda_0 I-A) ,$$

and is also the algebraic multiplicity of λ_0 as an eigenvalue of A.

If $r_-(\lambda_0) = 0$ $(r_+(\lambda_0) = 0)$, we say that λ_0 is a *positive definite* (*negative definite*) eigenvalue of A. Recall that a matrix A is H-*positive* if $[Ax,x] > 0$ for all $x \neq 0$, where $[x,y] = (Hx,y)$. In other words, A is H-positive if and only if the matrix HA is positive definite.

THEOREM 3.1 *Let* A *be H-selfadjoint, and let* λ_0 *be a simple real eigenvalue of* A. *Let* Γ *be any contour such that* λ_0 *is the only eigenvalue of* A *inside or on* Γ. *Then there exists an* $\varepsilon > 0$ *with the following property: For every H-positive matrix* A_0 *with* $||A_0|| < \varepsilon$, *the H-selfadjoint matrix* $A + A_0$ *has exactly* $r \overset{\text{def}}{=} r_+(\lambda_0) + r_-(\lambda_0)$ *eigenvalues inside* Γ *(counting multiplicities), and all of them are real. Moreover,* $r_+(\lambda_0)$ *of these eigenvalues (counting multiplicities) are greater than* λ_0, *and* $r_-(\lambda_0)$ *of them are less than* λ_0. *Furthermore, every eigenvalue* $\tilde{\lambda}$ *of* $A + A_0$ *inside* Γ *is simple and positive definite if* $\tilde{\lambda} > \lambda_0$, *or negative definite if* $\tilde{\lambda} < \lambda_0$.

PROOF. First observe that λ is an eigenvalue of A if and only if it is an eigenvalue of the hermitian pencil $\lambda H - HA$, in the sense that $\det(\lambda H-HA) = 0$. It will be convenient to prove the theorem in this context of hermitian pencils. Thus, writing $B = HA$ and $B_0 = HA_0$ it will be proved that if B_0 is positive definite (in the classical sense) and $||B_0|| < \varepsilon$ then the eigenvalues of linear polynomial $\lambda H - (B+B_0)$ have the properties indicated in the theorem. The conclusions of the theorem will then hold on replacing ε by $||H^{-1}||^{-1}\varepsilon$, i.e. $||B_0|| < ||H^{-1}||^{-1}\varepsilon$ will imply $||A_0|| < \varepsilon$.

Since B_0 is positive definite there is an invertible matrix S such that $B_0 = S^*S$ and then we may write

$$\lambda H - (B+B_0) = S^*\{\lambda H' - (B'+I)\}S$$

where $H' = S^{*-1}HS^{-1}$ and $B' = S^{*-1}BS^{-1}$. This congruence implies that when λ_0 is an eigenvalue of $\lambda H - B$ with associated parameters $r_+(\lambda_0)$, $r_-(\lambda_0)$ as in the statement of the theorem, then it is also an eigenvalue of $\lambda H' - B'$ with the same parameters $r_+(\lambda_0)$, $r_-(\lambda_0)$. Furthermore, with Γ defined as in the theorem, λ_0 is the only eigenvalue of $\lambda H' - B'$ inside or on Γ.

Theorem II.3.3 shows that $\lambda H' - B'$ admits a representation

$$\lambda H' - B' = U(\lambda)\operatorname{diag}[\mu_1(\lambda),\cdots,\mu_n(\lambda)]U(\lambda)^*$$

for all $\lambda \in \mathbb{R}$, where $U(\lambda)$ is an analytic matrix function with unitary values $(U(\lambda)U(\lambda)^* = I$ for $\lambda \in \mathbb{R})$; and $\mu_1(\lambda),\cdots,\mu_n(\lambda)$ are real-valued analytic functions of the real variable λ. Since λ_0 is a simple eigenvalue of $\lambda H - B$, and hence $\lambda H' - B'$, it follows that λ_0 is a simple zero of exactly r functions $\mu_{i_k}(\lambda)$, $k = 1,2,\cdots,r$ among $\mu_1(\lambda),\cdots,\mu_n(\lambda)$ and, moreover, exactly $r_+(\lambda_0)$ and $r_-(\lambda_0)$ of the r derivatives $\mu'_{i_k}(\lambda_0)$ are positive, and negative, respectively. It is supposed, for simplicity, that $\mu_1(\lambda) = \cdots = \mu_r(\lambda_0) = 0$ and $\mu'_j(\lambda_0)$ is positive for $j = 1,2,\cdots,r_+(\lambda_0)$, and negative for $j = r_+(\lambda_0)+1,\cdots,r$.

Choose an $\varepsilon > 0$ so that, for $||B_0|| < \varepsilon$, $\lambda H - (B+B_0)$, and hence $\lambda H' - (B'+I)$, has exactly r eigenvalues inside Γ. By continuity of the eigenvalues the same is true for $\lambda H' - (B'+\delta I)$ for any $\delta \in [0,1]$. Furthermore, defining $\mu_j(\lambda,\delta) = \mu_j(\lambda) - \delta$, we have

$$\lambda H' - (B'+\delta I) = U(\lambda)\mathrm{diag}[\mu_1(\lambda,\delta),\cdots,\mu_n(\lambda,\delta)]U(\lambda)^* .$$

Let $[\alpha,\beta]$ be the interval of $\mathbb{R}$ consisting of points inside and on Γ so that $\lambda_0 \in (\alpha,\beta)$. Fix j and consider the family of functions $\mu_j(\lambda,\delta)$ defined on $[\alpha,\beta]$; one for each $\delta \in [0,1]$. Clearly, $\mu_j(\lambda,0)$ has a simple zero at $\lambda = \lambda_0$ and for any $\delta \in [0,1]$, $\mu_j(\alpha,\delta) \neq 0$ and $\mu_j(\beta,\delta) \neq 0$. As $\mu_j(\alpha,\delta)$, $\mu_j(\beta,\delta)$ are continuous non-zero functions of $\delta \in [0,1]$ and $\mu_j(\alpha,0)\mu_j(\beta,0) < 0$ for $j = 1,\cdots,r$, also $\mu_j(\alpha,\delta)\mu_j(\beta,\delta) < 0$ for $j = 1,\cdots,r$ and $\delta \in [0,1]$. Thus, for each $\delta \in [0,1]$, $\mu_j(\lambda,\delta)$ has at least one zero in (α,β). The same applies for each j from 1 to r. Since the total number of zeros (counting multiplicities) of $\mu_j(\lambda,\delta)$, $j = 1,\cdots,r$ in the interval $[\alpha,\beta]$ does not exceed r, each function $\mu_j(\lambda,\delta)$ for a fixed j between 1 and r and fixed $\delta \in [0,1]$ has exactly one simple zero $\lambda_j(\delta)$ in (α,β). In particular, it follows that for any $\delta \in [0,1]$ the r eigenvalues of $\lambda H - (B+\delta B_0)$ which are inside Γ are all real.

Since $\lambda_i(\delta)$ is a simple zero of $\mu_j(\lambda,\delta)$, $j = 1,2,\cdots,r$, which depends continuously on $\delta \in [0,1]$, the derivative

$$\frac{d\mu_j}{d\lambda}(\lambda,\delta) \overset{\mathrm{det}}{=} \nu_j(\delta)$$

is either positive for all $\delta \in [0,1]$ or negative for all $\delta \in [0,1]$.

Since $\nu_j(0) > 0$ for $j = 1,\cdots,r_+(\lambda_0)$ and $\nu_j(0) < 0$ for $j = r_+(\lambda_0)+1, \cdots,r$, we conclude that for any $\delta \in [0,1]$ $\nu_j(\delta) > 0$ for $j = 1,\cdots,r_+(\lambda_0)$ and $\nu_j(\delta) < 0$ for $j = r_+(\lambda_0)+1,\cdots,r$.

Now let $\delta \in (0,1]$ and, since $\mu_j(\lambda_0,\delta) = \mu_j(\lambda_0) - \delta < 0$, it follows that $\lambda_j(\delta) > \lambda_0$ for $j = 1,2,\cdots,r_+(\lambda_0)$ and $\lambda_j(\delta) < \lambda_0$ for $j = r_+(\lambda_0)+1,\cdots,n$. Putting $\delta = 1$ the theorem is proved. □

Note that the ε appearing in the statement of Theorem 3.1 can be estimated. Using Theorem 2.2 in [20] it is found that one can use any ε satisfying

$$0 < \varepsilon < (\sup_{\lambda\in\Gamma}||\lambda H - HA)^{-1}||)^{-1}\cdot||H^{-1}||^{-1} .$$

A result similar to Theorem 3.1 holds for H-negative perturbations of A; in this case the words "greater than" and "less than" in the statement of the theorem must be interchanged.

For positive definite H the statement of Theorem 3.1 reduces to the following well-known fact (which is not difficult to prove using the variational properties of the eigenvalues of a hermitian matrix): Let A be an hermitian $n \times n$ matrix with eigenvalues $\lambda_1 \leq \cdots \leq \lambda_n$, then the eigenvalues $\mu_1 \leq \cdots \leq \mu_n$ of the hermitian matrix $A + A_0$ with positive definite A_0 satisfy the inequalities

$$\mu_i > \lambda_i , \quad i = 1,\cdots,n .$$

3.2 Hamiltonian Systems of Positive Type with Constant Coefficients

Consider the Hamiltonian system of differential equations with constant coefficients:

$$E\frac{dx}{dt} = i(H+H_0)x , \tag{3.1}$$

where $E = E^*$ is invertible, and $H = H^*$, $H_0 = H_0^*$. Formally, we shall regard the matrices H and H_0 as matrix functions with period ω (although they are constant in fact), in order to put the system (3.1) in the context of Hamiltonian systems with periodic coefficients. In the system (3.1) the matrix H_0 is viewed as a small perturbation of H; so $||H_0||$ is small in some sense. We are to study the behaviour of the eigenvalues of the monodromy matrix of (3.1) (called the *multiplicators* of the system (3.1)) as H_0 changes. The case when H_0 is positive definite is of special interest, and the system is then said to be of *positive type*.

Assume now that all solutions of the system

$$E\frac{dx}{dt} = iHx , \quad t \in \mathbb{R} \tag{3.2}$$

are bounded. As we know, (Theorem II.1.15) in this case the multiplicators of (3.2) are unimodular, and for every multiplicator λ_0 the quadratic form (Ex,x) is non-degenerate on $\mathrm{Ker}(\lambda_0 I-X)$, where $X = \exp(\omega i E^{-1}H)$ is the monodromy matrix of (3.2). We say that the multiplicator λ_0 has *positive multiplicity* $r_+(\lambda_0)$ and *negative multiplicity* $r_-(\lambda_0)$ if the quadratic form (Ex,x) defined on $\mathrm{Ker}(\lambda_0 I-X)$ has $r_+(\lambda_0)$ positive squares and $r_-(\lambda_0)$ negative squares in its canonical form. In particular, $r_+(\lambda_0) + r_-(\lambda_0)$ coincides with the dimension of $\mathrm{Ker}(\lambda_0 I-X)$, which in turn is equal to the multiplicity of λ_0 as a zero of $\det(\lambda I-X)$. The multiplicator λ_0 is called of *positive* (resp. *negative*) type if the quadratic form (Ex,x) is positive (resp. negative) definite on $\mathrm{Ker}(\lambda_0 I-X)$.

THEOREM 3.2 *Let* (3.1) *be a Hamiltonian system with constant coefficients of positive type, and assume that all solutions of the unperturbed system* (3.2) *are bounded. Then for* $||H_0||$ *small enough all solutions of* (3.1) *are also bounded; moreover, if* λ_0 *is a multiplicator of* (3.2) *with positive multiplicity* $r_+(\lambda_0)$ *and negative multiplicity* $r_-(\lambda_0)$, *then exactly* $r_+(\lambda_0)$ *(resp.* $r_-(\lambda_0)$*) multiplicators of* (3.1) *in a neighbourhood of* λ_0 *are of positive (resp. negative) type and situated on the unit circle in the positive (resp. negative) direction from* λ_0.

PROOF. As $X = \exp(\omega i E^{-1}H)$ is the monodromy matrix of (3.2) then, for any multiplicator λ_0 of (3.2) there is an eigenvalue μ_0 of the matrix $A = E^{-1}H\omega$ for which $\lambda_0 = e^{i\mu_0}$. Observe that A is E-selfadjoint and $E^{-1}H_0\omega$ is E-positive for H_0 positive definite; moreover, $\mathrm{Ker}(\lambda_0 I-X) = \mathrm{Ker}(\mu_0 I-A)$.

Now apply Theorem 3.1 to the E-positive perturbation $E^{-1}H_0\omega$ of $E^{-1}H\omega$ and, after mapping the perturbed real eigenvalues μ on the unit circle with the map $\lambda = e^{i\mu}$, the theorem is obtained. □

Note that positive definiteness of H_0 is essential in Theorem 3.2; indeed, we know from Theorem 2.1 that if λ_0 is not of positive or negative type, then there exists $H_0 = H_0^*$ as small as we wish such that the perturbed system $E\frac{dx}{dt} = i(H+H_0)x$ has an unbounded solution. Theorem 3.2 shows, in particular, that this situation is impossible for positive definite perturbations.

Let λ_0 be a multiplicator of the system (3.2) with bounded solutions, and let $r_+(\lambda_0)$ (resp. $r_-(\lambda_0)$) be the positive (resp. negative) multiplicity of λ_0. It is possible to regard λ_0 as $r_+(\lambda_0) + r_-(\lambda_0)$ equal multiplicators; $r_+(\lambda_0)$ of them being of positive type and $r_-(\lambda_0)$ of them being of negative type. With this convention, one can reformulate Theorem 3.2 in more informal terms as follows: a multiplicator of positive (resp. negative) type of the system (3.2) with bounded solutions moves counterclockwise (resp. clockwise) on the unit circle as the matrix H is perturbed by a positive definite matrix.

These rules of motion of multiplicators of Hamiltonian systems hold also in the case of (non-trivially) periodic coefficients. M.G. Krein proved the following result in a more general setting (see [26a]):

THEOREM 3.3 *Let*

$$E\frac{dx}{dt} = iH(t)x \tag{3.3}$$

be a Hamiltonian system ($E = E^*$ *is invertible,* $H(t) = (H(t))^*$ *is piecewise continuous with period* ω) *with only bounded solutions. Let* $H_0(t)$ *be a positive definite piecewise continuous matrix function with period* ω. *Then for small* $\varepsilon > 0$, *solutions of the Hamiltonian system*

$$E\frac{dx}{dt} = i(H(t) + \varepsilon H_0(t))x \tag{3.4}$$

are all bounded, and a multiplicator of positive (resp. negative) type of the system (3.4) *moves counterclockwise (resp. clockwise) on the unit circle as* ε *increases* $(0 \leq \varepsilon < \varepsilon_0)$.

The conditions on $H_0(t)$ in Theorem 3.3 can be relaxed somewhat. Namely, it is sufficient to require that $(H_0(t)x,x) \geq 0$ for all $0 \leq t \leq \omega$ and all $x \in \mathbb{C}^n$, and zero is the only vector x_0 for which $(H_0(t)x_0,x_0) \equiv 0$.

The proof of Theorem 3.3 is beyond the scope of this book.

CHAPTER 4

PERTURBATIONS OF INVARIANT MAXIMAL NEUTRAL SUBSPACES

During analysis of the Riccati equation in Chapter II.4 it was found that maximal M-invariant, H-neutral subspaces play an important part. Indeed, they are used to construct all hermitian solutions. Similarly, they also play an important role in the study of symmetric factorizations of nonnegative matrix functions in Chapters II.2 and II.3. The applications of perturbation theory made in this chapter allow us to analyse the dependence of extremal hermitian solutions of the Riccati equation on its coefficients in the first case, and in the second case, the dependence of the factors on the parent function is studied.

4.1 Continuity of Invariant Maximal Neutral Subspaces

Let A be H-selfadjoint and suppose that N is an A-invariant, maximal H-neutral subspace. In discussion of the continuous dependence of N on A and H we confine attention to the case when N is the unique subspace with the above properties. This case has been analysed in Theorem I.3.22. Accordingly, it is assumed that the real Jordan blocks of A are all of even size, and that the sign characteristic of (A,H) corresponding to any real eigenvalue is made up entirely of +1's, or of -1's. Let $\mathbb{N}$ be the set of all pairs (A,H) for which A has these properties and A is H-selfadjoint.

Theorem I.3.22 shows, in particular, that for $(A,H) \in \mathbb{N}$ there exists a unique A-invariant $\frac{n}{2}$-dimensional H-neutral subspace $N_u = N_u(A,H)$ (resp. $N_\ell = N_\ell(A,H)$) such that $\sigma(A|_{N_u})$ (resp. $\sigma(A|_{N_\ell})$) lies in the closed upper (resp. lower) half-plane.

THEOREM 4.1 *The subspaces* $N_u(A,H)$ *and* $N_\ell(A,H)$ *are continuous functions of* $(A,H) \in \mathbb{N}$.

The continuity of this statement is understood in terms of the gap metric (see the Appendix to Part III), i.e. given $(A_0,H_0) \in \mathbb{N}$, for every $\varepsilon > 0$ there exists $\delta > 0$ such that

$$||P_{N_u(A,H)} - P_{N_u(A_0,H_0)}|| < \varepsilon$$

for every pair $(A,H) \in \mathbb{N}$ such that

$$||A-A_0|| + ||H-H_0|| < \delta .$$

Here P_S is the orthogonal projector on the subspace S of $\mathbb{C}^n$. The continuity of $N_\ell(A,H)$ is understood in a similar way.

PROOF. We shall prove the continuity of N_u only (for N_ℓ the proof is completely.analogous). Let $(A_m,H_m) \in \mathbb{N}$, $m = 1,2,\cdots$ be a sequence such that $A_m \to A_0$, $H_m \to H_0$ and $(A_0,H_0) \in \mathbb{N}$. Consider the sequence of subspaces $N_\ell^{(m)} = N_\ell(A_m,B_m)$. By Theorem A.1 of the appendix, there exists a converging subsequence $N_\ell^{(m_k)} \to N_\ell^{(0)}$, where $N_\ell^{(0)}$ is some $\frac{n}{2}$-dimensional subspace of $\mathbb{C}^n$. By Theorem A.2 of the appendix,

$$N_\ell^{(0)} = \{x \in \mathbb{C}^n | \text{there exist } x_{m_k} \in N_\ell^{(m_k)}, \ k = 1,2,\cdots, \text{ such that } x_{m_k} \to x\}. \quad (4.1)$$

We shall check that in fact $N_\ell^{(0)} = N_\ell(A_0,H_0)$. Indeed, let $x,y \in N_\ell^{(0)}$. Using the characterization (4.1) pick $x_{m_k}, y_{m_k} \in N_\ell^{(m_k)}$ such that $x_{m_k} \to x$, $y_{m_k} \to y$. Then

$$(H_0x,y) = \lim_{k\to\infty}(H_{m_k}x_{m_k},y_{m_k}) = 0$$

as $N_\ell^{(m_k)}$ is H_{m_k}-neutral. So $N_\ell^{(0)}$ is H_0-neutral. Further, $A_{m_k}x_{m_k} \in N_\ell^{(m_k)}$ and $\lim_{k\to\infty}(A_{m_k}x_{m_k}) = A_0x$. Again by (4.1), $Ax \in N_\ell^{(0)}$, i.e. $N_\ell^{(0)}$ is A_0-invariant.

Now we prove that the spectrum of $A_0|_{N_\ell^{(0)}}$ lies in the closed lower half-plane. Pick a basis $\varphi_1,\cdots,\varphi_q$ in $N_\ell^{(0)}$. Then (using Corollary A.5 of the appendix and passing if neccessary to a subsequence of $N_\ell^{(m_k)}$)

there exists a basis $\varphi_{1,m_k},\cdots,\varphi_{q,m_k}$ in $N_\ell^{(m_k)}$, $i = 1,2,\cdots$ such that

$$\varphi_{i,m_k} \to \varphi_i \quad \text{as} \quad k \to \infty , \quad i = 1,\cdots,q . \tag{4.2}$$

Write $A_0|_{N_\ell^{(0)}}$ as a $q \times q$ matrix B_0 in the basis $\varphi_1,\cdots,\varphi_q$, and write $A_{m_k}|_{N_\ell^{(m_k)}}$ as a $q \times q$ matrix B_{m_k} in the basis $\varphi_{1,m_k},\cdots,\varphi_{q,m_k}$. Because of (4.2), we have

$$\lim_{k\to\infty} B_{m_k} = B_0 . \tag{4.3}$$

Now pick $\lambda_0 \in \mathbb{C}$ in the open upper half-plane. We claim that $(\lambda_0 I - B_0)$ is invertible. Suppose not, and let Γ be a small circle around λ_0 which does not intersect the real line, and such that $\det(\lambda I - B_0) \neq 0$ for $\lambda \in \Gamma$. Then the polynomial $\det(\lambda I - B_0)$ has a zero inside Γ (namely, $\lambda = \lambda_0$); hence the same is true for any polynomial of degree q whose coefficients are sufficiently close to those of $\det(\lambda I - B_0)$. Hence, by (4.3) $\det(\lambda I - B_{m_k})$ has a zero inside Γ for k large enough, but this contradicts the assumption $\mathrm{Im}\, \sigma(A_{m_k}|_{N_\ell^{(m_k)}}) \leq 0$. So $\lambda_0 I - B_0$ is invertible, and consequently $A_0|_{N_\ell^{(0)}}$ has no eigenvalues in the open upper half-plane.

So $N_\ell^{(0)}$ is $\frac{n}{2}$-dimensional, H_0-neutral, A_0-invariant and $\sigma(A_0|_{N_\ell^{(0)}})$ lies in the closed lower half-plane. By the uniqueness of such a subspace (Theorem I.3.22), $N_\ell^{(0)} = N_\ell(A_0, H_0)$.

We have proved that every converging subsequence $N_\ell^{(m_k)}$ necessarily converges to $N_\ell(A_0, H_0)$. Now the continuity of $N_\ell(A,H)$ follows by a standard argument. □

4.2 Analyticity of Invariant Maximal Neutral Subspaces

Now we shall study analyticity properties of the subspaces $N_u = N_u(A,H)$ and $N_\ell = N_\ell(A,H)$ introduced in the preceding section. Bearing in mind subsequent applications, we shall consider A and H as analytic functions of several real variables. It will turn out that N_u and

N_ℓ are analytic functions of these variables, provided A has no real eigenvalues.

Let Ξ be an open set in $\mathbb{R}^p$. A typical point in $\mathbb{R}^p$ will be denoted $\xi = \langle\xi_1,\cdots,\xi_p\rangle$. An $n \times n$ matrix function $A(\xi)$, $\xi \in \Xi$ is called *analytic* (on Ξ), if for every $\xi^o = \langle\xi_1^o,\cdots,\xi_p^o\rangle \in \Xi$ the function $A(\xi)$ admits a power series development

$$A(\xi) = \sum_{k_i \geq 0} A_{k_1 k_2 \cdots k_p} (\xi_1-\xi_1^o)^{k_1}(\xi_2-\xi_2^o)^{k_2}\cdots(\xi_p-\xi_p^o)^{k_p}, \quad \xi = (\xi_1,\cdots,\xi_p)$$

which converges absolutely in the open ball

$$\{\xi = (\xi_1,\cdots,\xi_p) \in \mathbb{R}^p \mid |\xi_i-\xi_i^o| < \alpha, \quad i = 1,\cdots,p\}$$

for some $\alpha > 0$. In particular, an analytic function has all partial derivatives $\dfrac{\partial^{k_1+\cdots+k_p}A(\xi)}{\partial^{k_1}\xi_1\partial^{k_2}\xi_2\cdots\partial^{k_p}\xi_p}$ (where k_i are nonnegative integers) which in turn are also analytic functions in Ξ.

A family of subspaces $L(\xi) \subset \mathbb{C}^n$ for $\xi \in \Xi$ is called *analytic* (on Ξ) if the orthogonal projector on $L(\xi)$ is analytic. We consider here matrix valued analytic functions of several *real* variables. So $A(\xi) = [a_{ij}(\xi)]_{i,j=1}^n$, where the complex valued scalar functions $a_{ij}(\xi)$ are analytic on Ξ. This means that both the real part and the imaginary part of $a_{ij}(\xi)$ are analytic functions on Ξ.

The following lemma will be useful. Note that the projectors $P(\xi)$ introduced there are not necessarily orthogonal.

LEMMA 4.2 *Let* $P(\xi)$, $\xi \in \Xi$ *be an analytic projector valued function. Then* $L(\xi) = \operatorname{Im} P(\xi)$ *is an analytic family of subspaces, i.e. the orthogonal projector on* $L(\xi)$ *is analytic.*

PROOF. Pick $\xi_0 \in \Xi$, and consider a $k \times k$ non-zero minor of $P(\xi_0)$, where $k = \dim \operatorname{Im} P(\xi_0)$. For simplicity of notation assume that this minor is in the left upper corner of $P(\xi_0)$:

$$P(\xi) = \begin{bmatrix} P_{11}(\xi) & * \\ P_{12}(\xi) & * \end{bmatrix},$$

where $P_{11}(\xi)$ (resp. $P_{12}(\xi)$) is of size $k \times k$ (resp. $(n-k) \times k$), and $P_{11}(\xi_0)$ is invertible. Then, for $\xi \in \Xi$ close enough to ξ_0, $P_{11}(\xi)$ is invertible as well, and

$$L(\xi) = \operatorname{Im}\begin{bmatrix} I \\ Z(\xi) \end{bmatrix},$$

where $Z(\xi) = P_{12}(\xi)(P_{11}(\xi))^{-1}$. The orthogonal projector on $L(\xi)$ is now given by the formula

$$\begin{bmatrix} I & (Z(\xi))^* \\ Z(\xi) & Z(\xi)(Z(\xi))^* \end{bmatrix} \begin{bmatrix} (I+(Z(\xi))^* Z(\xi))^{-1} & 0 \\ 0 & [I+Z(\xi)(Z(\xi))^*]^{-1} \end{bmatrix}$$

(it is convenient to check this bearing in mind that a projector P is orthogonal if and only if $P = P^*$). The analyticity of $L(\xi)$ on a neighbourhood of ξ_0 follows from this representation. Since $\xi_0 \in \Xi$ was arbitrary, $L(\xi)$ is analytic on Ξ. □

Consider now $n \times n$ analytic matrix functions $A(\xi)$, $H(\xi)$ defined on Ξ with the properties that $A(\xi)$ is $H(\xi)$-selfadjoint and $(A(\xi),H(\xi)) \in \mathbb{N}$ for all $\xi \in \Xi$ (see the preceding section for the definition of $\mathbb{N}$). Does it follow that the subspaces $N_u(\xi) = N_u(A(\xi),H(\xi))$ and $N_\ell(\xi) = N_\ell(A(\xi),H(\xi))$ are analytic on Ξ? The following example shows that, in contrast to the continuity properties of N_u and N_ℓ, the answer is, in general, no.

EXAMPLE 4.1 For any $t \in \mathbb{R}$ define

$$A(t) = \begin{bmatrix} 0 & i \\ it^2 & 0 \end{bmatrix}; \quad H(t) = \begin{bmatrix} -t^2 & i \\ -i & 1 \end{bmatrix}.$$

Then $H(t)A(t) = (A(t))^* H(t)$, i.e. $A(t)$ is $H(t)$-selfadjoint, and $\sigma(A(t)) = \{it,-it\}$. So $\sigma(A(t)) \cap \mathbb{R} = \emptyset$ for $t \neq 0$, and $A(0)$ has the only eigenvalue 0, and has the Jordan form $\begin{bmatrix} 0 & 1 \\ 0 & 0 \end{bmatrix}$. Thus, $(A(t),H(t)) \in \mathbb{N}$ for all $t \in \mathbb{R}$. Further, the subspace $N_u(t)$ (resp. $N_\ell(t)$) is spanned by the vector $\langle 1,|t|\rangle$ (resp. $\langle 1,-|t|\rangle$), so the orthogonal projectors on $N_u(t)$ and $N_\ell(t)$ are

$$\frac{1}{1+t^2}\begin{bmatrix} 1 & |t| \\ |t| & t^2 \end{bmatrix}, \quad \frac{1}{1+t^2}\begin{bmatrix} 1 & -|t| \\ -|t| & t^2 \end{bmatrix},$$

respectively, and are not analytic at $t = 0$. □

However, if $A(\xi)$ has no real eigenvalues, then analyticity of $N_u(\xi)$ and $N_\ell(\xi)$ follows without difficulty:

THEOREM 4.3 *Let* $(A(\xi),H(\xi)) \in \mathbb{N}$, *where* $A(\xi)$ *and* $H(\xi)$ *are analytic functions on* $\Xi \subset \mathbb{R}^p$. *Assume that for every* $\xi \in \Xi$, *the matrix* $A(\xi)$

has no real eigenvalues. Then the A-*invariant* $\frac{n}{2}$ *-dimensional* H-*neutral subspaces* $N_\ell(\xi) = N_\ell(A(\xi),H(\xi))$ *and* $N_u(\xi) = N_u(A(\xi),H(\xi))$ *are analytic on* Ξ.

PROOF. Pick $\xi^o \in \Xi$, and let Γ be a closed contour such that all eigenvalues of $A(\xi^o)$ in the open upper half-plane are inside Γ, and all other eigenvalues of $A(\xi^o)$ (which lie in the open lower half-plane) are outside Γ. Let U be a neighbourhood of ξ^o such that $A(\xi)$ has no eigenvalues on Γ itself whenever $\xi \in U$ (such a U always exists, because the eigenvalues are continuous functions of ξ on Ξ). By definition, $N_u(\xi)$ is the sum of the root subspaces of $A(\xi)$ corresponding to the eigenvalues inside Γ. In other words, for any $\xi \in U$

$$N_u(\xi) = \operatorname{Im} P_\Gamma(\xi) ,$$

where

$$P_\Gamma(\xi) = \frac{1}{2\pi i} \int_\Gamma (\lambda I - A(\xi))^{-1} d\lambda$$

defines an analytic projector-valued function on Ξ. Hence by Lemma 4.2 $N_u(\xi)$ is an analytic family of subspaces. For $N_\ell(\xi)$ the proof of analyticity is similar. □

4.3 Extremal Solutions of the Algebraic Riccati Equation

Consider the algebraic Riccati equation

$$XDX + XA + A^*X - C = 0 , \tag{4.4}$$

where A, C, D are given $n \times n$ complex matrices with the properties that $C^* = C$, D is non-negative definite and

$$\operatorname{rank}[D, AD, \cdots, A^{n-1}D] = n .$$

The equation with exactly the same conditions A, C, D was studied in Chapter II.4. We have seen in Section II.4.6 that if (4.4) has a hermitian solution at all, then it has extremal hermitian solutions X_+ and X_-, in the sense that for every hermitian solution X of (4.4) and every $x \in \mathbb{C}^n$ the inequalities

$$(X_-x,x) \leq (Xx,x) \leq (X_+x,x)$$

hold. The maximal (resp. minimal) hermitian solution X_+ (resp. X_-) is characterized by the property that $Re\,\sigma(A+DX_+) \geq 0$ (resp. $Re\,\sigma(A+DX_-) \leq 0$).

Introduce the set $\mathcal{R}$ of all triples of $n \times n$ matrices (A,C,D) such that $C^* = C$, D is non-negative definite and $\mathrm{rank}[D,AD,\cdots,A^{n-1}D] = n$. Since $\mathcal{R}$ is a subset of $\mathbb{R}^{3n^2}$, any metric on $\mathbb{R}^{3n^2}$ induces a metric on $\mathcal{R}$; so one can speak about continuous functions defined on $\mathcal{R}$, etc.

THEOREM 4.4 *The extremal hermitian solutions* X_+ *and* X_- *of* (4.4) *are continuous functions of the matrices* $(A,C,D) \in \mathcal{R}$.

PROOF. Put (as in Chapter II.4)

$$M = i\begin{bmatrix} A & D \\ C & -A^* \end{bmatrix}, \quad H = \begin{bmatrix} -C & A^* \\ A & D \end{bmatrix},$$

and assume that H is invertible. Theorem II.4.8 shows that $M|_{L_\pm}$ is similar to $i(A+DX_\pm)$, where

$$L_\pm = \mathrm{Im}\begin{bmatrix} I \\ X_\pm \end{bmatrix} \tag{4.5}$$

are n-dimensional H-neutral subspaces. Since M is H-selfadjoint, $Im\,\sigma(M|_{L_+}) = Im\,\sigma(A+DX_+) \geq 0$, and the sign characteristic of (M,H) consists of the same signs for each real eigenvalue of M (Theorem II.4.3), we can apply Theorem 4.1 and deduce that L_+ is a continuous function of $(A,C,D) \in \mathcal{R}$. In other words, the orthogonal projector P_{L_+} on L_+ is a continuous function of $(A,C,D) \in \mathcal{R}$. The orthogonal projector on L_+ is given by the formula (cf. the proof of Lemma 4.2)

$$P_{L_+} = \begin{bmatrix} I & X_+^* \\ X_+ & X_+X_+^* \end{bmatrix}\begin{bmatrix} (I+X_+^*X_+)^{-1} & 0 \\ 0 & (I+X_+X_+^*)^{-1} \end{bmatrix}, \tag{4.6}$$

from which the continuity of X_+ as a function of $(A,C,D) \in \mathcal{R}$ follows immediately. The proof of continuity of X_- is similar. □

For the analytic dependence of extremal solutions on the coefficients, the following result holds, and can be proved using the same line of argument as the proof of Theorem 4.4, provided Theorem 4.3 is used in place of Theorem 4.1.

THEOREM 4.5 *Let* $D(\xi)$, $A(\xi)$, $C(\xi)$ *be analytic* $n \times n$ *matrix functions of* $\xi \in \Xi$, *where* Ξ *is an open connected set in* $\mathbb{R}^p$. *Assume the following conditions hold:*

(i) $(C(\xi))^* = C(\xi)$ *for all* $\xi \in \Xi$;

(ii) $(D(\xi)x,x) \geq 0$ *for all* $\xi \in \Xi$ *and* $x \in \mathbb{C}^n$;

(iii) $\mathrm{rank}[D(\xi),A(\xi)D(\xi),\cdots,(A(\xi))^{n-1}D(\xi)] = n$ *for all* $\xi \in \Xi$;

(iv) *the matrix*

$$M(\xi) = i\begin{bmatrix} A(\xi) & D(\xi) \\ C(\xi) & -(A(\xi))^* \end{bmatrix}$$

has no real eigenvalues for all $\xi \in \Xi$. *Then the maximal and minimal hermitian solutions of the algebraic Riccati equation*

$$X(\xi)D(\xi)X(\xi) + X(\xi)A(\xi) + (A(\xi))^*X(\xi) - C(\xi) = 0$$

are analytic functions in Ξ.

In the next section we shall use this result to deduce analyticity properties of the solution of an optimal control problem.

4.4 Application to the Optimal Control Problem

Consider the linear system

$$\begin{cases} \dfrac{dx(t)}{dt} = Fx(t) + Gu(t)\ ; \quad x(0) = x_0\ ; \\ y(t) = Rx(t)\ , \end{cases} \tag{4.7}$$

where G, F, R are real matrices of sizes $n \times m$, $n \times n$ and $p \times n$ respectively, and $x(t)$, $u(t)$, $y(t)$ are complex vector valued functions of dimensions n, m and p respectively. As in Section II.4.1, consider the following optimal control problem: find $u = u(x)$ such that the integral

$$E = \int_0^\infty [u^T(t)V_1u(t) + y^T(t)V_2y(t)]dt \tag{4.8}$$

is minimal, where V_1 and V_2 are constant real positive definite matrices of sizes $m \times m$ and $p \times p$ respectively, and $y(t)$ is found from the system (4.7) with $u = u(x)$. It will be shown in the next theorem that the optimal control and the minimal value of E are analytic in an appropriate sense.

THEOREM 4.6 *Assume that*

$$\operatorname{rank}[G,FG,\cdots,F^{n-1}G] = \operatorname{rank}\begin{bmatrix} R \\ RF \\ \cdot \\ \cdot \\ \cdot \\ RF^{n-1} \end{bmatrix} = n\ , \tag{4.9}$$

and that V_1 *and* V_2 *of* (4.8) *are positive definite. Then the solution* $u = u(t)$ *of the optimal control problem exists and is given by the formula*

$$u(t) = Zx(t) \tag{4.10}$$

where Z *is an* $m \times n$ *matrix independent of* t *which depends analytically on the entries of* F, G, R, V_1 *and* V_2. *The value of the integral* (4.8) *under the optimal control* (4.10) *is also an analytic function of the entries of* F, G, R, V_1, V_2.

Consider more carefully the use of the term "analytic" in this theorem. Define the set Λ of all quintets of real matrices $\{G,F,R,V_1,V_2\}$, where the sizes of G, F, R, V_1, V_2 are $n \times m$, $n \times n$, $p \times n$, $m \times m$, $p \times p$ respectively, (4.9) is satisfied, and V_1 and V_2 are positive definite. The set Λ is an open subset in the set Λ_0 of all quintets of matrices $\{G,F,R,V_1,V_2\}$ with the sizes as above and such that V_1 and V_2 are real and symmetric. Indeed, the condition (4.9) means that some $n \times n$ submatrices in each of $[G,FG,\cdots,F^{n-1}G]$ and $\begin{bmatrix} R \\ RF \\ \vdots \\ RF^{n-1} \end{bmatrix}$ are invertible. Clearly, this property holds also for matrices G', F', R' which are sufficiently close to G, F, R respectively. Further, any real symmetric matrix sufficiently close to a positive definite matrix V, is also positive definite. In turn, the set Λ_0 may be identified with $\mathbb{R}^q$, where $q = (m+n+p)n + \frac{m(m+1)}{2} + \frac{p(p+1)}{2}$ (here mn, n^2, np, $\frac{m(m+1)}{2}$, $\frac{p(p+1)}{2}$ is the number of independent entries in G, F, R, V_1, V_2 respectively, for $\{G,F,R,V_1,V_2\} \in \Lambda_0$). So Λ is an open set in $\mathbb{R}^q$, and Theorem 4.6 states that the matrix Z and the value of integral (4.8) under the optimal control are analytic functions on Λ, as defined in Section 4.2.

In fact, the solution of the optimal control problem of the theorem is given by the formula

$$u_0(x) = -V^{-1}G^T X_+ x(t) \ , \tag{4.11}$$

where X_+ is the maximal hermitian solution of the algebraic Riccati equation

$$XGV_1^{-1}G^T X - XF - F^T X - R^T V_2 R = 0 \ , \tag{4.12}$$

and the value of integral (4.8) under the optimal control $u_0(x)$ is $x_0^T X_+ x_0$ (see [28]; also Section II.4.1).

For the proof of Theorem 4.6 we need the following lemma.

LEMMA 4.7 *If* $\{G,F,R,V_1,V_2\} \in \Lambda$, *then the matrix*

$$A = \begin{bmatrix} -F & GV_1^{-1}G^T \\ R^T V_2 R & F^T \end{bmatrix}$$

has no pure imaginary eigenvalues.

PROOF. Suppose $i\alpha$, $\alpha \in \mathbb{R}$ is an eigenvalue of A:

$$\begin{bmatrix} -F & GV_1^{-1}G^T \\ R^T V_2 R & F^T \end{bmatrix} \begin{bmatrix} z_1 \\ z_2 \end{bmatrix} = i\alpha \begin{bmatrix} z_1 \\ z_2 \end{bmatrix}, \tag{4.13}$$

where $z_1, z_2 \in \mathbb{C}^n$ and $||z_1||^2 + ||z_2||^2 > 0$. Premultiplying (4.13) by $[z_2^*, z_1^*]$, we get

$$-z_2^* F z_1 + z_1^* R^T V_2 R z_1 + z_2^* G V_1^{-1} G^T z_2 + z_1^* F^T z_2 = i\alpha(z_2^* z_1 + z_1^* z_2) \ .$$

Separating real and imaginary parts, we obtain

$$z_1^* R^T V_2 R z_1 + z_2^* G V_1^{-1} G^T z_2 = 0 \ ,$$

and consequently, $Rz_1 = 0$ and $G^T z_2 = 0$. Now (4.13) yields $-Fz_1 = i\alpha z_1$ and $F^T z_2 = i\alpha z_2$. Hence

$$\begin{bmatrix} R \\ RF \\ \cdot \\ \cdot \\ \cdot \\ RF^{n-1} \end{bmatrix} z_1 = 0 \quad \text{and} \quad \begin{bmatrix} G^T \\ G^T F^T \\ \cdot \\ \cdot \\ \cdot \\ G^T (F^T)^{n-1} \end{bmatrix} z_2 = 0 \ ,$$

which in view of (4.9) implies $z_1 = z_2 = 0$, a contradiction with our assumption. □

Proof of Theorem 4.6. Consider the algebraic Riccati equation (4.12), and put $D = GV_1^{-1}G^T$; $A = -F$; $C = R^TV_2R$. Clearly, $C = C^*$ and D is nonnegative definite. Also

$$\operatorname{rank}[D, AD, \cdots, A^{n-1}D] = n \ . \tag{4.14}$$

To check this, observe first that (4.9) implies

$$\operatorname{Im} G + F(\operatorname{Im} G) + \cdots + F^{n-1}(\operatorname{Im} G) = \mathbb{C}^n \ .$$

It is claimed that $\operatorname{Im} G = \operatorname{Im}(GV_1^{-1}G^T)$. The inclusion of $\operatorname{Im}(GV_1^{-1}G^T)$ in $\operatorname{Im} G$ is evident. To check the reverse inclusion write $V_1^{-1} = V_3V_3^*$ for some invertible $m \times m$ matrix V_3 and let $y \in \operatorname{Im} G$. Then $y = GV_3x$ for some $x \in \mathbb{C}^n$. Subtracting from x its orthogonal projection on $\operatorname{Ker} GV_3$ it may be assumed that x is orthogonal to $\operatorname{Ker} GV_3$. But then $x \in \operatorname{Im}(GV_3)^* = \operatorname{Im}(V_3^*G^T)$, so $x \in V_3^*G^Tz$ and

$$y = GV_3V_3^*G^Tz \in \operatorname{Im}(GV_1^{-1}G^T) \ ,$$

as required.

Now we obtain

$$\operatorname{Im}(GV_1^{-1}G^T) + F(\operatorname{Im}(GV_1G^T)) + \cdots + F^{n-1}(\operatorname{Im}(GV_1^{-1}G^T)) = \mathbb{C}^n$$

which is equivalent to (4.14).

Combining Lemma 4.7 and Theorem 4.5, we find that the maximal hermitian solution of (4.12) is an analytic function of the entries of G, F, R, V_1, V_2, provided (4.9) holds and V_1 and V_2 are positive definite. Application of formula (4.11) and the formula $x_0^TX_+x_0$ for the value of (4.8) under the optimal control completes the proof of the theorem. □

4.5 Continuity of Canonical Factorization of Nonnegative Rational Matrix Functions

In this section we shall apply Theorem 4.1 to factorization of nonnegative rational matrix functions. The results and remarks from Chapter II.3 will be used extensively.

Let $W(\lambda)$ be a nonnegative definite rational matrix function which is equal to I at infinity. The unique factorization $W(\lambda) = L_*(\lambda)L(\lambda)$ described in Theorem II.3.12 in which the rational matrix functions $L(\lambda)$ and $L(\lambda)^{-1}$ have no poles in the open upper half-plane, and $L(\infty) = I$, will be called the *canonical factorization* of $W(\lambda)$. We prove in this section that the canonical factorization depends continuously on $W(\lambda)$. More exactly, the following holds:

THEOREM 4.8 *Let* $W(\lambda)$ *be a nonnegative definite rational matrix function with* $W(\infty) = I$, *and let* $W(\lambda) = L_*(\lambda)L(\lambda)$ *be its canonical factorization. Then* $L(\lambda)$ *depends continuously on* $W(\lambda)$ *in the following sense. Let* $\theta = (A,B,C,I;\mathbb{C}^{2k},\mathbb{C}^n)$ *and* $\theta_L = (A_L,B_L,C_L,I;\mathbb{C}^k,\mathbb{C}^n)$ *be minimal nodes with transfer functions* $W(\lambda)$ *and* $L(\lambda)$, *respectively. Then for every* $\varepsilon > 0$ *there exists a* $\delta > 0$ *such that, whenever* $W_0(\lambda)$ *is a nonnegative definite rational matrix function with minimal node* $(A_0,B_0,C_0,I;\mathbb{C}^{2k},\mathbb{C}^n)$, *where* $||A_0-A|| + ||B_0-B|| + ||C_0-C|| < \delta$, *and with canonical factorization* $W_0(\lambda) = L_{0*}(\lambda)L_0(\lambda)$, *the function* $L_0(\lambda)$ *has a minimal node* $(A_{L_0},B_{L_0},C_{L_0},I;\mathbb{C}^k,\mathbb{C}^n)$ *satisfying*

$$||A_{L_0}-A_L|| + ||B_{L_0}-B_L|| + ||C_{L_0}-C_L|| < \varepsilon .$$

As remarked after the proof of Theorem II.3.11 the main matrix A in a minimal node for $W(\lambda)$ has even size, say $2k \times 2k$, and the main matrix in a minimal node of $L(\lambda)$ is then $k \times k$.

Since any two minimal nodes with the same transfer function are similar, it is easily seen that the property described in Theorem 4.8 does not depend on the choice of θ and θ_L, i.e. this is indeed a property of the functions $W(\lambda)$ and $L(\lambda)$.

PROOF. Let $A^\times = A - BC$, and let M (resp. N) be the $A^\times$-invariant (resp. A-invariant) H-neutral maximal H-nonnegative subspace such that $Im\ (A^\times|_M) \lesssim 0$ (resp. $Im\ \sigma(A|_N) \lesssim 0$); by Theorems II.3.11 and I.3.22 the subspaces M and N exist and are unique. Further, Theorem II.3.7 ensures that M and N are direct complements to each other. Let π be the projector on M along N; then $L(\lambda)$ is the transfer function of $pr_\pi(\theta)$. Write A, B, C in the block matrix form with respect to the decomposition $\mathbb{C}^{2k} = N \dotplus M$:

$$A = \begin{bmatrix} A_{11} & A_{12} \\ A_{21} & A_{22} \end{bmatrix}; \quad B = \begin{bmatrix} B_1 \\ B_2 \end{bmatrix}; \quad C = [\ C_1, C_2\ ] ;$$

then $pr_\pi(\theta) = (A_{22},B_2,C_2,I;M,\mathbb{C}^n)$. Here A_{ij}, $i,j = 1,2$, B_1, B_2, C_1, C_2 are $k \times k$ matrices representing the corresponding linear transformation in fixed bases $\varphi_1,\cdots,\varphi_k$ in M and $\psi_1,\cdots,\psi_k$ in N.

Since both θ_L and $pr_\pi(\theta)$ are minimal nodes with the same transfer function, they are similar; so it is sufficient to prove the theorem for $\theta_L = pr_\pi(\theta)$, i.e. identifying $\mathbb{C}^k$ with M and assuming $A_L = A_{22}$, $B_L = B_2$, $C_L = C_2$.

Assume now that the assertion of Theorem 4.8 is not true. So there exist sequences of $m \times m$ matrices A_{0q}, $n \times m$ matrices C_{0q}, $m \times n$ matrices B_{0q}, $q = 1,2,\cdots$ such that the rational function $W_{0q}(\lambda) \overset{def}{=} I + C_{0q}(\lambda I - A_{0q})^{-1}B_{0q}$ is nonnegative definite,

$$\lim_{q\to\infty}\{||A_{0q}-A|| + ||B_{0q}-B|| + ||C_{0q}-C||\} = 0 , \tag{4.15}$$

but every minimal node $\theta_{L_q} = (A_{L_q},B_{L_q},C_{L_q},I;\mathbb{C}^k,\mathbb{C}^n)$ of the rational matrix function $L_q(\lambda)$ from the canonical factorization $W_{0q}(\lambda) = L_{q*}(\lambda)L_q(\lambda)$ satisfies the inequality

$$||A_{L_q}-A_{22}|| + ||B_{L_q}-B_2|| + ||C_{L_q}-C_2|| \geq \varepsilon_0 , \tag{4.16}$$

where $\varepsilon_0 > 0$ is some fixed number.

For every $q = 1,2,\cdots$ let H_{0q_*} be the hermitian invertible matrix such that $H_{0q}A_{0q} = A^*_{0q}H_{0q}$; $H_{0q}B_{0q} = C^*_{0q}$; $C_{0q} = B^*_{0q}H_{0q}$. We claim that $\lim_{q\to\infty} H_{0q} = H$, where H is the hermitian invertible matrix such that $HA = A^*H$; $HB = C^*$; $C = B^*H$. Indeed, the linear system of equation

$$XA_{0q} = A^*_{0q}X , \quad XB_{0q} = C^*_{0q} , \quad C_{0q} = B^*_{0q}X$$

has a unique solution $X = H_{0q}$ (see the appendix to Chapter II.3). For the same reason, the linear system

$$YA = A^*Y ; \quad YB = C^* ; \quad C = B^*Y$$

also has a unique solution $Y = H$. But it is clear that the solution of a linear system depends continuously on the coefficients of the system, as long as the solution is unique. So indeed

$$\lim_{q\to\infty} H_{0q} = H .$$

Let M_q (resp. N_q) be the $(A_{0q}-B_{0q}C_{0q})$-invariant (resp. A_{0q}-invariant) H_{0q}-neutral maximal H_{0q}-nonnegative subspace such that $Im\,\sigma((A_{0q}-B_{0q}C_{0q})|_{M_q}) \leq 0$ (resp. $Im\,\sigma(A_{0q}|_{N_q}) \leq 0$). By Theorem 4.1, $M_q \to M$ and $N_q \to N$ as $q \to \infty$. Now apply Theorem A.2 of the Appendix to find bases $\varphi_{q1},\cdots,\varphi_{qk}$ in M_q and $\psi_{q1},\cdots,\psi_{qk}$ in N_q such that

$$\lim_{q\to\infty} \varphi_{qi} = \varphi_i \ ; \quad \lim_{q\to\infty} \psi_{qi} = \psi_i \ , \quad i = 1,2,\cdots,k \ . \tag{4.17}$$

Write A_{0q}, B_{0q}, C_{0q} as matrices with respect to the decomposition $\mathbb{C}^{2k} = N_q \dotplus M_q$, where $\varphi_{q1},\cdots,\varphi_{qk}$ (resp. $\psi_{q1},\cdots,\psi_{qk}$) is the chosen basis in M_q (resp. N_q); thus,

$$A_{0q} = \begin{bmatrix} A_{11q} & A_{12q} \\ A_{21q} & A_{22q} \end{bmatrix} ; \quad B_{0q} = \begin{bmatrix} B_{1q} \\ B_{2q} \end{bmatrix} ; \quad C_{0q} = [\ C_{1q}, C_{2q}\ ] \ .$$

Now (4.15) and (4.17) imply $A_{22q} \to A_{22}$, $B_{2q} \to B_2$, $C_{2q} \to C_2$, which contradicts (4.16), because $(A_{22q}, B_{2q}, C_{2q}, I; M_q, \mathbb{C}^{2k})$ is a minimal node with the transfer function $L_q(\lambda)$. The proof of Theorem 4.8 is completed. □

Combining Theorem 4.1 and 4.3 with Theorem II.2.6 and the remarks after this theorem, we obtain the following result on factorization of nonnegative matrix polynomials.

THEOREM 4.9 *Let* $L(\lambda)$ *be a monic nonnegative* $n \times n$ *matrix polynomial. Then there exists a unique monic* $n \times n$ *matrix polynomial* $M_+(\lambda)$ *(resp.* $M_-(\lambda)$*) such that* $L(\lambda) = (M_+(\bar{\lambda}))^* M_+(\lambda)$ *(resp.* $L(\lambda) = (M_-(\bar{\lambda}))^* M_-(\lambda)$*) and* $M_+(\lambda)$ *(resp.* $M_-(\lambda)$*) is invertible for every* λ *in the open lower (resp. upper) half-plane.*

Moreover, the coefficients of $M_+(\lambda)$ *and* $M_-(\lambda)$ *depend continuously on the coefficients of* $L(\lambda)$*, and if* $L(\lambda)$ *has no real eigenvalues, then the coefficients of* $M_+(\lambda)$ *and* $M_-(\lambda)$ *depend analytically on the coefficients of* $L(\lambda)$.

CHAPTER 5

PERTURBATIONS WHICH PRESERVE JORDAN STRUCTURE

In Chapter III.1 the "general perturbations" of pairs (A,H), in which A is H-selfadjoint, are of a broad class and, in fact, stability results have been obtained only in special cases. For example, in Chapter III.1 the unperturbed matrices are finally restricted to those which are similar to a real diagonal matrix, and in Chapter III.3 the perturbations are restricted to H-positive matrices. In this chapter a different restriction is imposed on the perturbations; namely, those which preserve the real Jordan structure of the unperturbed matrix A. Thus, A and the perturbed matrices are to have the same partial multiplicities associated with real eigenvalues, even though the real eigenvalues themselves are not fixed. Observe that if $H = I$, this analysis will include the case of general hermitian perturbations.

Perturbations which preserve the real Jordan structure allow us to compare the sign characteristics of the unperturbed and perturbed matrices, and one of the main results of the chapter is that the sign characteristic is stable under such perturbations. Similar results are obtained for H-unitary matrices, and the case of real matrices is developed separately.

5.1 Stability of the Sign Characteristic

In this section the basic definition concerning perturbations which preserve the real Jordan structure of an H-selfadjoint matrix A is made, and the main result establishing the stability of the sign characteristic under such perturbations is stated and proved.

For any complex matrix A with a real eigenvalue λ_0, the λ_0-*structure preserving neighbourhood* of A consists of all $n \times n$ matrices A_1 such that $||A-A_1||$ is small enough, A_1 has exactly one real eigenvalue

λ_1 in a neighbourhood of λ_0 and, in addition, the partial multiplicities of $\lambda I - A_1$ at λ_1 (i.e. the sizes of Jordan blocks with eigenvalue λ_1 in the Jordan form of A_1) are the same as those of $I\lambda - A$ at λ_0.

The following is the stability theorem for the sign characteristic.

THEOREM 5.1 *Let the* $n \times n$ *matrix* A *be* H*-selfadjoint. Let* λ_0 *be a real eigenvalue of* A. *Then there exists a* λ_0*-structure preserving neighbourhood* $\mathcal{U}_A$ *of* A *and a neighbourhood* $\mathcal{U}_H$ *of* H *such that, if* $A_1 \in \mathcal{U}_A$ *is* H_1*-selfadjoint for some hermitian invertible matrix* $H_1 \in \mathcal{U}_H$, *then the sign characteristics of* (A,H) *at* λ_0 *and the sign characteristic of* (A_1,H_1) *at the eigenvalue close to* λ_0 *are the same.*

By the sign characteristic of (A,H) at λ_0 we mean that part of the sign characteristic of (A,H) corresponding to the eigenvalue λ_0 of A.

For the proof of Theorem 5.1 we need some preliminary results. For a given subspace $\mathcal{L} \subset \mathbb{C}^n$, denote by $P_{\mathcal{L}}$ the orthogonal projector on $\mathcal{L}$. Two increasing sequences of subspaces of $\mathbb{C}^n$

$$\mathcal{M}_1 \subset \mathcal{M}_2 \subset \cdots \subset \mathcal{M}_k \quad \text{and} \quad \mathcal{N}_1 \subset \mathcal{N}_2 \subset \cdots \subset \mathcal{N}_k$$

are said to be δ*-close* if $||P_{\mathcal{M}_j} - P_{\mathcal{N}_j}|| < \delta < 1$ for $j = 1,2,\cdots,k$. Note that in this case $\dim \mathcal{M}_j = \dim \mathcal{N}_j$, and $P_{\mathcal{M}_j}(\mathcal{N}_j) = \mathcal{M}_j$, $P_{\mathcal{N}_j}(\mathcal{M}_j) = \mathcal{N}_j$ for each j.

LEMMA 5.2 *Given* $\varepsilon > 0$, *there exists a* δ *such that, for any* δ*-close sequence of subspaces* $\{\mathcal{M}_j\}_{j=1}^k$ *and* $\{\mathcal{N}_j\}_{j=1}^k$ *in* $\mathbb{C}^n$ *there is an invertible* $n \times n$ *matrix* S *(depending on the sequences) for which* $S\mathcal{M}_j = \mathcal{N}_j$, $j = 1,2,\cdots,k$ *and* $||I-S|| < \varepsilon$.

PROOF. We use induction on k. Put $S_1 = I - (P_{\mathcal{M}_1} - P_{\mathcal{N}_1})$ and observe that $S_1\mathcal{M}_1 = \mathcal{N}_1$. Then $||P_{\mathcal{M}_1} - P_{\mathcal{N}_1}|| < \delta < 1$ implies $||S_1 - I|| < \delta$ and this is the case $k = 1$. Note also that $||S_1^{-1} - I|| < \delta/(1-\delta)$. Furthermore, $S_1 P_{\mathcal{M}_j} S_1^{-1}$ is a projector on $S_1\mathcal{M}_j$ and therefore

$$\begin{aligned} ||P_{S_1\mathcal{M}_j} - P_{\mathcal{N}_j}|| &\leq ||S_1 P_{\mathcal{M}_j} S_1 - P_{\mathcal{N}_j}|| \\ &\leq ||S_1 P_{\mathcal{M}_j}(S_1^{-1} - I)|| + ||(S_1 - I)P_{\mathcal{M}_j}|| + ||P_{\mathcal{M}_j} - P_{\mathcal{N}_j}|| \\ &< (1+\delta)\frac{\delta}{1-\delta} + \delta + \delta = \delta\left(\frac{3-\delta}{1-\delta}\right) . \end{aligned}$$

By the induction hypothesis there is a $\tilde{\delta} > 0$ and an invertible linear transformation $\tilde{S} : N_k \to N_k$ such that $\tilde{S}(S_1 M_j) = N_j$ for $j = 2,\cdots,k$ and $||I-\tilde{S}|| < \frac{1}{2}\varepsilon$ provided that

$$||P_{S_1 M_j} - P_{N_j}|| < \tilde{\delta}\ , \quad j = 2,\cdots,k\ . \tag{5.1}$$

Define the linear transformation $S : \mathbb{C}^n \to \mathbb{C}^n$ by: $S|_{M_1} = \tilde{S}S_1|_{M_1}$ and $S|_{M_1^\perp} = I$. Then, given (5.1) we have

$$||I-S|| = ||I-S|_{M_1}|| \leq ||(I-\tilde{S})S_1|_{M_1}|| + ||I-S_1|| < \frac{1}{2}\varepsilon(1+\delta) + \delta\ .$$

Thus, it is sufficient to choose $\delta > 0$ so that

$$\delta\left[\frac{3-\delta}{1-\delta}\right] < \tilde{\delta} \quad \text{and} \quad \delta < \min\{\varepsilon/(2+\varepsilon),1\}\ . \ \square$$

A closer inspection of the proof of Lemma 5.2 reveals that, in fact, one can choose $\delta = \frac{\varepsilon}{12^{n-1}}$ (provided $\varepsilon \leq \frac{1}{2}$). If the sequences of subspaces are of length not exceeding k, then one can take $\delta = \frac{\varepsilon}{12^{k-1}}$.

The following corollary is an immediate consequence of Lemma 5.2.

COROLLARY 5.3 *Let* $L_1 \subset L_2$ *be subspaces in* $\mathbb{C}^n$, *and let* $\eta_1,\cdots,\eta_k$ *be a basis in some direct complement to* L_1 *in* L_2. *Then for every* $\varepsilon > 0$ *there exists a* $\delta > 0$ *such that, if* $L_1' \subset L_2'$ *are subspaces in* $\mathbb{C}^n$ *with* $||P_{L_i'} - P_{L_i}|| < \delta$, $i = 1,2$, *then for some basis* $\eta_1',\cdots,\eta_k'$ *in a direct complement to* L_1' *in* L_2', *the inequalities* $||\eta_j'-\eta_j|| < \varepsilon$, $j = 1,\cdots,k$ *hold.*

LEMMA 5.4 *Let* X *be an* $n \times m$ *complex matrix. Then there exists a constant* $K > 0$ *and a neighbourhood* U_X *of* X *such that*

$$||P_{\mathrm{Ker}\,Y} - P_{\mathrm{Ker}\,X}|| \leq K||X-Y|| \tag{5.2}$$

for every $Y \in U_X$ *satisfying* rank Y = rank X.

PROOF. Assume first that X is right invertible, i.e. the rows of X are linearly independent. For brevity, write $P_{\mathrm{Ker}\,X} = P_X$. There exists a right inverse X^I of X such that $\mathrm{Im}\,X^I = \mathrm{Im}(I-P_X)$, and then $X^I X = I - P_X$ (indeed, both sides are projectors with the same kernel and the same image). The matrices Y with the property

$$||Y-X|| < \frac{1}{2}||X^I||^{-1} \tag{5.3}$$

are also right invertible, and a right inverse Y^I of such a Y is given by the easily verified formula: $Y^I = X^I Z$ where $Z = \sum_{j=0}^{\infty} ((X-Y)X^I)^j$. Note that Z is invertible and $||Z|| \leq 2$; $||I-Z|| \leq 2||X^I||\ ||X-Y||$.

Now define the (not necessarily orthogonal) projector $R_Y = I - Y^I Y$. We have

$$\begin{aligned} ||P_X - R_Y|| &= ||X^I X - Y^I Y|| = ||X^I X - X^I Z Y|| \\ &\leq ||X^I||\ ||X - ZY|| \leq ||X^I||\{||I-Z||\ ||X|| + ||Z||\ ||X-Y||\} \\ &\leq ||X^I||\{2||X^I||\ ||X-Y||\ ||X|| + 2||X-Y||\}\ , \end{aligned}$$

and hence

$$||P_X - R_Y|| \leq K||X-Y||\ ,$$

for $K = 2||X^I||^2||X|| + 2||X^I||$ and for Y satisfying (5.3).

We now use the fact that $||P_X - P_Y|| \leq ||Q_X - Q_Y||$, where Q_X (resp. Q_Y) is an arbitrary projector on $\text{Ker } X$ (resp. $\text{Ker } Y$) (see Proposition A.3 in the Appendix). Hence $||P_X - P_Y|| \leq ||P_X - R_Y||$; so the lemma is proved when X is right invertible.

Now consider the case when X is not right invertible, and let r be the dimension of a complementary subspace N to $\text{Im } X$ in $\mathbb{C}^n$. Consider the linear transformation

$$\tilde{X} : \mathbb{C}^m \dotplus \mathbb{C}^r \to \mathbb{C}^n$$

defined by $\tilde{X}(x+y) = Xx + Zy$; $x \in \mathbb{C}^m$, $y \in \mathbb{C}^r$, where $Z : \mathbb{C}^r \to N$ is some invertible linear transformation. Clearly, $\text{Im } \tilde{X} = \mathbb{C}^n$, so $\tilde{X}$ is right invertible. Also $\text{Ker } \tilde{X} = \text{Ker } X$. Let $P_{\tilde{X}}$ be the orthogonal projector on $\text{Ker } \tilde{X}$. Using the part of Lemma 5.4 already proved, we find a constant $\tilde{K}$ and a neighbourhood $\mathcal{U}_X$ of $\tilde{X}$ such that

$$||P_{\tilde{X}} - P_{\tilde{Y}}|| \leq K||\tilde{X} - \tilde{Y}|| \tag{5.4}$$

for every linear transformation $\tilde{Y} : \mathbb{C}^m \dotplus \mathbb{C}^r \to \mathbb{C}^n$ which belongs to $\mathcal{U}_{\tilde{X}}$. Note that equality $\text{rank } \tilde{Y} = \text{rank } \tilde{X}$ holds automatically for $\mathcal{U}_{\tilde{X}}$ small enough, because then $\tilde{Y}$ will also be right invertible, together with $\tilde{X}$ (see the first part of the proof). Apply (5.4) for $\tilde{Y}$ of the form $\tilde{Y}(x+y) =$

$= Yx + Zy$; $x \in \mathbb{C}^m$, $y \in \mathbb{C}^r$, where Y is a matrix sufficiently close to X (so that $\widetilde{Y} \in \mathcal{U}_{\widetilde{X}}$), and $\dim \operatorname{Ker} Y = \dim \operatorname{Ker} X$. It is easily seen that $\operatorname{Ker} \widetilde{Y} \subset \mathbb{C}^m$. Indeed,

$$\operatorname{rank} \widetilde{Y} = \operatorname{rank} \widetilde{X} = \operatorname{rank} X = \operatorname{rank} Y ,$$

and since $\operatorname{Ker} Y \subset \operatorname{Ker} \widetilde{Y}$ we have, in fact, $\operatorname{Ker} \widetilde{Y} = \operatorname{Ker} Y$ and therefore $\operatorname{Ker} \widetilde{Y} \subset \mathbb{C}^m$. Also $P_X = (P_{\widetilde{X}})|_{\mathbb{C}^m}$ and $P_Y = (P_{\widetilde{Y}})|_{\mathbb{C}^m}$. So (5.2) follows from (5.4). □

Proof of Theorem 5.1. Let $\mathcal{U}_A$ be a sufficiently small real λ_0-structure preserving neighbourhood of A, and let λ_1 be the eigenvalue of $A_1 \in \mathcal{U}_A$ which is close to λ_0. Let γ be the largest partial multiplicity of $I\lambda - A$ at λ_0, and denote by Ψ_i, $i = 1,\cdots,\gamma$, the subspace spanned by all the eigenvectors of $I\lambda - A$ corresponding to λ_0 which generate a Jordan chain of A with length not less than i. Let $\Psi_i^{(1)}$, $i = 1,\cdots,\gamma$, be the similarly defined subspace for $I\lambda - A_1$ and its eigenvalue λ_1.

Put $E_\alpha = \operatorname{Ker}(I\lambda_0 - A)^\alpha$ and for $\alpha = 1,2,\cdots,\gamma$ choose a basis $\eta_{\alpha 1},\cdots,\eta_{\alpha,k_\alpha}$ in a direct complement to $E_{\alpha-1}$ in E_α. Then for $i = 1,2,\cdots,\gamma$ the vectors

$$\varphi_{ij} = (\lambda_0 I - A)^{i-1}\eta_{ij} , \quad j = 1,\cdots,k_i$$

form a basis in Ψ_i. We claim that it is possible to choose bases $\varphi_{i1}^{(1)},\cdots,\varphi_{i,k_i}^{(1)}$ in $\Psi_i^{(1)}$ such that $\varphi_{ij}^{(1)}$ is as close as we wish to φ_{ij} ($j = 1,\cdots,k_i$, $i = 1,\cdots,\gamma$) provided the neighbourhood $\mathcal{U}_A$ is small enough.

Indeed, it follows from Lemma 5.4 that for $\alpha = 1,2,\cdots,\gamma$ the subspaces $E_\alpha^{(1)} = \operatorname{Ker}(I\lambda_1 - A)^\alpha$ can be made arbitrarily close to E_α (in the sense that $||P_{E_\alpha} - P_{E_\alpha^{(1)}}||$ is as close to zero as we wish provided $\mathcal{U}_A$ is small enough). From Corollary 5.3 we find that there exist bases $\eta_{\alpha 1}^{(1)},\cdots,\eta_{\alpha,k_\alpha}^{(1)}$ in a direct complement to $E_{\alpha-1}^{(1)}$ in $E_\alpha^{(1)}$ such that $\eta_{\alpha j}^{(1)}$ is arbitrarily close to $\eta_{\alpha j}$ ($j = 1,\cdots,k_\alpha$; $\alpha = 1,\cdots,\gamma$). Now we can put

$$\varphi_{ij}^{(1)} = (\lambda_1 I - A_1)^{i-1}\eta_{ij}^{(1)}$$

to produce bases $\varphi_{i1}^{(1)},\cdots,\varphi_{i,k_i}^{(1)}$ in $\Psi_i^{(1)}$ ($i = 1,\cdots,\gamma$), where $\varphi_{ij}^{(1)}$ is arbitrarily close to φ_{ij} for $j = 1,\cdots,k_i$, $i = 1,\cdots,\gamma$.

Since we have found bases φ_{ij} in Ψ_i and $\varphi_{ij}^{(1)}$ in $\Psi_i^{(1)}$ which are close enough, the assertion of Theorem 5.1 is easily obtained by using Theorem I.3.16. Indeed, let $f_i(x,y)$ and G_i be the bilinear form and selfadjoint linear transformation respectively defined in that theorem for A and H. By $f_i^{(1)}(x,y)$ and $G_i^{(1)}$ we denote the analogously defined quantities for A_1 and H_1. Then, if H_1 is close enough to H and the bases $\varphi_{ij}^{(1)}$ are close enough to φ_{ij}, the matrix representation of $f_i^{(1)}$ in the basis $\varphi_{i1}^{(1)},\cdots,\varphi_{i,k_i}^{(1)}$ will be arbitrarily close to the matrix representation of f_i in the basis $\varphi_{i1},\cdots,\varphi_{i,k_{i1}}$ $(i = 1,\cdots,\gamma)$. By Theorem I.3.16(ii), the same is true for the matrix representations of $G_i^{(1)}$ and G_i, and now Theorem I.3.16(iv) ensures that the sign characteristic of (A,H) at λ_0 coincides with the sign characteristic of (A_1,H_1) at λ_1. □

5.2 Stability of Unitary Similarity

Theorem 5.1 allows us to deduce some stability properties of unitary similarity. Recall that an H_1-selfadjoint matrix A_1 is unitary similar to an H_2-selfadjoint matrix A_2 if $A_1 = S^{-1}A_2S$ and $H_1 = S^*H_2S$ for some invertible matrix S. In particular, A_1 and A_2 are similar, but in general the similarity of A_1 and A_2 alone is not sufficient to ensure their (H_1,H_2)-unitary similarity (although it is sufficient if H_1 is positive definite or negative definite). The next theorem shows, in particular, that if (A_1,H_1) and (A_2,H_2) are unitarily similar, then the same is true for (B_1,H_1) and (B_2,H_2) provided B_i is sufficiently close and similar to A_i for $i = 1$ and $i = 2$. Roughly speaking, similarity with the additional requirement of closeness ensures unitary similarity.

THEOREM 5.5 *Let* A_i *be* H_i*-selfadjoint matrices,* $i = 1,2$, *and suppose that* (A_1,H_1) *and* (A_2,H_2) *are unitarily similar. Then for some neighbourhoods* $U(A_i)$ *and* $U(H_i)$ *of* A_i *and* H_i, *respectively,* $i = 1,2$, *the following property holds: If (for* $i = 1$ *and* 2) *we have* $(B_iG_i) \in U(A_i) \times U(H_i)$, B_i *is* G_i*-selfadjoint, and* B_i *is similar to* A_i, *then* (B_1,G_1) *and* (B_2,G_2) *are unitarily similar.*

PROOF. Applying Theorem 5.1 for each real eigenvalue of A_i it is found that the sign characteristic of (B_i,G_i) coincides with the sign characteristic of (A_i,H_i) provided only that B_i is close enough to A_i, and G_i to H_i for $i = 1$ and 2. Since, by Theorem I.3.6, the sign characteristic of (A_1,H_1) and the sign characteristic of (A_2,H_2) coincide, the same is true for the sign characteristics of (B_i,G_i), $i = 1,2$. Appeal again to Theorem I.3.6 to deduce that (B_1,G_1) and (B_2,G_2) are unitarily similar. □

The proof of Theorem 5.5 shows that the condition that B_i be similar to A_i can be somewhat weakened. Namely, one can replace this condition by the hypotheses that B_1 and B_2 are similar, and for every real eigenvalue λ_0 of A_i, the matrix B_i is in some real λ_0-structure preserving neighbourhood of A_i for $i = 1,2$. The latter requirement cannot be removed, as the following example shows.

EXAMPLE 5.1 Let

$$A_1 = A_2 = \begin{bmatrix} 0 & 0 \\ 0 & 0 \end{bmatrix} ; \; H_1 = \begin{bmatrix} 1 & 0 \\ 0 & -1 \end{bmatrix} ; \; H_2 = \begin{bmatrix} -1 & 0 \\ 0 & 1 \end{bmatrix} ; \; B_1 = B_2 = \begin{bmatrix} \delta & 0 \\ 0 & 0 \end{bmatrix} ; \; \delta > 0 .$$

All the conditions with $G_i = H_i$, except for the similarity of B_i and A_i of Theorem 5.5 are satisfied. However, (B_1,H_1) and $B_2,H_2)$ are not unitarily similar. Note that B_i does not belong to any 0-structure preserving neighbourhood of A_i. □

The following particular case of Theorem 5.5 is noteworthy.

COROLLARY 5.6 *Let* A *be an* H-*selfadjoint matrix. Then every* H-*selfadjoint matrix which is similar to* A *and sufficiently close to* A *is* H-*unitarily similar to* A.

Observe that when G is positive definite, G-selfadjoint matrices A_1 and A are G-unitarily similar if and only if they are similar. The same is true for negative definite G. However, in general, it is true only under the additional assumption that A_1 is sufficiently close to A.

Theorems I.3.6 and Corollary 5.6 combine to prove the following fact.

THEOREM 5.7 *The set* $\Sigma(A)$ *of all* H-*selfadjoint matrices which are similar to a fixed* H-*selfadjoint matrix* A *is a disconnected union of a finite number of equivalence classes of* H-*unitarily similar matrices, each class is a connected set and consists of all* $B \in \Sigma(A)$ *with like sign characteristic.*

It is not difficult to express the number of different equivalence classes in $\Sigma(A)$ in terms of the Jordan structure of A and the sign characteristic of (A,H). Namely, let J be the Jordan form of A, and let (A,H) and $(J,P_{\varepsilon,J})$ be $(H,P_{\varepsilon,J})$-unitarily similar, i.e. $J = T^{-1}AT$,

$P_{\varepsilon,J} = T^*HT$ for some invertible T (cf. Theorem I.3.3).

Then for every set of signs ε' such that $\operatorname{sig} P_{\varepsilon',J} = \operatorname{sig} P_{\varepsilon,J}$ there exists an invertible matrix T' such that $P_{\varepsilon',J} = T'^*HT$. Put $A' = T'JT'^{-1}$; then A' is H-selfadjoint. So for every set of sign ε' such that $\operatorname{sig} P_{\varepsilon',J} = \operatorname{sig} P_{\varepsilon,J}$ we constructed an H-selfadjoint matrix $A' = A(\varepsilon')$. By Theorem I.3.6, $A(\varepsilon')$ and $A(\varepsilon")$ are H-unitarily similar if and only if the sets of signs ε' and $\varepsilon"$ are equivalent (i.e. one is obtained from another by permutation of signs corresponding to equal Jordan blocks), and for every equivalence class in $\Sigma(A)$ there is a representative $A(\varepsilon')$ in this class for some ε'. Thus, the number of such equivalence classes in $\Sigma(A)$ coincides with the number of non-equivalent systems of signs ε' such that $\operatorname{sig} P_{\varepsilon',J} = \operatorname{sig} P_{\varepsilon,J}$. Let $\lambda_1,\cdots,\lambda_s$ be all the different real eigenvalues of A, with corresponding sizes of Jordan blocks $\alpha_{i1},\cdots,\alpha_{i,k_i}$, $i = 1,\cdots,s$. Then

$$\operatorname{sig} P_{\varepsilon,J} = \frac{1}{2} \sum_{i,j} [1-(-1)^{\alpha_{ij}}]\varepsilon_{ij}$$

where ε_{ij} are the signs in ε. So the number of different equivalence classes in $\Sigma(A)$ is equal to the number of all systems of signs $\varepsilon' = (\varepsilon'_{ij})$, which are pairwise non-equivalent and satisfy

$$\sum_{i,j} [1-(-1)^{\alpha_{ij}}]\varepsilon'_{ij} = \sum_{i,j} [1-(-1)^{\alpha_{ij}}]\varepsilon_{ij} . \tag{5.5}$$

5.3 Special Cases of Unitary Similarity

Here, we use the results of the preceding section to describe unitary similarity for special classes of H-selfadjoint matrices.

Let A be H-selfadjoint and denote by $\Gamma(A,H)$ the number of H-unitary similarity classes in the set $\Sigma(A)$ of all matrices B which are H-selfadjoint and which are also similar to A. By Theorem 5.7, $\Gamma(A,H)$ is a finite number. The important case when $\Gamma(A,H) = 1$ means that if A is similar to an H-selfadjoint matrix B, then the similarity matrix can be chosen H-unitary; because the relations $HB = B^*H$, $A = S^{-1}BS$ for invertible S imply (provided $\Gamma(A,H) = 1$) the existence of a U such that $A = U^{-1}BU$ and $U^*HU = H$.

First, consider the case when $\operatorname{sig} H = n - 2$ (i.e. H has exactly 1 negative eigenvalue). Taking into account equality (5.5) and the results described in Section I.3.7, we obtain the following results.

THEOREM 5.8 *Let* sig $H = n - 2$, *and let* A *be* H-*selfadjoint.*
(i) *If* $\sigma(A)$ *is real and all elementary divisors of* $I\lambda - A$ *are linear, then* $\Gamma(A,H)$ *is equal to the number of different eigenvalues of* A;
(ii) *If* $\sigma(A)$ *is real, one elementary divisor of* $I\lambda - A$ *is quadratic and the rest are linear, then* $\Gamma(A,H) = 2$;
(iii) *If* $\det(I\lambda - A)$ *has non-real zero, then* $\Gamma(A,H) = 1$;
(iv) *If* $\sigma(A)$ *is real, one elementary divisor of* $I\lambda - A$ *is cubic, and the rest are linear, then* $\Gamma(A,H) = 1$.

For the case when H has 2 negative eigenvalues, we have the following table of values of $\Gamma(A,H)$ corresponding to the 12 cases listed in Section I.3.7. We denote by k the number of different real eigenvalues of A.

Case	$\Gamma(A,H)$
1	$\frac{k(k-1)}{2} + \ell$, where ℓ is the number of different multiple (with multiplicity ≥ 2) eigenvalues of A.
2	4, if the quadratic elementary divisors of $I\lambda - A$ belong to different eigenvalues; 3, if the quadratic elementary divisors belong to the same eigenvalue.
3	1
4	2
5	2
6	k, if the elementary divisor of degree 3 is the only elementary divisor of $I\lambda - A$ belonging to the same eigenvalue; $k + 1$, otherwise.
7	the same as case 6
8	$2k$
9	k
10	2
11	1
12	1

Now consider matrices which are definite with respect to H (ref. Section I.3.8). Combining Theorems I.3.14 and I.3.6 we obtain the following corollary.

COROLLARY 5.9 *If* A *is H-positive, then* $\Gamma(A,H) = 1$. *Moreover, an H-selfadjoint matrix* B *is H-unitarily similar to* A *if and only if* $\dim(\lambda I-A) = \dim(\lambda I-B)$ *for all real* λ. *In this case* B *is also H-positive.*

Using Theorem I.3.15 and equality (5.5) one can easily check that Corollary 5.9 holds if "positive" is replaced by "nonnegative".

5.4 Continuous Dependence of the Canonical Form

In Section 5.1 we proved that the sign characteristic is stable under small structure preserving perturbations. In this section we shall show that, in fact, the canonical form is also stable. For this, we shall need to assume that the Jordan structure is preserved for every eigenvalue of the H-selfadjoint matrix A, and not only for a fixed real eigenvalue.

Let $A(t)$, $t \in [0,1]$ be a continuous family of $n \times n$ matrices. We say that $A(t)$ has *fixed Jordan structure* if the following conditions are satisfied:

1) the number of different eigenvalues of $A(t)$ is independent of $t \in [0,1]$;
2) the sizes of Jordan blocks with a continuous eigenvalue $\lambda_0(t)$, $t \in [0,1]$ in the Jordan form of $A(t)$ are independent of $t \in [0,1]$. (In view of (1) the notion of a continuous eigenvalue of $A(t)$ is not ambiguous.)

Let A be H-selfadjoint. Let $\lambda_1, \cdots, \lambda_{a+b}$ be all the different eigenvalues of A such that $\lambda_1, \cdots, \lambda_a$ are in the open upper half-plane; $\lambda_{a+1}, \cdots, \lambda_{a+b}$ are real; $\lambda_{a+b+i} = \bar{\lambda}_i$. Let $m_{i1}, \cdots, m_{i,k_i}$, $i = 1, \cdots, 2a+b$ be the sizes of Jordan blocks with eigenvalue λ_i in the Jordan form of A. A Jordan basis of A in $\mathbb{C}^n$

$$x_{10}^{(i)}, \cdots, x_{1,m_{i1}-1}^{(i)} \; ; \; \cdots \; ; \; x_{k_i 0}^{(i)}, \cdots, x_{k_i, m_{ik_i}-1}^{(i)} \; ; \qquad i = 1, \cdots, 2a+b \tag{5.6}$$

(so that $x_{pj}^{(i)}$, $j = 0, \cdots, m_{ip}-1$ is a Jordan chain of A corresponding to λ_i, for $p = 1, \cdots, k_i$) is called *selfadjoint* if

$$[Hx^{(i_1)}_{j_1\ell_1}, x^{(i_2)}_{j_2\ell_2}] = \begin{cases} \varepsilon(i_1,j_1), & \text{if } a+1 \le i_1 = i_2 \le a+b \text{ and } j_1 = j_2 \text{ and} \\ & \ell_1 + \ell_2 = m_{i_1 j_1} - 1 \\ 1, & \text{if } 1 \le i_1 \le \alpha;\ i_2 = i_1 + a + b \text{ and} \\ & j_1 = j_2 \text{ and } \ell_1 + \ell_2 = m_{i_1 j_1} - 1 \\ 0, & \text{in all other cases.} \end{cases}$$

Here $\varepsilon(i_1,j_1) = \pm 1$ depending on i_1 and j_1 only (and not depending on ℓ_1 and ℓ_2).

The proof of Theorem I.3.3 shows that for every H-selfadjoint matrix A there exists a selfadjoint Jordan basis. The following result shows that this basis can be chosen to depend continuously on a parameter, provided A and H are continuous and A has a fixed Jordan structure.

THEOREM 5.10 *Let* $A(t)$, $H(t)$, $t \in [0,1]$ *be continuous families of matrices such that* $H(t) = H(t)^*$ *is invertible, and* $A(t)$ *is* $H(t)$*-selfadjoint for all* $t \in [0,1]$. *Assume that* $A(t)$ *has a fixed Jordan structure; let* $\lambda_1(t),\cdots,\lambda_a(t)$ *and* $\lambda_{a+1}(t),\cdots,\lambda_{a+b}(t)$ *be all the different eigenvalues of* $A(t)$ *in the open upper half-plane and on the real line, respectively; and let* $m_{i1},\cdots,m_{i,k_i}$ *be the sizes the Jordan blocks with eigenvalue* $\lambda_i(t)$ *in the Jordan form of* $A(t)$, $i = 1,2,\cdots,2a+b$ (*it is assumed that* $\lambda_{a+b+i}(t) = \overline{\lambda_i(t)}$, $i = 1,\cdots,a$). *Then there exist continuous vector functions in* $\mathbb{C}^n$

$$x^{(i)}_{10}(t),\cdots,x^{(i)}_{1,m_{i1}-1}(t);\cdots;x^{(i)}_{k_i 0}(t),\cdots,x^{(i)}_{k_i,m_{i,k_i}-1}(t)\ ;\ i = 1,\cdots,2a+b \quad (5.7)$$
$$t \in [0,1]$$

such that for each $t \in [0,1]$ *the vectors* (5.7) *form a selfadjoint Jordan basis for* $A(t)$ *in* $\mathbb{C}^n$.

In the proof of Theorem 5.10 we shall use extensively the notions and results given in the Appendix to Part III.

PROOF. We shall take advantage of the proof of Theorem I.3.3 and the notation introduced there. Note that the subspaces $X_i(t)$ are continuous, i.e. images of a continuous projector valued function (in fact, $X_i(t) = \operatorname{Im} Q_i(t)$, where $Q_i(t)$ is the projector $\frac{1}{2\pi i}\int_{\Gamma_i(t)}(\lambda I - A(t))^{-1}d\lambda$ for some suitable contour $\Gamma_i(t)$; evidently $Q_i(t)$ is continuous).

Consider a fixed i, $1 \le i \le a$. By Corollary A.8 in the Appendix, we find a continuous vector function $a_1(t) \in X_i'(t)$ such that $(A(t)-\lambda_i(t)I)^m a_1(t) = 0$, $(A(t)-\lambda_i(t)I)^{m-1} a_1(t) \ne 0$. To find $b_1 \in X_i''$ we have to solve the following equation:

$$((Q_i''(t))^* H(t)(A(t)-\lambda_i(t)I)^{m-1} a_1(t) \ , \ b_1) = 1 \ , \tag{5.8}$$

where

$$Q_i''(t) = \frac{1}{2\pi i} \int_{\Gamma_i''(t)} (\lambda I - A(t))^{-1} d\lambda \ ,$$

and $\Gamma_i''(t)$ is a contour such that $\bar{\lambda}_i$ is inside $\Gamma_i''(t)$, and all other eigenvalues of $A(t)$ are outside $\Gamma_i''(t)$. System (5.8) has a continuous solution; again by Corollary A.8 in the Appendix. Indeed, the continuous $1 \times n$ matrix

$$g(t) \stackrel{\text{def}}{=} [(Q_i''(t))^* H(t)(A(t)-\lambda_i(t)I)^{m-1} a_1(t)]^T$$

has rank 1 for every $t \in [0,1]$; so there exists a continuous vector function $\tilde{b}(t)$ such that $g(t)\tilde{b}(t) \ne 0$, $t \in [0,1]$. Then put $b(t) = (g(t)\tilde{b}(t))^{-1}\tilde{b}(t)$ to satisfy (5.8).

Then we construct continuous vector functions $a_m(t),\cdots,a_1(t)$; $b_m(t),\cdots,b_1(t)$ as in the proof of Theorem I.3.3. Define

$$c_1(t) = a_1(t) + \sum_{j=2}^{m} \alpha_j(t) a_j(t) \ ,$$

where the numbers $\alpha_j(t)$ are such that

$$[c_1(t), b_{m-1}(t)] = \cdots = [c_1(t), b_1(t)] = 0 \ .$$

This amounts to the following relations (in view of (I.3.6), (I.3.7)):

$$[a_1(t), b_{m-1}(t)] + \alpha_2(t) = 0 \ ;$$

$$[a_1(t), b_{m-2}(t)] + \alpha_2(t)[a_2(t), b_{m-1}(t)] + \alpha_3(t) = 0 \ ;$$

$$[a_1(t), b_1(t)] + \alpha_2(t)[a_2(t), b_2(t)] + \cdots + \alpha_m(t) = 0 \ .$$

Evidently, these equations define $\alpha_i(t)$ as continuous functions of $t \in [0,1]$.

Put $N_1(t) = \mathrm{Span}\{c_1(t),\cdots,c_m(t),b_1(t),\cdots,b_m(t)\}$. Clearly, the matrix $Z(t) \overset{def}{=} [c_1(t),\cdots,c_m(t),b_1(t),\cdots,b_m(t)]$ is of size $n \times 2m$ is continuous for $t \in [0,1]$ and has linearly independent columns for all $t \in [0,1]$. Let $x_1(t),\cdots,x_p(t)$ be a continuous basis in $\mathrm{Ker}(Z(t)^*H(t))$ (such a basis exists in view of Theorem A.6 in the Appendix). Clearly, $\mathrm{Span}\ x_1(t),\cdots,x_p(t)$ coincides with the H-orthogonal complement $N_1(t)^{[\perp]} = \{x \in \mathbb{C}^n \mid (H(t)x,y) = 0 \text{ for all } y \in N_1(t)\}$. Write $\hat{A}(t) = A(t)|_{N_1(t)^{[\perp]}}$ and $\hat{H}(t) = PH(t)|_{N_1(t)^{[\perp]}}$, where P is the orthogonal projector on $N_1(t)^{[\perp]}$, as $p \times p$ continuous matrices in the basis $x_1(t),\cdots,x_p(t)$, and repeat the same argument for $\hat{A}(t)$ and $\hat{H}(t)$ until the non-real eigenvalues of $A(t)$ are exhausted.

Now consider a fixed $X_i(t)$, where $i = a+1,\cdots,a+b$, so that λ_i is real. Again, let m be such that $(A(t)-\lambda_i(t)I)^m|_{X_i(t)} = 0$ but $(A(t)-\lambda_i(t)I)^{m-1}|_{X_i(t)} \neq 0$. Let $Q_i(t)$ be the orthogonal (in the sense of the original scalar product in $\mathbb{C}^n$) projector on $X_i(t)$ and define the continuous (on t) linear transformation $F(t) : X_i(t) \to X_i(t)$ by $F(t) = Q_i(t)H(t)(A(t) - \lambda_i(t)I)^{m-1}$. Since $\lambda_i(t)$ is real, $F(t)$ is selfadjoint. Moreover, $F(t) \neq 0$. So there exists a continuous vector function $a_1(t) \in X_i(t)$, $t \in [0,1]$ such that

$$(F(t)a_1(t),a_1(t)) \neq 0 , \quad t \in [0,1] .$$

(One can prove this fact using Theorem A.6 and Corollary A.9 of the Appendix; Corollary A.9 is applicable, because the number of positive (resp. negative) eigenvalues of $F(t)$ coincides with the number of signs $+1$ (resp. -1) corresponding to the Jordan blocks of $A(t)$ with eigenvalue $\lambda_i(t)$ and maximal size in the sign characteristic of $(A(t),H(t))$. This number is constant in view of Theorem 5.1.) Normalizing $a_1(t)$ we can assume that

$$(F(t)a_1(t),a_1(t)) = \varepsilon , \quad \varepsilon = \pm 1 , \quad t \in [0,1] ,$$

or

$$(A(t)-\lambda_i(t)I)^{m-1}a_1(t),a_1(t) \ = \varepsilon . \tag{5.9}$$

Let $a_j(t) = (A(t)-\lambda_i(t)I)^{j-1}a_1(t)$, $j = 1,\cdots,m$. Verify:

$$[a_j(t),a_k(t)] = \begin{cases} \varepsilon & \text{for} \quad j+k = m+1 \\ 0 & \text{for} \quad j+k > m+1 . \end{cases}$$

Put

$$b_1(t) = a_1(t) + \sum_{j=2}^{m} \alpha_j(t)a_j(t) ;$$

$$b_j(t) = (A(t)-\lambda_i(t)I)^{j-1}b_1(t) ; \quad j = 1,\cdots,m ,$$

where $\alpha_j(t)$ are such that

$$[b_1(t),b_1(t)] = \cdots = [b_1(t),b_{m-1}(t)] = 0 .$$

It is easily seen that $\alpha_j(t)$ are uniquely determined and continuous. Now $b_1(t),\cdots,b_m(t)$ is a part of a selfadjoint canonical set of Jordan chains. Put

$$N(t) = \mathrm{Span}\{b_1(t),\cdots,b_m(t)\}$$

and repeat the same process for $N(t)^{[\perp]}$. □

The following corollary is just a restatement of Theorem 5.10 in different terms.

COROLLARY 5.11 *Let* $A(t)$, $H(t)$ *be as in Theorem* 5.10. *Then there exists a continuous invertible* $n \times n$ *matrix function* $S(t)$, $t \in [0,1]$ *such that*

$$A(t) = S(t)^{-1}J(t)S(t) ; \quad H(t) = S(t)^{*}P_{\varepsilon,J}S(t) , \tag{5.10}$$

where $(J(t),P_{\varepsilon,J})$ *is the canonical form of* $(A(t),H(t))$.

Indeed, let $x_1(t),\cdots,x_n(t)$ be a continuous selfadjoint Jordan basis of $A(t)$; which exists in view of Theorem 5.10. Let $S(t)$ be the invertible $n \times n$ matrix whose columns are $x_1(t),\cdots,x_n(t)$. Then (5.10) holds.

We point out two more corollaries from Theorem 5.10.

COROLLARY 5.12 *Let* H *be a hermitian invertible* $n \times n$ *matrix and let* $A(t)$, $B(t)$, $t \in [0,1]$ *be continuous families of* H-*selfadjoint matrices. Suppose that for every* $t \in [0,1]$, $A(t)$ *and* $B(t)$ *are* H-*unitarily similar. If the Jordan structure of* $A(t)$, *as well as the Jordan structure of* $B(t)$, *is fixed, then there exists a continuous* H-*unitary matrix function*

$S(t)$, $t \in [0,1]$ *such that*

$$A(t) = S(t)^{-1}B(t)S(t) .$$

PROOF. By Corollary 5.11, write:

$$A(t) = S_A(t)^{-1}J(t)S_A(t) \; ; \quad H = (S_A(t))^* P_{\varepsilon,J} S_A(t) ,$$

where $(J(t),P_{\varepsilon,J})$ is the canonical form of $(A(t),H)$, and $S_A(t)$ is a continuous invertible matrix function. (Note that $P_{\varepsilon,J}$ is fixed in view of the fixed Jordan structure of $A(t)$). Similarly,

$$B(t) = S_B(t)^{-1}J(t)S_B(t) \; ; \quad H = (S_B(t))^* P_{\varepsilon,J} S_B(t) ,$$

where $S_B(t)$ is a continuous invertible matrix function, and $(J(t),P_{\varepsilon,J})$ is the canonical form of $(B(t),H)$ (in view of the H-unitary similarity between $B(t)$ and $A(t)$, the canonical form is the same for $(A(t),H)$ and $(B(t),H)$). Put $S(t) = S_B(t)^{-1}S_A(t)$ to satisfy the requirements of Corollary 5.12. □

Note that the condition that the Jordan structures of $A(t)$ and $B(t)$ are fixed, is essential in Corollary 5.12, even when H is positive definite, as the following example shows.

EXAMPLE 5.2 Let

$$A(t) = e^{-\frac{1}{t^2}} \begin{bmatrix} \cos\frac{2}{t} & \sin\frac{2}{t} \\ \sin\frac{2}{t} & -\cos\frac{2}{t} \end{bmatrix} , \quad t \neq 0 \; ; \quad A(0) = 0 \; ;$$

$$B(t) = \begin{bmatrix} e^{-\frac{1}{t^2}} & 0 \\ 0 & -e^{-\frac{1}{t^2}} \end{bmatrix} , \quad t \neq 0 \; ; \quad B(0) = 0 .$$

Then $A(t)$ and $B(t)$ are continuous and even infinitely differentiable, for all real t (including zero). For each $t \in \mathbb{R}$ the hermitian matrices $A(t)$ and $B(t)$ have the same eigenvalues and therefore are I-unitarily similar. However, any unitary matrix $S(t)$ for which the equality $A(t) = S(t)B(t)S^{-1}(t)$, $t \neq 0$ holds is of the form

$$S(t) = \begin{bmatrix} \alpha(t)\cos\frac{1}{t} & \beta(t)\sin\frac{1}{t} \\ \alpha(t)\sin\frac{1}{t} & -\beta(t)\cos\frac{1}{t} \end{bmatrix}$$

where $\alpha(t)$ and $\beta(t)$ are unimodular numbers. Evidently, $S(t)$ does not admit a continuous extension to the point $t = 0$. □

COROLLARY 5.13 *Let* H *be hermitian and invertible, and let* $A(t)$, $t \in [0,1]$ *be a continuous family of* H*-selfadjoint matrices. Suppose* $A(t)$ *is* H*-unitarily similar to* $A(0)$, *for every* $t \in [0,1]$. *Then there exists a continuous* H*-unitary matrix function* $S(t)$, $t \in [0,1]$ *such that*

$$A(t) = S(t)^{-1}A(0)S(t)\ , \quad t \in [0,1]\ .$$

The proof is immediate: Apply Corollary 5.12 with $B(t) \equiv A(0)$.

The results of this section as well as their proofs, are also valid for other classes of matrix functions; for instance, for C^p-functions, where p is a positive integer or $p = \infty$.

5.5 Analytic Dependence of the Canonical Form

In this section we shall state a result analogous to Theorem 5.10 for analytic matrix functions. An $n \times n$ matrix function $A(t)$ defined for $a < t < b$ (here a and b are real numbers) is called *(real) analytic on* (a,b) if in a neighbourhood of each point t_0, $A(t)$ can be represented by a power series:

$$A(t) = \sum_{j=0}^{\infty} A_j(t-t_0)^j\ .$$

Analytic vector functions are defined similarly.

THEOREM 5.14 *Let* $A(t)$, $H(t)$ *be* $n \times n$ *analytic matrix functions on an interval* $(a,b) \in \mathbb{R}$, *for which* $H(t) = H(t)^*$ *is invertible, and* $A(t)$ *is* $H(t)$*-selfadjoint for all* $t \in (a,b)$. *Assume that* $A(t)$ *has a fixed Jordan structure; let* $\lambda_1(t),\cdots,\lambda_a(t)$ *and* $\lambda_{a+1}(t),\cdots,\lambda_{a+b}(t)$ *be all the different eigenvalues of* $A(t)$ *in the open upper half-plane and on the real line, respectively; and let* $m_{i1},\cdots,m_{i,k_i}$ *be the sizes of Jordan blocks with eigenvalue* $\lambda_i(t)$ *in the Jordan form of* $A(t)$, $i = 1,2,\cdots,2a+b$ *(here* $\lambda_{a+b+i}(t) = \overline{\lambda_i(t)}$, $i = 1,\cdots,a$). *Then there exist analytic vector functions on* (a,b):

$$x_{10}^{(i)}(t),\cdots,x_{1,m_{i1}-1}^{(i)}(t);\cdots;x_{k_i 0}^{(i)}(t),\cdots,x_{k_i,m_{i,k_i}-1}^{(i)}(t)\ ,\quad i = 1,\cdots,2a+b\ ;\quad t \in (a,b)$$

which form a selfadjoint Jordan basis of $A(t)$ *in* $\mathbb{C}^n$ *for every* $t \in (a,b)$.

Theorem 5.14 is proved exactly as Theorem 5.10, using analytic analogues of Theorem A.6 and Corollaries A.8, A.9 (see Section A.3 of the Appendix).

Theorem 5.14 can also be reformulated in terms of matrices which produce the unitary similarity of $(A(t),H(t))$ to its canonical form, as follows (this is an analogue of Corollary 5.11):

COROLLARY 5.15 *Let* $A(t)$, $H(t)$ *be as in Theorem* 5.14. *Then there exists an analytic invertible* $n \times n$ *matrix function* $S(t)$ *on the interval* (a,b) *such that*

$$A(t) = S(t)^{-1}J(t)S(t)\ ;\quad H(t) = (S(t))^*P_{\varepsilon,J}S(t)\ ,\quad t \in (a,b)$$

where $(J(t),P_{\varepsilon,J})$ *is the canonical form of* $(A(t),H(t))$.

5.6 H-Unitary Matrices

The results developed in Sections 5.1-5.5 admit natural analogues for H-unitary matrices. One can obtain these analogues via the Cayley transform: If A is H-selfadjoint, and $\omega \notin \sigma(A)$ is a non-real number, then the matrix $U = (A-\bar{\omega}I)(A-\omega I)^{-1}$ is H-unitary; conversely, if U is an H-unitary matrix and $\alpha \notin \sigma(U)$ is a unimodular number, then $A = i(U+\alpha I)(U-\alpha I)^{-1}$ is H-selfadjoint (see Chapter I.2). However, Theorems 5.10 and 5.14 require caution: Let $U(t)$, $H(t)$ be continuous (resp. analytic) families of $n \times n$ matrices for $t \in [0,1]$ (resp. $a < t < b$) such that $U(t)$ is $H(t)$-unitary for each t and $U(t)$ has a fixed Jordan structure. Then, in order to make the Cayley transform work, one must choose a continuous (resp. analytic) unimodular number $\alpha(t)$ such that $\alpha(t) \notin \sigma(U(t))$ for all t. It is easy to see that such a choice is always possible.

As an example, we present some unitary analogues of the results in Sections 5.1-5.5.

THEOREM 5.16 *Let* U *be H-unitary and let* λ_0 *be a unimodular eigenvalue of* U. *Then there exists a* λ_0*-structure preserving neighbourhood* $\mathcal{U}_U$ *of* U *and a neighbourhood* $\mathcal{U}_H$ *of* H *such that if* $U_1 \in \mathcal{U}_U$ *is* H_1*-unitary for some hermitian invertible matrix* $H_1 \in \mathcal{U}_H$, *then the sign characteri-*

stic of (U,H) *at* λ_0 *and the sign characteristic of* (U_1,H_1) *at the unimodular eigenvalue close to* λ_0 *are the same.*

This theorem is obtained from Theorem 5.1. A λ_0-structure preserving neighbourhood of an H-unitary matrix corresponding to its unimodular eigenvalue λ_0, is defined as for a λ_0-structure preserving neighbourhood of an H-selfadjoint matrix corresponding to its real eigenvalue λ_0 (see Section 5.1).

The following analogue of Corollary 5.12 will be used in the sequel.

COROLLARY 5.17 *Let* $H = H^*$ *be invertible, and let* $U(t)$, $V(t)$, $t \in [0,1]$ *be continuous families of* H-*unitary matrices. Suppose that for every* $t \in [0,1]$, $U(t)$ *and* $V(t)$ *are* H-*unitarily similar; and suppose that the Jordan structure of* $U(t)$, *as well as the Jordan structure of* $V(t)$, *is fixed. Then there exists a continuous* H-*unitary function* $S(t)$, $t \in [0,1]$ *such that* $U(t) = S(t)^{-1}V(t)S(t)$.

5.7 Connected Components of Selfadjoint Matrices With Like Real Jordan Structure

Denote by $\mathcal{A}$ the set of all pairs (A,H), where H is an $n \times n$ hermitian invertible matrix and A is H-selfadjoint. We introduce a set of parameters which will determine the Jordan structure of an $n \times n$ matrix. Let S be an ordered set of numbers, consisting of:

(i) a nonnegative integer r;

(ii) positive integers $s_1,\cdots,s_r$;

(iii) a set of positive integers $m_{i1} \geq \cdots \geq m_{i,s_i}$ for every $i = 1,\cdots,r$.

Denote by $\mathcal{S}$ the (finite) collection of all ordered sets S with the structure described in (i), (ii), and such that $\sum_{i=1}^{r} \sum_{j=1}^{s_i} m_{ij} \leq n$ (the number n is fixed; so $\mathcal{S} = \mathcal{S}(n)$). For every $S \in \mathcal{S}$ denote by $\mathcal{A}_S$ the set of pairs $(A,H) \in \mathcal{A}$ such that A has r distinct real eigenvalues $\lambda_1 < \cdots < \lambda_r$ (which may depend on A), and the integers $m_{i1} \geq \cdots \geq m_{is_i}$ are the sizes of Jordan blocks with eigenvalue λ_i in the Jordan form of $A (i = 1,\cdots,r)$. Here

$$S = \{r;s_1,\cdots,s_r;m_{11},\cdots,m_{1s_1},m_{21},\cdots,m_{2s_2},\cdots,m_{rs_r}\} .$$

For every $(A,H) \in \mathcal{A}_S$ the sign characteristic of (A,H) is assumed to satisfy the following normalization condition: if $m_{ij_1} = m_{ij_2} \neq 0$ and $j_1 < j_2$ then the corresponding signs satisfy $\varepsilon_{ij_1} \geq \varepsilon_{ij_2}$.

It turns out that the connected components in A_S are distinguished by the sign characteristic.

THEOREM 5.18 *The set* A_S *is a disconnected union of the arcwise connected sets* $A_{S,\varepsilon}$, *where* $A_{S,\varepsilon}$ *consists of all pairs* (A,H) *with fixed sign characteristic* $\varepsilon = (\varepsilon_{ij})_{i=1,\cdots,r}^{j=1,\cdots,s_i}$.

PROOF. Theorem 5.1 ensures that each $A_{S,\varepsilon}$ is open in A_S. This implies that if two pairs $(A,H),(B,G) \in A_S$ are connected by a continuous path, then they belong to the same $A_{S,\varepsilon}$.

So it remain to prove that each $A_{S,\varepsilon}$ is connected.

Let $(A,H),(B,G) \in A_{S,\varepsilon}$. We shall prove that there is a continuous path in $A_{S,\varepsilon}$ connecting (A,H) and (B,G). Let $L_+(A)$, $L_0(A)$, $L_-(A)$ be the spectral subspaces of A corresponding to the parts of $\sigma(A)$ lying in the open upper half-plane, the real line, and the open lower half-plane respectively. Similarly, introduce subspaces $L_+(B)$, $L_0(B)$, $L_-(B)$. Passing, if necessary, to the canonical forms of (A,H) and (B,G) and using Theorem I.3.3, we shall assume that $L_j(A) = L_j(B)$, $j = +,-,0$, and the subspaces $L_+(A)$, $L_0(A)$, $L_-(A)$ are mutually orthogonal (with respect to the usual scalar product in $\mathbb{C}^n$).

Write A, H, B, G as block matrices with respect to the decomposition

$$\mathbb{C}^n = L_+(A) \oplus L_0(A) \oplus L_-(A) :$$

$$A = \operatorname{diag}[A_+, A_0, A_-] \ ; \quad B = \operatorname{diag}[B_+, B_0, B_-] \ ;$$

$$H = \begin{bmatrix} 0 & 0 & H_+ \\ 0 & H_0 & 0 \\ H_+^* & 0 & 0 \end{bmatrix} ; \quad G = \begin{bmatrix} 0 & 0 & G_+ \\ 0 & G_0 & 0 \\ G_+^* & 0 & 0 \end{bmatrix}$$

Since A is H-selfadjoint and B is G-selfadjoint, we find that A_0 is H_0-selfadjoint, B_0 is G_0-selfadjoint and $H_+A_- = A_+^*H_+$, $G_+B_- = B_+^*G_+$.

Let $H_+(t)$, $t \in [0,1]$ be a continuous path of invertible linear transformations $L_-(A) \to L_+(A)$ such that $H_+(0) = H_+$; $H_+(1) = G_+$. Let

$A_+(t)$, $t \in [0,1]$ be a continuous path of matrices with the properties that $\sigma(A_+(t))$ is in the open upper half-plane for all $t \in [0,1]$; $A_+(0) = A_+$; $A_+(1) = B_+$. (Such a path is easily constructed by considering the Jordan normal forms or upper triangular forms of A_+ and B_+). Put

$$B(t) = \mathrm{diag}[A_+(t), B_0, H_+(t)^{-1}A_+(t)^*H_+(t)] \ ;$$

$$G(t) = \begin{bmatrix} 0 & 0 & H_+(t) \\ 0 & G_0 & 0 \\ H_+(t)^* & 0 & 0 \end{bmatrix}.$$

Then $(B(t),G(t)) \in A_{S,\varepsilon}$ for all $t \in [0,1]$; $(B(1),G(1)) = (B,G)$;

$$B(0) = \begin{bmatrix} A_+ & 0 & 0 \\ 0 & B_0 & 0 \\ 0 & 0 & A_- \end{bmatrix} ; \quad G(0) = \begin{bmatrix} 0 & 0 & H_+ \\ 0 & G_0 & 0 \\ H_+^* & 0 & 0 \end{bmatrix}$$

So it remains to construct a continuous path between (A_0,H_0) and (B_0,G_0) in $A_{S,\varepsilon}$ (considered as matrices in some orthonormal basis in the possibly smaller space $L_0(A)$).

Using Theorem I.3.3, we can replace (A_0,H_0) and (B_0,G_0) by their canonical forms $(J,P_{\varepsilon,J})$ and $(J',P_{\varepsilon',J'})$ respectively. Arrange the blocks in J and J' in the non-decreasing order of eigenvalues; and arrange blocks corresponding to the same eigenvalue in the order of non-increasing sizes. So we have:

$$J = \mathrm{diag}[J_1,\cdots,J_r] \ ; \quad J_i = \mathrm{diag}[J_{i1},\cdots,J_{is_i}] \ ,$$

where $\sigma(J_i) = \{\lambda_i\}$; $\lambda_1 < \cdots < \lambda_r$; the size of J_{ik} is m_{ik}; $m_{i1} \geq m_{i2} \geq \cdots \geq m_{is_i}$. Also

$$J' = \mathrm{diag}[J_1',\cdots,J_r'] \ ; \quad J_i' = \mathrm{diag}[J_{i1}',\cdots,J_{is_i}'] \ ,$$

where $\sigma(J_i') = \{\mu_i\}$; $\mu_1 < \cdots < \mu_r$; and the size of J_{ik}' is m_{ik}; $m_{i1} \geq \cdots \geq m_{is_i}$. Observe that since (A_0,H_0), (B_0,G_0) belong to the same $A_{S,\varepsilon}$ the numbers r, q_i, m_{ik} are the same for J and J'. Also the sign characteristic is the same, so $P_{\varepsilon,J} = P_{\varepsilon',J'} \overset{\text{def}}{=} P$. Now let $\lambda_i(t)$, $i = 1,\cdots,r$ be continuous functions on $t \in [0,1]$ such that

$$\lambda_1(t) < \cdots < \lambda_r(t) \ , \quad t \in [0,1] \ ;$$

$\lambda_i(0) = \lambda_i$; $\lambda_i(1) = \mu_i$. Put $J(t) = \mathrm{diag}[J_1+(\lambda_1(t)-\lambda_1)I,\cdots,J_r+(\lambda_r(t)-\lambda_r)I]$. Then $(J(t),P) \in A_{S,\varepsilon}$; $(J(0),P) = (J,P)$; $(J(1),P) = (J',P)$, and the theorem is proved. □

An analogue of Theorem 5.18 holds also if the indefinite scalar product is kept fixed, as we show in the following theorem. Given $S \in \mathcal{S}$, let $A_S(H) = \{A \mid (A,H) \in A_S\}$; define also $A_{S,\varepsilon}(H) = \{A \mid (A,H) \in A_{S,\varepsilon}\}$, where $\varepsilon = (\varepsilon_{ij})_{i=1,\cdots,r}^{j=1,\cdots,s_i}$ is a corresponding set of signs.

THEOREM 5.19 *The set* $A_S(H)$ *is a disconnected union of the arcwise connected sets* $A_{S,\varepsilon}(H)$.

PROOF. Again, by Theorem 5.1, we have only to prove that every set $A_{S,\varepsilon}(H)$ is arcwise connected.

Let $A,B \in A_{S,\varepsilon}(H)$. We shall show that there exists a continuous path in $A_{S,\varepsilon}(H)$ connecting A and B. As in the proof of Theorem 5.18, introduce the subspaces $L_j(A)$, $L_j(B)$, $j = +,-,0$. Let $x_1(A),\cdots,x_p(A)$ be a basis in $L_0(A)$ in which the pair $(A|_{L_0(A)},P_0(A)H|_{L_0(A)})$ has a canonical form (here $P_0(A)$ is the orthogonal projector on $L_0(A)$). Let $y_1(A),\cdots,y_q(A)$ and $z_1(A),\cdots,z_q(A)$ be bases in $L_+(A)$ and $L_-(A)$ respectively such that

$$(Hy_i(A),z_j(A)) = \begin{cases} 0 & \text{if} \quad i \neq j \\ 1 & \text{if} \quad i = j \end{cases}$$

(and of course $(Hy_i(A),y_j(A)) = (Hz_i(A),z_j(A)) = 0$). The existence of such bases follows from the canonical form. Similarly construct bases $x_1(B),\cdots,x_p(B)$; $y_1(B),\cdots,y_q(B)$; $z_1(B),\cdots,z_q(B)$ in $L_0(B)$, $L_+(B)$, $L_-(B)$, respectively (we assume, in view of the condition $A,B \in A_{S,\varepsilon}(H)$ that the matrix of $P_0(A)H|_{L_0(A)}$ in the basis $x_j(A)$, $j = 1,\cdots,p$ and that of $P_0(B)H|_{L_0(B)}$ in the basis $x_j(B)$, $j = 1,\cdots,p$ are identical). Let T be the invertible $n \times n$ matrix defined by the rule $T(x_i(A)) = x_i(B)$; $i = 1,\cdots,p$; $T(y_i(A)) = y_i(B)$; $T(z_i(A)) = z_i(B)$, $i = 1,\cdots,q$. Then $(Hx,y) = (HTx,Ty)$ for all $x,y \in \mathbb{C}^n$, and therefore T is H-unitary: $H = T^*HT$. Put $B_1 = T^{-1}BT$. Then B_1 is H-selfadjoint and $B_1 \in A_{S,\varepsilon}(H)$. In view of Theorem I.3.9 it remains to show that B_1 is arcwise connected with A in $A_{S,\varepsilon}(H)$.

Observe that $L_i(B_1) = L_i(A)$; $i = +,0,-$ (in the selfevident notation). Without loss of generality we can assume that $(A,H) = (J,P_{\varepsilon,J})$ is in the canonical form. Then the subspaces $L_+(A)$, $L_0(A)$, $L_-(A)$ are orthogonal (in the usual scalar product in $\mathbb{C}^n$).

We follow now the line of argument used in the proof of Theorem 5.18. Write with respect to the decomposition $\mathbb{C}^n = L_+(A) \oplus L_0(A) \oplus L_-(A)$:

$$B_1 = \mathrm{diag}[B_{1+}, B_{10}, B_{1-}] \ ; \quad A = \mathrm{diag}[J_+, J_0, J_-] \quad (\text{here } J_- = \bar{J}_+) \ ;$$

$$H = \begin{bmatrix} 0 & 0 & P_1 \\ 0 & P_0 & 0 \\ P_1 & 0 & 0 \end{bmatrix} .$$

Let $A_+(t)$, $t \in [0,1]$ be a continuous path of matrices such that $\sigma(A_+(t))$ is in the open upper half-plane for all $t \in [0,1]$; $A_+(0) = J_+$; $A_+(1) = B_{1+}$. Put

$$B_1(t) = \mathrm{diag}[A_+(t), B_{10}, P_1^{-1} A_+(t)^* P_1] \ .$$

Then $B_1(1) = B_1$; $B_1(t)$ is H-selfadjoint for all $t \in [0,1]$; $B_1(t) \in A_{S,\varepsilon}(H)$; and $B_1(0) = \mathrm{diag}[J_+, B_{10}, J_-]$. So in order to find a connection between B_1 and A, it is sufficient to find a connection between B_{10} and J_0. To this end pass to the canonical form (J_0', P_0) of (B_{10}, P_0) using Theorem I.3.9 and then apply the argument used in the last part of the proof of Theorem 5.18. □

Theorems 5.18 and 5.19 admit H-unitary analogues, which can be proved by using the Cayley transform. In this case two sets

$$S_1 = \{r; s_1, \cdots, s_r; m_{11}, \cdots, m_{1s_1}; \cdots; m_{r1}, \cdots, m_{rs_r}\} \in S$$

and

$$S_2 = \{r; p_1, \cdots, p_r; n_{11}, \cdots, n_{1p_1}; \cdots; n_{r1}, \cdots, n_{rp_r}\} \in S$$

should be identified if there is a cyclic permutation $\pi : \{1, \cdots, r\} \to \{1, \cdots, r\}$ such that $p_{\pi(i)} = s_i$, $n_{\pi(i),q} = m_{iq}$ for all $i = 1, \cdots, r$ and $q = 1, \cdots, s_i$.

5.8 The Real Case

If A is H-selfadjoint and both A and H are real matrices, then many of the results of this chapter still apply provided the notions of unitary similarity and H-unitary similarity are replaced by real unitary similarity and real H-unitary similarity, respectively. Also, the real Jordan structure of A has to be used rather than the Jordan form for complex matrices. All the results of Sections 5.1-5.6 (with the exception of Theorem 5.7) hold also in the real case.

Theorem 5.7 as stated does not hold in the real case, because the classes of real H-unitary similarity are not necessarily connected (Theorem I.5.5). However, the set $\Sigma_r(A)$ of all real H-selfadjoint matrices (with fixed H) which are similar to a fixed real H-selfadjoint matrix A, is a disconnected union of a finite number of real H-unitary similarity classes. Each class consists of all $B \in \Sigma_r(A)$ with like sign characteristic.

The results of Section 5.7 do not hold in the real case. Let us present the description of connected components in the set of real selfadjoint matrices with like Jordan structure.

Let S be an ordered set of numbers, as in Section 5.7. Consider the set A_S^r of all pairs of real $n \times n$ matrices $(A,H) \in A_S$. As in the complex case, A_S^r is a disconnected union of the sets $A_{S,\varepsilon}^r$ consisting of all pairs $(A,H) \in A_S^r$ with fixed sign characteristic ε. But now the sets $A_{S,\varepsilon}$ are not necessarily connected, as the following theorem shows.

THEOREM 5.20 *Assume that all eigenvalues of* A *are real. If the Jordan form* J *of* A *has a block of odd size, then the set* $A_{S,\varepsilon}^r$ *is connected; otherwise* $A_{S,\varepsilon}^r$ *has exactly two connected components.*

PROOF. Let J have a block of odd size, let $(A,H),(B,G) \in A_{S,\varepsilon}^r$, and let $(J,P_{\varepsilon,J})$, $(J',P_{\varepsilon',J'})$ be the (real) canonical forms of (A,H) and (B,G) respectively. By Theorem I.5.5, there exist continuous paths from (A,H) to $(J,P_{\varepsilon,J})$ and from (B,G) to $(J',P_{\varepsilon',J'})$ in $A_{S,\varepsilon}^r$. Now one can connect between $(J,P_{\varepsilon,J})$ and $(J',P_{\varepsilon',J'})$ in $A_{S,\varepsilon}^r$ as in the proof of Theorem 5.18.

Assume now the Jordan form J of A does not have a block of odd size. By Theorem I.5.5 the class of real unitarily similar pairs containing (A,H) has two connected components. Choose (B,G) to be unitarily similar to (A,H) but lying in the other connected component. We claim that (A,H) and (B,G) are not connected in $A_{S,\varepsilon}^r$. Assume the contrary, and let $(A(t),H(t)) \in A_{S,\varepsilon}^r$, $t \in [0,1]$ be a continuous function such that $A(0) = A$, $H(0) = H$, $A(1) = B$, $H(1) = G$. Let $\lambda_1(t) < \cdots < \lambda_r(t)$ be the (real) distinct eigenvalues of $A(t)$. Clearly, $\lambda_i(t)$ are continuous functions of t

(as zeros of the polynomial $\det(\lambda I-A(t))$ with coefficients continuous in t). Let $E_i(t)$ be the root subspace of $A(t)$ corresponding to the eigenvalue $\lambda_i(t)$. Then $\mathbb{C}^n = E_1(t) \dotplus \cdots \dotplus E_r(t)$. Since $(A(t),H(t)) \in A^r_{S,\varepsilon}$, the dimension of $E_i(t)$ does not depend on $t \in [0,1]$, and therefore $E_i(t) = \operatorname{Ker}[(A(t)-\lambda_i(t)I)^n]$ is a continuous family of subspaces (see Proposition A.4 of the Appendix to Part III). Let $U(t)$, $t \in [0,1]$ be the real matrix defined as follows: $U(t)x = (\lambda_i(0)-\lambda_i(t))x$, for $x \in E_i(t)$. It is easily seen (using, for example, the existence of continuous bases in $E_1(t),\cdots,E_r(t)$; see Theorem A.6 of the Appendix) that $U(t)$ is continuous. Also $\widetilde{A}(t) \overset{\text{def}}{=} A(t) + U(t)$ is H-selfadjoint; $\widetilde{A}(t)$ is similar to $A(0)$ and the sign characteristics of $(\widetilde{A}(t),H(t))$ and of $(A(0),H(0))$ coincide. By the real version of Theorem I.3.6, the pair $(\widetilde{A}(t),H(t))$ is unitarily similar to $(A(0),H(0))$. But $(\widetilde{A}(1),H(1)) = (B,G)$; so (A,H) and (B,G) are connected in the unitary similarity class containing (A,H), a contradiction with the choice of (B,G).

So the set $A^r_{S,\varepsilon}$ has at least two connected components. It remains to prove that $A^r_{S,\varepsilon}$ has exactly two connected components. Actually the connected components $A^+_{S,\varepsilon}$ and $A^-_{S,\varepsilon}$ can be described as follows: $A^+_{S,\varepsilon}$ (resp. $A^-_{S,\varepsilon}$) consists of all pairs $(A,H) \in A^r_{S,\varepsilon}$ such that $A = T^{-1}JT$ and $H = T^*P_{\varepsilon,J}T$ for some real canonical form $(J,P_{\varepsilon,J})$ and some real T with $\det T > 0$ (resp. $\det T < 0$). Indeed, the connectedness of $A^+_{S,\varepsilon}$ and $A^-_{S,\varepsilon}$ can be proved as in Theorem 5.18, using the fact that the set of real matrices with fixed size and positive (resp. negative) determinant is connected (see Lemma I.5.6). □

APPENDIX TO PART III

SUBSPACES IN FINITE DIMENSIONAL COMPLEX SPACE

This appendix contains well-known material about a metric in the set of all subspaces of a finite dimensional vector space. Basic properties of this metric space, continuous and analytic families of subspaces are discussed.

A.1 The Metric Space of Subspaces

Let $\mathbb{C}^n$ be the n-dimensional complex space with the standard scalar product (.), and norm $||x|| = \sqrt{(x,x)}$. The norm of an $n \times n$ matrix A (considered as a linear transformation from $\mathbb{C}^n$ to $\mathbb{C}^n$) is defined by

$$||A|| = \max_{x \in \mathbb{C}^n \setminus \{0\}} \frac{||Ax||}{||x||} .$$

For subspaces $S_1, S_2 \subset \mathbb{C}^n$ define the gap $\theta(S_1,S_2)$ by

$$\theta(S_1,S_2) = ||P_{S_1} - P_{S_2}|| ,$$

where P_{S_i} is the orthogonal projector on S_i, $i = 1,2$. Then the function $\theta(S_1,S_2)$ satisfies all the requirements of a metric:

(i) $\theta(S_1,S_2) \geq 0$ and $\theta(S_1,S_2) = 0$ if and only if $S_1 = S_2$;

(ii) $\theta(S_1,S_2) = \theta(S_2,S_1)$;

(iii) $\theta(S_1,S_2) \leq \theta(S_1,S_3) + \theta(S_3,S_1)$.

Consider the metric space $\mathbb{G}_m$ of all subspaces in $\mathbb{C}^n$ of fixed dimension m $(0 \leq m \leq n)$ with the metric $\theta(.,.)$.

THEOREM A.1 *The set* $\mathbb{G}_m$ *is a complete compact metric space, and is open in the set of all subspaces in* $\mathbb{C}^n$.

Recall that compactness of $\mathbb{G}_m$ means that for every sequence

$L_1, L_2, \cdots$ of subspaces in $\mathbb{G}_m$ there exists a converging subsequence $L_{i_1}, L_{i_2}, \cdots$, i.e. such that

$$\lim_{k\to\infty} \theta(L_{i_k}, L_0) = 0$$

for some $L_0 \in \mathbb{G}_m$. Completeness of $\mathbb{G}_m$ means that every sequence of subspaces L_i, $i = 1,2,\cdots$ such that $\lim_{i,j\to\infty} \theta(L_i, L_j) = 0$, converges.

PROOF. Let us prove first the compactness of $\mathbb{G}_m$. To this end consider the set $\mathbb{S}_m$ of all orthonormal systems $u = \{u_k\}_{k=1}^m$ consisting of m vectors $u_1,\cdots,u_m$ in $\mathbb{C}^n$.

For $u = \{u_k\}_{k=1}^m$, $v = \{v_k\}_{k=1}^m \in \mathbb{S}_m$ define

$$\delta(u,v) = \left[\sum_{k=1}^m ||u_k - v_k||^2\right]^{\frac{1}{2}} .$$

It is easily seen that $\delta(u,v)$ is a metric in $\mathbb{S}_m$. So we may treat $\mathbb{S}_m$ as a metric space. For each $u = \{u_k\}_{k=1}^m \in S_m$ define $A_m u = \mathrm{Span}\{u_1,\cdots,u_m\} \in \mathbb{G}_m$. This way we obtain a map $A_m : \mathbb{S}_m \to \mathbb{G}_m$ of metric spaces $\mathbb{S}_m$ and $\mathbb{G}_m$.

We prove now that the map A_m is continuous. Indeed, let $L \in \mathbb{G}_m$ and let $v_1,\cdots,v_m$ be an orthonormal basis in L. Pick some $u = \{u_k\}_{k=1}^m \in \mathbb{S}_m$ (which is taken to be in a neighbourhood of $v = \{v_k\}_{k=1}^m \in \mathbb{S}_m$). For v_i, $i = 1,\cdots,m$ we have (where $M = A_m u$ and P_N stands for the orthogonal projector on the subspace N):

$$\begin{aligned} ||(P_M - P_L)v_i|| &= ||P_M v_i - v_i|| \leq ||P_M(v_i - u_i)|| + ||u_i - v_i|| \\ &\leq ||P_M|| \cdot ||v_i - u_i|| + ||u_i - v_i|| \leq 2\delta(u,v) , \end{aligned}$$

and therefore for $x = \sum_{i=1}^m \alpha_i v_i \in L$ with $||x|| = 1$ we have

$$||(P_M - P_L)x|| \leq 2\delta(u,v) \sum_{i=1}^m |\alpha_i| . \tag{A.1}$$

Now, since $||x|| = \sum_{i=1}^m |\alpha_i|^2 = 1$, we obtain that $|\alpha_i| \leq 1$ and $\sum_{i=1}^m |\alpha_i| \leq m$, and (A.1) gives

$$||(P_M - P_L)|_L|| \leq 2m\delta(u,v) . \tag{A.2}$$

Now fix some $y \in L^\perp$, $||y|| = 1$. We wish to evaluate $P_M y$. For every $x \in L$, write

$$(x,P_M y) = (P_M x,y) = ((P_M - P_L)x,y) + (x,y) = ((P_M - P_L)x,y) ,$$

and

$$|(x,P_M y)| \leq 2m||x||\delta(u,v) \tag{A.3}$$

by (A.2). On the other hand, write

$$P_M y = \sum_{i=1}^{m} \alpha_i u_i ,$$

then for every $z \in L^{\perp}$,

$$(z,P_M y) = (z, \sum_{i=1}^{m} \alpha_i(u_i - v_i)) + (z, \sum_{i=1}^{m} \alpha_i v_i) = (z, \sum_{i=1}^{m} \alpha_i(u_i - v_i))$$

and

$$|(z,P_M y)| \leq ||z|| \; |\sum_{i=1}^{m} \alpha_i(u_i - v_i)| \leq ||z||m \max_{1 \leq i \leq m} |\alpha_i| \; ||u_i - v_i|| .$$

But $||y|| = 1$ implies that $\sum_{i=1}^{m} |\alpha_i|^2 \leq 1$, so $\max_{1 \leq i \leq m} |\alpha_i| \leq 1$. Consequently,

$$|(z,P_M y)| \leq ||z||m\delta(u,v) . \tag{A.4}$$

Combining (A.3) and (A.4) we find that

$$|(t,P_M y)| \leq 3m\delta(u,v)$$

for every $t \in \mathbb{C}^n$, $||t|| = 1$. Thus,

$$||P_M y|| \leq 3m\delta(u,v) . \tag{A.5}$$

Now we can easily prove the continuity of A_m. Pick an $x \in \mathbb{C}^n$ with $||x|| = 1$. Then, using (A.2), (A.5), we have:

$$||(P_M - P_L)x|| \leq ||(P_M - P_L)P_L x|| + ||P_M(x - P_L x)|| \leq 5m\delta(u,v) ,$$

so

$$\theta(M,L) = ||P_M - P_L|| \leq 5m\delta(u,v) ,$$

which obviously implies the continuity of A_m.

It is easily seen that $\$_m$ is compact; since $A_m : \$_m \to \mathbb{G}_m$ is a continuous map onto $\mathcal{G}_m$, the latter is also compact.

Let us prove the completeness of $\mathcal{G}_m$. Let $L_1, L_2, \cdots$ be a Cauchy sequence in $\mathcal{G}_m$, i.e. $\theta(L_i, L_j) \to 0$ as $i,j \to 0$. By compactness, there exists a subsequence L_{i_k} such that $\lim_{k\to\infty} \theta(L_{i_k}, L) = 0$ for some $L \in \mathcal{G}_m$. But then it is easily seen that in fact $L = \lim_{i\to\infty} L_i$.

To prove that $\mathcal{G}_m$ is open in the set of all subspaces in $\mathbb{C}^n$ it is sufficient to check that for subspaces $L, M \subset \mathbb{C}^n$, the inequality $\theta(L,M) < 1$ implies that $L \cap M^\perp = L^\perp \cap M = \{0\}$ (these equalities ensure that $\dim L = \dim M$). Assume the contrary; for instance, assume that $L \cap M^\perp \neq \{0\}$, and let $x \in L \cap M^\perp$ with $||x|| = 1$. Then

$$||(P_L - P_M)x|| = ||x|| = 1 ,$$

so $\theta(L,M) \geq 1$, a contradiction. □

In particular, Theorem A.1 implies that the metric space of all subspaces in $\mathbb{C}^n$ is a disconnected union of the metric spaces $\mathcal{G}_m$, $m = 0,1,\cdots,n$. One can show that each $\mathcal{G}_m$ is connected.

The limit of subspaces can be conveniently described in terms of the limits of vectors from these subspaces, as follows.

THEOREM A.2 *Let* $S_1, S_2, \cdots$ *be a converging sequence of* m-*dimensional subspaces in* $\mathbb{C}^n$, *and* $\theta(S_p, S) \to 0$ *for some* m-*dimensional subspace* S. *Then* S *consists of exactly those vectors* $x \in \mathbb{C}^n$ *for which there exists a sequence of vectors* $x_p \in S_p$, $p = 1,2,\cdots$ *which converges to* x.

PROOF. Let $x \in S$. Then

$$||P_{S_p}x - x|| = ||P_{S_p}x - P_S x|| \leq \theta(S_p, S)||x|| \to 0 ;$$

so the sequence $x_p = P_{S_p}x \in S_p$ converges to x.

Conversely, let $x_p \in S_p$, $p = 1,2,\cdots$ be a converging sequence and $\lim_{p\to\infty} x_p = x$. Then

$$\begin{aligned} ||P_S x - x|| &\leq ||P_S x - P_{S_p}x|| + ||P_{S_p}x - x_p|| + ||x_p - x|| \\ &\leq \theta(S, S_p)||x|| + 2||x - x_p|| \to 0 \quad \text{as} \quad p \to \infty . \end{aligned}$$

So $P_S x = x$, and $x \in S$. □

We have defined the gap between two subspaces as the norm of difference of the orthogonal projectors on these subspaces. It turns out that if one takes projectors which are not necessarily orthogonal, the norm of the difference can only increase:

PROPOSITION A.3 *Let* M, L *be subspaces in* $\mathbb{C}^n$. *Then for any projectors* P_1 *(resp.* P_2*) such that* $\operatorname{Im} P_1 = M$ *(resp.* $\operatorname{Im} P_2 = L$*), the inequality*

$$\theta(L,M) \leq ||P_1-P_2||$$

holds.

PROOF. Denote by S_L (resp. S_M) the unit sphere in L (resp. M). Given $x \in \mathbb{C}^n$ and a set $Y \subset \mathbb{C}^n$, denote $d(x,Y) = \inf_{y\in Y} ||x-y||$.

For every $x \in S_L$ we have:

$$||x-P_2x|| = ||(P_1-P_2)x|| \leq ||P_1-P_2|| \; ;$$

therefore

$$\sup_{x\in S_L} d(x,M) \leq ||P_1-P_2|| \; .$$

Analogously, $\sup_{x\in S_M} d(x,L) \leq ||P_1-P_2||$; so

$$\max\{\rho_L,\rho_M\} \leq ||P_1-P_2|| \; , \tag{A.6}$$

where $\rho_L = \sup_{x\in S_L} d(x,M)$; $\rho_M = \sup_{x\in S_M} d(x,L)$.

So the proof of this proposition will be completed if we show that

$$||P_M-P_L|| \leq \max\{\rho_L,\rho_M\} \; , \tag{A.7}$$

where P_M (resp. P_L) is the orthogonal projector on L (resp. M). (In fact there is an equality in (A.7), because (A.6) holds, in particular, for $P_1 = P_L$, $P_2 = P_M$).

We now prove (A.7). Observe that $\rho_L = \sup_{x\in S_L} ||(I-P_M)x||$; $\rho_M = \sup_{x\in S_M} ||(I-P_L)x||$. Consequently, for every $x \in \mathbb{C}^n$ we have

$$||(I-P_L)P_Mx|| \leq \rho_M||P_Mx|| \; ; \quad ||(I-P_M)P_Lx|| \leq \rho_L||P_Lx|| \; . \tag{A.8}$$

Now

$$||P_M(I-P_L)x||^2 = ((I-P_L)P_M(I-P_L)x,(I-P_L)x)$$
$$\leq ||(I-P_L)P_M(I-P_L)x||\ ||(I-P_L)x|| \ .$$

Hence by (A.8)

$$||P_M(I-P_L)x||^2 \leq \rho_M||P_M(I-P_L)x||\ ||(I-P_L)x|| \ ,$$

$$||P_M(I-P_L)x|| \leq \rho_M||(I-P_L)x|| \ . \tag{A.9}$$

On the other hand, using the relation

$$P_M - P_L = P_M(I-P_L) - (I-P_M)P_L$$

and the orthogonality of P_M, we obtain:

$$||(P_M-P_L)x||^2 = ||P_M(I-P_L)x||^2 + ||(I-P_M)P_Lx||^2 \ .$$

Combining with (A.8) and (A.9) gives:

$$||(P_M-P_L)x||^2 \leq \rho_M^2||(I-P_L)x||^2 + \rho_L^2||P_Lx||^2 \leq \max\{\rho_M^2,\rho_L^2\}||x||^2 \ .$$

So (A.7) follows. □

A.2 Continuous Families of Subspaces

Let $L(t) \subset ¢^n$, $t \in [0,1]$ be a subspace-valued function, i.e. for each $t \in [0,1]$, $L(t)$ is a subspace in $¢^n$. The function $L(t)$ is called *continuous* if the orthogonal projector on $L(t)$ is a continuous matrix valued function on $t \in [0,1]$ (in other words, continuity is understood in the gap metric). In this case we say that $L(t)$ is a continuous family of subspaces on $[0,1]$.

Examples of continuous families of subspaces are provided by the following proposition.

PROPOSITION A.4 *Let* $B(t)$ *be a continuous* $m \times n$ *complex matrix function on* $[0,1]$ *such that* rank $B(t) \overset{\text{def}}{=} p$ *is independent of* t *on* $[0,1]$. *Then* Ker $B(t)$ *and* Im $B(t)$ *are continuous families of subspaces on* $[0,1]$.

PROOF. Take $t_0 \in [0,1]$. There exists a non-zero minor of size

$p \times p$ of $B(t_0)$. For simplicity of notation assume that this minor is in the upper left corner of $B(t_0)$. By continuity the $p \times p$ minor in the upper left corner of $B(t)$ is also non-zero, provided t belongs to some neighbourhood $\mathcal{U}_0$ of t_0. So (here we use the assumption that rank $B(t)$ is independent of t)

$$\operatorname{Im} B(t) = \operatorname{Span}\{b_1(t),\cdots,b_p(t)\} , \quad t \in \mathcal{U}_0 , \tag{A.10}$$

where $b_i(t)$ is the i-th column of $B(t)$. Let $b_{ij}(t)$ be the (i,j)-th entry in $B(t)$; and let $D(t) = [b_{ij}(t)]_{i=p+1;j=1}^{i=m;j=p}$; $C(t) = [b_{ij}(t)]_{i,j=1}^{p}$. Then the matrix

$$P(t) = \begin{bmatrix} I_p & 0 \\ D(t)C(t)^{-1} & 0 \end{bmatrix}, \quad t \in \mathcal{U}_0$$

is a continuous projector with $\operatorname{Im} P(t) = \operatorname{Im} B(t)$. Hence $P(t)$ is uniformly continuous on $\mathcal{U}_1$, where $\mathcal{U}$ is a neighbourhood of t_0 in $[0,1]$ such that $\bar{\mathcal{U}}_1 \subset \mathcal{U}_0$. By Proposition A.3, the orthogonal projector on $\operatorname{Im} B(t)$ is also uniformly continuous on $\mathcal{U}_1$.

The statement concerning $\operatorname{Ker} B(t)$ can be reduced to that already considered because $\operatorname{Ker} B(t)$ is the orthogonal complement to $\operatorname{Im}(B(t))^*$ (note that $B(t)^*$ is continuous in t if $B(t)$ is). □

In particular, we obtain an important case:

COROLLARY A.5 *Let* $P(t)$ *be a continuous projector valued function on* $[0,1]$. *Then* $\operatorname{Im} P(t)$ *and* $\operatorname{Ker} P(t)$ *are continuous families of subspaces on* $[0,1]$.

We have to show that rank $P(t)$ is constant if the projector function $P(t)$ is continuous. But this follows from Proposition A.3 and the fact that the set of subspaces of fixed dimension is open in the set of all subspaces in $\mathbb{C}^n$ (Theorem A.1).

We shall need the following result on the existence of continuous bases in a continuous family of subspaces.

THEOREM A.6 *Let* $B(t)$ *be a continuous* $m \times n$ *matrix function on* $[0,1]$ *with* rank $B(t) \overset{\text{def}}{=} p$ *independent of* t. *Then there exists a continuous basis* $x_1(t),\cdots,x_{n-p}(t)$ *in* $\operatorname{Ker} B(t)$, *and a continuous basis* $y_1(t),\cdots,y_p(t)$ *in* $\operatorname{Im} B(t)$.

For the proof of this result we need the following lemma.

LEMMA A.7 *Let* $M(t) \subset \mathbb{C}^n$ *be a continuous family of subspaces on* $[0,1]$. *Then there exists an invertible continuous* $n \times n$ *matrix function* $S(t)$ *on* $[0,1]$ *such that*

$$S(t)M(0) = M(t) , \quad t \in [0,1] .$$

PROOF. Let $0 = t_0 < t_1 < t_2 < \cdots < t_{p-1} < t_p = 1$ be points with the property that

$$||P_{M(t_i)} - P_{M(\eta)}|| < 1 \quad \text{for} \quad t_i \leq \eta \leq t_{i+1} , \quad i = 0,\cdots,p-1 .$$

For each $i = 0,\cdots,p-1$, the matrix $S_i(\eta)$, $t_i \leq \eta \leq t_{i+1}$, defined by $S_i(\eta) = I - (P_{M(t_i)} - P_{M(\eta)})$ maps $M(t_i)$ on $M(\eta)$, is invertible and $S_i(t_i) = I$ (cf. the proof of Lemma 5.2). Now put

$$S(t) = S_i(t)\cdots S_1(t_2)S_0(t_1) \quad \text{for} \quad t_i \leq t \leq t_{i+1}$$

to satisfy the conditions of Lemma A.7. □

Proof of Theorem A.6. By Proposition A.4, Ker $B(t)$ is a continuous family of subspaces. By Lemma A.7 there exists an invertible continuous matrix function $S(t)$, $t \in [0,1]$ such that

$$S(t)\operatorname{Ker} B(0) = \operatorname{Ker} B(t)$$

for $t \in [0,1]$. Pick a basis $x_1,\cdots,x_{n-p}$ in Ker $B(0)$, and put

$$x_i(t) = S(t)x_i , \quad i = 1,\cdots,n-p$$

to obtain a continuous basis in Ker $B(t)$. For Im $B(t)$ the proof is similar. □

The following corollaries of Theorem A.6 will often be useful.

COROLLARY A.8 *Let* $B(t)$ *be a continuous* $m \times n$ *matrix function on* $[0,1]$ *such that* rank $B(t)$ *is constant (i.e. does not depend on* $t \in [0,1]$) *and* $B(t) \neq 0$. *Then there exists a continuous vector function* $x(t)$ *on* $[0,1]$ *such that* $B(t)x(t) \neq 0$ *for all* $t \in [0,1]$.

PROOF. Use the property:

$$\mathbb{C}^n = \operatorname{Ker} B(t) \oplus \operatorname{Im}(B(t))^* \tag{A.11}$$

and Theorem 6 to find a continuous non-zero vector $x(t) \in \operatorname{Im}(B(t))^*$. In view of (A.11), $B(t)x(t) \neq 0$ for all $t \in [0,1]$. □

COROLLARY A.9 *Let* $F(t)$ *be a continuous non-zero hermitian* $n \times n$ *matrix function on* $[0,1]$. *Assume that the number of positive eigenvalues of* $F(t)$ *(counting multiplicities), as well as the number of zero eigenvalues of* $F(t)$ *and the number of negative eigenvalues of* $F(t)$ *(also counting multiplicities), do not depend on* $t \in [0,1]$. *Then there exists a continuous vector function* $x(t) \in \mathbb{C}^n$, $t \in [0,1]$ *such that*

$$(F(t)x(t),x(t)) \neq 0 \quad \textit{for all} \quad t \in [0,1] .$$

PROOF. Assume for definiteness, that $F(t)$ has a positive eigenvalue. Put

$$Q(t) = \frac{1}{2\pi i} \int_{\Gamma(t)} (\lambda I - F(t))^{-1} d\lambda ,$$

where $\Gamma(t)$ is a closed contour such that the positive eigenvalues of $F(t)$ are inside $\Gamma(t)$, and zero and negative eigenvalues of $F(t)$ are outside $\Gamma(t)$. Clearly, $Q(t)$ is a continuous matrix function on $[0,1]$. Moreover, the conditions on $F(t)$ ensure that the rank of $Q(t)$ is constant (i.e. independent of t on $[0,1]$). Also $Q(t) \neq 0$. Using Theorem A.6, pick a continuous non-zero vector $x(t)$ in $\operatorname{Im} Q(t)$. Then clearly,

$$(F(t)x(t),x(t)) > 0 \quad \text{for all} \quad t \in [0,1] . \quad \square$$

We shall need also the following result on continuous families of subspaces with definiteness properties with respect to an indefinite scalar product.

THEOREM A.10 *Let* $H = H^*$ *be an invertible complex* $n \times n$ *matrix, and let* $M(t)$ *be a continuous family of subspaces on* $[0,1]$ *such that the quadratic form* (Hx,x) *is positive definite on* $M(t)$ *for all* $t \in [0,1]$. *Then there exists a basis* $v_1(t),\cdots,v_k(t)$ *in* $M(t)$ *such that* $v_i(t)$ *are continuous functions on* $[0,1]$, *and*

$$(Hv_i(t),v_j(t)) = \begin{cases} 1 , & i = j \\ 0 , & i \neq j \end{cases} . \tag{A.12}$$

Moreover, let $u_1,\cdots,u_k$ *(resp.* $w_1,\cdots,w_k$*) be a basis in* $M(0)$ *(resp. in* $M(1)$*) such that*

$$(Hu_i,u_j) = (Hw_i,w_j) = \delta_{ij} , \quad (1 \leq i,j \leq k) \tag{A.13}$$

Then the continuous basis $v_1(t),\cdots,v_k(t)$ *in* $M(t)$ *with the property* (A.12) *can be chosen so that* $v_i(0) = u_i$, $v_i(1) = w_i$, $i = 1,\cdots,k$.

PROOF. By Theorem A.6 choose a continuous basis $x_1(t),\cdots,x_k(t)$ in $M(t)$, $t \in [0,1]$. Applying to the positive definite matrix $[(Hx_i(t),x_j(t))]_{i,j=1}^k$ Lagrange's method for reduction to the sum of squares (see, e.g. [30b] or [14]), we obtain a continuous invertible $k \times k$ matrix $T(t)$, $t \in [0,1]$ such that the matrix $[(HT(t)x_i(t),T(t)x_j(t))]_{i,j=1}^k$ is just the identity matrix. Now put $v_i(t) = T(t)x_i(t)$, $i = 1,\cdots,k$.

Let now $\{u_j\}_{j=1}^k$ and $\{w_j\}_{j=1}^k$ be a basis in $M(0)$ and $M(1)$ respectively with the property (A.13). Let $U_0 : M(0) \to M(0)$ be the linear transformation such that $U_0v_j(0) = u_j$, $j = 1,\cdots,k$, written as the $k \times k$ matrix with respect to the basis $v_1(0),\cdots,v_k(0)$. As this basis is orthonormal in the positive definite scalar product $[x,y] = (Hx,y)$ on $M(0)$, the matrix U_0 is easily seen to be unitary. Analogously, let $U_1 : M(1) \to M(1)$ be the linear transformation such that $U_1v_j(1) = w_j$, $j = 1,\cdots,k$, written as the $k \times k$ unitary matrix with respect to the basis $v_1(1),\cdots,v_k(1)$. The set of all $k \times k$ unitary matrices is arcwise connected; so there is a continuous path $U(t)$, $t \in [0,1]$ of $k \times k$ unitary matrices such that $U(0) = U_0$, $U(1) = U_1$. Identifying $U(t)$ with a linear transformation $M(t) \to M(t)$ written in the basis $v_1(t),\cdots,v_k(t)$ and putting $v_j'(t) = U(t)v_j(t)$ we obtain a continuous basis $v_1'(t),\cdots,v_k'(t)$ in $M(t)$, $t \in [0,1]$ with the properties that $(Hv_i'(t),v_j'(t))$ is equal to 1 if $i = j$ and zero otherwise, and $v_i'(0) = u_i$, $v_i'(1) = w_i$, $i = 1,\cdots,k$. □

The results of this section, as well as their proofs are valid also for the class of C^p-functions, where p is a positive integer, or $p = \infty$.

A.3 Analytic Families of Subspaces

Let $L(t)$ be a subspace valued function defined for $a < t < b$, where a, b are real. We say that $L(t)$ is *analytic* if the orthogonal projector $P(t)$ on $L(t)$ is an analytic function of the real variable $t \in (a,b)$, i.e. the real and pure imaginary parts of each entry in the matrix $P(t)$ are analytic functions of t.

Results analogous to those in Section A.2 hold also in the context of analytic functions of the real variable (but the proofs in general are different from those in Section A.2). As an example we shall present the analogues of Proposition A.4 and Theorem A.6.

PROPOSITION A.11 *Let* $B(t)$ *be an analytic* $m \times n$ *complex matrix function on* (a,b) *such that* $\operatorname{rank} B(t) \overset{\text{def}}{=} p$ *is independent of* t *on* (a,b). *Then* $\operatorname{Ker} B(t)$ *and* $\operatorname{Im} B(t)$ *are analytic families of subspaces on* (a,b).

PROOF. As in the proof of Proposition A.4, we find that, in a neighbourhood of each $t_0 \in (a,b)$, we have $\operatorname{Im} P(t) = \operatorname{Im} B(t)$, where $P(t)$ is an analytic projector in this neighbourhood of the form

$$P(t) = \begin{bmatrix} I & 0 \\ X(t) & 0 \end{bmatrix}.$$

One checks easily that the orthogonal projector $P_0(t)$ on $\operatorname{Im} P(t) = \operatorname{Im} B(t)$ is given by the formula

$$\begin{bmatrix} Y(t) & Y(t)X(t)^* \\ X(t)Y(t) & X(t)Y(t)X(t)^* \end{bmatrix}$$

where $Y(t) = (I+X(t)^*X(t))^{-1}$. Since $P_0(t)$ is analytic in the *real* variable t in a neighbourhood of t_0, and $t_0 \in (a,b)$ was arbitrary, the proposition follows for $\operatorname{Im} B(t)$. The case of $\operatorname{Ker} B(t)$ follows from the fact that $\operatorname{Ker} B(t)$ is the orthogonal complement of $\operatorname{Im} B(t)^*$. □

THEOREM A.12 *Let* $A(t)$ *be an analytic* $m \times n$ *matrix function of the real variable* $t \in (a,b)$. *Let* $r = \max_{t \in (a,b)} \operatorname{rank} A(t)$. *Then there exist analytic (in* (a,b)*) vector valued functions* $x_1(t),\cdots,x_r(t) \in \mathbb{C}^m$ *and analytic (in* (a,b)*) vector valued functions* $y_{r+1}(t),\cdots,y_n(t) \in \mathbb{C}^n$ *with the following properties:*

(i) $x_1(t),\cdots,x_r(t)$ *are linearly independent for every* $t \in (a,b)$;

(ii) $y_{r+1}(t),\cdots,y_n(t)$ *are linearly independent for every* $t \in (a,b)$;

(iii)
$$\operatorname{Span}\{x_1(t),\cdots,x_r(t)\} = \operatorname{Im} A(t)$$

and

$$\operatorname{Span}\{y_{r+1}(t),\cdots,y_n(t)\} = \operatorname{Ker} A(t)$$

for every $t \in (a,b)$ *except for a set of isolated points which consists exactly of those* $t_0 \in (a,b)$ *for which* $\operatorname{rank} A(t_0) < r$. *For such exceptional* t_0 *the inclusions*

$$\operatorname{Span}\{x_1(t_0),\cdots,x_r(t_0)\} \supset \operatorname{Im} A(t_0)$$

and

$$\operatorname{Span}\{y_{r+1}(t_0),\cdots,y_n(t_0)\} \subset \operatorname{Ker} A(t_0)$$

hold.

For the proof of Theorem A.12 (originally proved in [43]) we refer to Chapter S6 in [19b].

The analogues of Corollaries A.8 and A.9 for analytic matrix functions of a real variable follow from Theorem A.12 exactly in the same way as Corollaries A.8 and A.9 follow from Theorem A.6.

NOTES TO PART III

Section 1.1, as well as the first part of Section 1.3, is taken from [19c]. The results of Sections 1.2 and the second part of Section 1.3 are due mainly to Krein [26a](see also [15]). The presentation of Sections 1.7, 1.8, 1.9 is based on [15]. The results of Sections 1.4 and 1.5 are probably presented for the first time.

Sections 2.2 and 2.4 are taken from [19c]. In the Section 2.3 the results and notions from [36] are used. The main part of Theorem 2.7 (the sufficiency condition) was proved in [26a] for the real case. The necessity of this condition appeared in [15] (also the real case). The complex case was considered in [11]. Results on selfadjoint equations in Section 2.5 appear in [32]. Generalizations of the results in Section 2.5 for infinite dimensional Hamiltonian equations, together with an extensive bibliography, are found in [26b].

Results of Chapter 3 in a more general setting were obtained in [26a]. We give another proof which is better adjusted to the ideas of this book.

Sections 4.1 - 4.4 are based on the paper [40a]. Lemma 4.7 is taken from [49]. The results of Theorems 4.8 and 4.9 in such form appear probably for the first time. For perturbations and stability of invariant (not necessarily maximal) neutral subspaces see [39].

The first two sections of Chapter 5 are based on the results of papers [19a, 19c]. Sections 5.4, 5.5, 5.7, 5.8 probably contain new results. Example 5.2 is Example II.5.3 in [25]. Lemma 5.4 is taken from [18] (Theorem 13.1).

The Appendix contains well-known material some of which is true also in the infinite dimensional case (see [17a, 25]).

PART IV

CONNECTED COMPONENTS OF DIFFERENTIAL EQUATIONS

This part of the book begins with a description of the connected components in the set of all matrices which are H-selfadjoint, have real spectrum, and are stably diagonable. These results are then applied to the study of the connected components of differential and difference equations with constant hermitian coefficients and stably bounded solutions. The last, and most substantial, chapter is dedicated to an exposition, with complete proofs of well-known results of Gelfand and Lidskii [15], with extensions by Coppel and Howe [11], concerning the connected components of linear Hamiltonian systems with periodic coefficients and stably bounded solutions.

The first chapter of Part IV can be viewed as the final installment in the development of the theory of matrices in the presence of an indefinite scalar product.

CHAPTER 1

CONNECTED COMPONENTS OF STABLY DIAGONABLE MATRICES

Since the set of all H-selfadjoint stably diagonable matrices with real spectrum is open in the set of all H-selfadjoint matrices, it consists of certain connected components. These connected components are to be studied in this chapter. It turns out that the sign characteristic carries the full information required to differentiate these connected components. Note that this problem is peculiar to the indefinite scalar product spaces, for if H is positive definite there is just one connected component.

The material of this chapter forms a foundation for the rest of Part IV, where it is used to investigate the structure of connected components of different classes of differential and difference equations.

1.1 H-Selfadjoint Stably r-Diagonable Matrices

Let $H = H^*$ be an invertible complex $n \times n$ matrix. Recall that an H-selfadjoint matrix A is called stably r-diagonable if A is similar to a real diagonal matrix and this property holds also for every matrix A' which is sufficiently close to A and which is H'-selfadjoint for some hermitian matrix H' sufficiently close to H. Denote by $S_r(H)$ the class of all H-selfadjoint stably r-diagonable matrices. We know (Theorem III.1.4) that $A \in S_r(H)$ if and only if all eigenvalues of A are real and definite. The latter means that the quadratic form (Hx,x) is either positive definite or negative definite on the subspace $\mathrm{Ker}(\lambda_0 I-A)$ for every eigenvalue λ_0 of A. So there is a unique sign associated with each eigenvalue λ_0 of a matrix $A \in S_r(H)$ (which coincides with the sign of the quadratic form (Hx,x), $x \in \mathrm{Ker}(\lambda_0 I-A)$).

Let A be an H-selfadjoint stably r-diagonable matrix. We now define the *index* of A as follows. Let $(\alpha_0,\alpha_1),(\alpha_1,\alpha_2),\cdots,(\alpha_{p-1},\alpha_p),\alpha_0 = -\infty$, $\alpha_p = \infty$, be consecutive intervals on the real line such that every interval (α_i,α_{i+1}) contains the largest possible number of eigenvalues of A having the same sign in the sign characteristic (so adjacent intervals contain eigenvalues with opposite signs). Let n_i be the sum of multiplicities of the eigenvalues of A lying in (α_{i-1},α_i), multiplied by (-1) if the sign of these eigenvalues is negative. Thus, the sign of n_i coincides with the sign (in the sign characteristic of (A,H)) of eigenvalues belonging to (α_{i-1},α_i). The sequence $\{n_1,\cdots,n_p\}$ will be called the *index* of A and will be denoted $\text{H-ind}_r(A)$. It is easily seen that the index does not depend on the choice of α_i (subject to the condition mentioned above). Observe the following properties of $\text{H-ind}_r(A)$:

$$n_i n_{i+1} < 0 , \quad i = 1,\cdots,p-1 ; \tag{1.1}$$

$$\sum_{i=1}^{p} |n_i| = n ; \tag{1.2}$$

$$\sum_{i=1}^{p} n_i = \text{sig } H . \tag{1.3}$$

Note that if $n_1,\cdots,n_p$ are integers with the properties (1.1)-(1.3), then there is a matrix $A \in S_r(H)$ such that $\{n_1,\cdots,n_p\}$ is the index of A. Indeed, put

$$K = \text{diag}[I_{n_1}, 2I_{n_2},\cdots,pI_{n_p}] ,$$

and

$$Q = \text{diag}[\varepsilon_1 I_{n_1},\varepsilon_2 I_{n_2},\cdots,\varepsilon_p I_{n_p}] ,$$

where I_j is the $j \times j$ unit matrix and $\varepsilon_j = \text{sgn } n_j$. Clearly, K is Q-selfadjoint. Moreover, because of (1.3), $\text{sig } Q = \text{sig } H$. So there exists an invertible S such that $H = S^* QS$. Now $A = S^{-1}KS$ is H-selfadjoint, stably r-diagonable and

$$\text{H-ind}_r(A) = \{n_1,\cdots,n_p\} .$$

The set $S_r(H)$ of all H-selfadjoint stably r-diagonable matrices is open, and therefore splits into open connected components. All such components may be described as follows.

THEOREM 1.1 *All matrices from* $S_r(H)$ *with the same index form a connected component in* $S_r(H)$, *and each connected component in* $S_r(H)$ *has such a form.*

PROOF. Let $A,B \in S_r(H)$ with canonical forms

$$J_A = \mathrm{diag}[\alpha_1,\cdots,\alpha_n]\ ,\quad P_{\varepsilon,J_A} = \mathrm{diag}[\delta_1,\cdots,\delta_n] \quad \text{and}$$

$$J_B = \mathrm{diag}[\beta_1,\cdots,\beta_n]\ ,\quad P_{\varepsilon,J_B} = \mathrm{diag}[\zeta_1,\cdots,\zeta_n]\ , \text{ respectively}$$

(so $\alpha_i,\beta_i \in \mathbb{R}$ and $\delta_i,\zeta_i = \pm 1$). We assume $\alpha_1 \leq \cdots \leq \alpha_n$; $\beta_1 \leq \cdots \leq \beta_n$.

Suppose first H-ind$_r$(A) = H-ind$_r$(B). Then $P_{\varepsilon,J_A} = P_{\varepsilon,J_B}$, and $J(t) = \mathrm{diag}[t\alpha_1+(1-t)\beta_1,\cdots,t\alpha_n+(1-t)\beta_n]$, $t \in [0,1]$ is a continuous path of P_{ε,J_A}-selfadjoint stably r-diagonable matrices connecting J_A and J_B. Let $A = S^{-1}J_AS$, $H = S^*P_{\varepsilon,A}S$ be the unitary similarity relations. Then $A(t) = S^{-1}J_A(t)S$, $t \in [0,1]$ is a continuous path of matrices from $S_r(H)$, and A(1) is similar to B and has the same sign characteristic. By Theorem I.3.6, A(1) and B are H-unitarily similar. Since the classes of H-unitary similarity are connected (see Theorem I.3.9), A(1) and B belong to the same connected component of $S_r(H)$. So the same is true for A and B.

Suppose now that H-ind$_r$(A) is not equal to H-ind$_r$(B). We shall prove that A and B belong to different connected components of $S_r(H)$. Assume the contrary. Then there exists a continuous path A(t), $t \in [0,1]$ from A to B in $S_r(H)$. Let

$$t_0 = \inf\{t \in [0,1] \mid \text{H-ind}_r(A(t)) \neq \text{H-ind}_r(A)\}\ .$$

Let $\mu_1 < \cdots < \mu_r$ be the different eigenvalues of $A(t_0)$. We claim that for some μ_i, the form (Hx,x), $x \in \mathrm{Ker}(\mu_iI-A(t_0))$ is not definite. Indeed, if all the forms (Hx,x), $x \in \mathrm{Ker}(\mu_iI-A(t_0))$ were definite, then the same is true for the forms (Hx,x), $x \in \mathrm{Im}\ \frac{1}{2\pi i}\int_{\Gamma_i}(\lambda I-A(t))^{-1}d\lambda$, where Γ_i is a small contour around μ_i, and t belongs to some neighbourhood $\mathcal{U}$ of t_0 (see Theorem III.1.3 and remarks thereafter). But then, by definition of the index, we have H-ind$_r$(A(t)) = H-ind$_r$(A(t_0)), $t \in \mathcal{U}$, a contradiction with the choice of t_0. By Theorem III.1.4, $A(t_0)$ is not stably r-diagonable, which contradicts the choice of the path $A(t)$, $t \in [0,1]$. □

More generally, consider the connected components of the set S_r of pairs (A,H), where A is an H-selfadjoint stably r-diagonable $n \times n$ matrix. Given integers $n_1,\cdots,n_p$ such that $n_i n_{i+1} < 0$, $i = 1,\cdots,p$, and $\sum_{i=1}^{p} |n_i| = n$, it is easily seen that the set of all pairs $(A,H) \in S_r$ for which the index $H\text{-ind}_r(A)$ is equal to $\{n_1,\cdots,n_p\}$ (then necessarily $\sum_{i=1}^{p} n_i = \text{sig } H$) is not empty. Analogously to Theorem 1.1 such sets are precisely the connected components of S_r:

THEOREM 1.2 *All pairs* $(A,H) \in S_r$ *with the same index* $H\text{-ind}_r(A)$ *form a connected component in* S_r, *and each connected component in* S_r *has such a form.*

PROOF. Let H_i-selfadjoint matrices A_i, $i = 1,2$ be such that A_i is stably r-diagonable and $H_1\text{-ind}_r(A_1) = H_2\text{-ind}_r(A_2) = \{n_1,\cdots,n_p\}$. Then, in particular, $\text{sig } H_1 = \text{sig } H_2$. So there exists an invertible matrix S such that $H_2 = S^* H_1 S$. Put $A_2' = S^{-1}A_1 S$. Then A_2' is H_2-selfadjoint, and since $\sigma(A_2') = \sigma(A_1)$, also $H_2\text{-ind}_r(A_2') = H_2\text{-ind}_r(A_1)$. By Theorem 1.1 there is a continuous path of matrices from $S_r(H_2)$ connecting A_2 and A_2'. Also there is a continuous path $F(t)$, $t \in [0,1]$ of invertible matrices connecting S and I. Put $A(t) = F(t)^{-1}A_1F(t)$, $H(t) = F(t)^*H_1F(t)$ to obtain a continuous path of matrices $A(t) \in S_r(H(t))$ such that $A(0) = A_2'$; $A(1) = A_1$. So the set of all pairs $(A,H) \in S_r$ with $H\text{-ind}_r(A) = \{n_1,\cdots,n_p\}$ is connected.

Now let $(A_1,H_1),(A_2,H_2) \in S_r$, and assume that there exists a continuous path $(A(t),H(t)) \in S_r$ such that $A(0) = A_1$; $H(0) = H_1$; $A(1) = A_2$; $H(1) = H_2$. Then clearly, $\text{sig } H_1 = \text{sig } H_2$. Also $H_1\text{-ind}_r(A_1) = H_2\text{-ind}_r(A_2)$, and this is proved in the same way as in Theorem 1.1. □

1.2 H-Unitary Stably u-Diagonable Matrices

Let $H = H^*$ be an invertible complex $n \times n$ matrix. Recall that an H-unitary matrix U is called stably u-diagonable if in some basis in $\mathbb{C}^n$ U is a diagonal matrix with unimodular eigenvalues and this property holds also for every matrix U' which is sufficiently close to U and which is G-unitary for some hermitian invertible matrix G sufficiently close to H. Denoting by $S_u(H)$ the set of all H-unitary stably u-diagonable matrices, it follows from Theorem III.1.7, that $U \in S_u(H)$ if and only if U is similar to a diagonal matrix with unimodular eigenvalues and all eigenvalues of U are definite. The latter statement means that the signs in the sign characteristic of (U,H) which corresponds to the Jordan blocks with the same eigenvalue, are either all +1's or all -1's. So there is a sign corresponding to each eigenvalue of $U \in S_u(H)$ (this sign coincides with the sign in the sign characteristic of (U,H) corresponding to this eigenvalue).

We will describe the structure of the connected components in $S_u(H)$. To this end we introduce the index for the stably u-diagonable H-unitary matrix U (cf. the definition of the index for stably r-diagonalizable H-selfadjoint matrices). Let $(\alpha_0,\alpha_1),(\alpha_1,\alpha_2),\cdots,(\alpha_{p-1},\alpha_p)$, $\alpha_p = \alpha_0$, be consecutive intervals on the unit circle such that every interval (α_i,α_{i+1}) contains the largest possible number of eigenvalues of U of the same sign (so adjacent intervals contain eigenvalues of opposite sign). In particular, p is even or else $p = 1$, and the latter case may occur only if H is positive definite or negative definite. Let v_i $(i = 1,\cdots,p)$ be the sum of multiplicities of the eigenvalues of U lying in (α_{i-1},α_i), multiplied by (-1) if the sign of these eigenvalues is negative (so the sign of v_i coincides with the sign of eigenvalues belonging to (α_{i-1},α_i)). The sequence $\{v_1,\cdots,v_p\}$, as well as any sequence $\{v_i,v_{i+1},\cdots,v_p,v_1,\cdots,v_{i-1}\}$, $i = 2,\cdots,p$ (obtained from $\{v_1,\cdots,v_p\}$ by a cyclic permutation) will be called the *index* of U and denoted $H\text{-ind}_u(U)$. It is easily seen that the index of U does not depend on the choice of α_i (subject to the above conditions), provided that one takes into account the possible cyclic permutation of $H\text{-ind}_u(U)$. Observe the following properties of $H\text{-ind}_u(U)$:

$$\text{if } p > 1, \text{ then } v_i v_{i+1} < 0\,, \quad i = 1,\cdots,p \quad (v_{p+1} = v_1 \text{ by definition}) \tag{1.4}$$

$$\sum_{i=1}^{p} |v_i| = n\,; \tag{1.5}$$

$$\sum_{i=1}^{p} v_i = \text{sig } H\,; \tag{1.6}$$

and note that (1.4) implies $p = 1$ or p is even. Let $\{v_1,\cdots,v_p\}$ be any sequence of non-zero integers with properties (1.4)-(1.6). It is easily seen that the set of all matrices from $S_u(H)$ whose index is $\{v_1,\cdots,v_p\}$ is not empty.

THEOREM 1.3 *All matrices from* $S_u(H)$ *whose indices are obtained from each other by a cyclic permutation form a connected component in* $S_u(H)$, *and every connected component in* $S_u(H)$ *has this form.*

PROOF. Let $U_1, U_2 \in S_u(H)$; let

$$K_1 = \mathrm{diag}[e^{i\theta_1}, e^{i\theta_2}, \cdots, e^{i\theta_n}] \ ; \quad P_{\varepsilon_1,J} = \mathrm{diag}[\zeta_1, \zeta_2, \cdots, \zeta_n] \ , \quad \zeta_i = \pm 1 \ ,$$

be the canonical form of the pair (U_1, H), and let

$$K_2 = \mathrm{diag}[e^{i\pi_1}, e^{i\pi_2}, \cdots, e^{i\pi_n}] \ ; \quad P_{\varepsilon_2,J} = \mathrm{diag}[\eta_1, \eta_2, \cdots, \eta_n] \ , \quad \eta_i = \pm 1$$

be the canonical form of the pair (U_2, H). We assume that

$$\theta_1 \leq \theta_2 \leq \cdots \leq \theta_n \ ; \quad \pi_1 \leq \pi_2 \leq \cdots \leq \pi_r \ ; \quad \theta_n - \theta_1 < 2\pi \ ; \quad \pi_n - \pi_1 < 2\pi \ . \qquad (1.7)$$

Suppose now that $H\text{-ind}_u(U_1)$ is equal to $H\text{-ind}_u(U_2)$. This means that, after some cyclic permutation of terms on the main diagonal in K_2, and the same permutation of terms in $P_{\varepsilon_2,J}$, we obtain $K_3 = \mathrm{diag}[e^{i\rho_1}, e^{i\rho_2}, \cdots, e^{i\rho_n}]$ and $P_{\varepsilon_1,J}$, respectively. Put

$$K(t) = \mathrm{diag}[e^{i\sigma_1(t)}, \cdots, e^{i\sigma_n(t)}] \ ; \quad \sigma_j(t) = (1+t)\theta_j + t\rho_j \ ; \quad j = 1, \cdots, n \ .$$

Now write $U_1 = S^{-1}K_1S$, $H = S^*P_{\varepsilon,J}S$ for some invertible matrix S. Then $U(t) = S^{-1}K(t)S$, $t \in [0,1]$ is a continuous path of H-unitary matrices from $S_u(H)$ connecting U_1 and $S^{-1}K_3S$. Also the H-unitary matrices $S^{-1}K_3S$ and U_2 are similar and have the same sign characteristic. By Theorem I.3.6, $S^{-1}K_3S$ and U_2 are H-unitarily similar. Since the class of H-unitary similarity of H-unitary matrices is connected (Theorem I.3.9), we can find a continuous path from $S^{-1}K_3S$ to U_2 in this class. Clearly, this path belongs to $S_u(H)$.

Now suppose that $H\text{-ind}_u(U_1)$ is not equal to $H\text{-ind}_u(U_2)$. Assume there exists a continuous path $U(t)$, $t \in [0,1]$, of H-unitary matrices in $S_u(H)$ connecting U_1 and U_2. As in the proof of Theorem 1.1, we pick

$$t_0 = \inf\{t \in [0,1] \mid H\text{-ind}_u(U(t)) \text{ is not equal to } H\text{-ind}_u(U_1)\} \ ,$$

and show that for all t in a sufficiently small neighbourhood of t_0,

$H\text{-ind}_u(U(t))$ is equal to $H\text{-ind}_u(U(t_0))$; a contradiction. □

Finally, we consider the connected components of the set S_u of all pairs (U,H), such that U is an $n \times n$ matrix belonging to $S_u(H)$. Again, given integers $v_1,\cdots,v_p$ with the properties (1.4) and (1.5), the set of all pairs $(U,H) \in S_u$ such that $H\text{-ind}_u(U)$ is equal to $\{v_1,\cdots,v_p\}$ is not empty.

THEOREM 1.4 *All pairs* $(U,H) \in S_u$ *whose indices* $H\text{-ind}_u(U)$ *are obtained from each other by a cyclic permutation form a connected component in* S_u. *Every connected component in* S_u *has such a form.*

PROOF. Assume $(U_1,H_1),(U_2,H_2) \in S_u$ are such that $H_1\text{-ind}_u(U_1) = \{v_1,\cdots,v_p\}$ is obtained from $H_2\text{-ind}_u(U_2)$ by a cyclic permutation. Then $\sum_{i=1}^{p} v_i = \text{sig}\, H_1 = \text{sig}\, H_2$. Let S be an invertible matrix such that $H_2 = S^*H_1S$, and put $U_3 = S^{-1}U_1S$. Let $S(t)$, $t \in [0,1]$ be a continuous path of invertible matrices with $S(0) = S$; $S(1) = I$. Put

$$U_3(t) = S^{-1}(t)U_1S(t) \; ; \quad H_2(t) = S^*(t)H_1S(t) \; .$$

Then $U_3(t)$ is $H_2(t)$-unitary; $\text{sig}\, H_2(t) = \text{sig}\, H_1$; and $U_3(t)$ is stably u-diagonable and the index $H_2(t)\text{-ind}_u(U_3(t))$ is equal to $H_1\text{-ind}_u(U_1)$. Further, $(U_3(0),H_2(0)) = (U_3,H_2)$; $(U_3(1),H_2(1)) = (U_1,H_1)$. By Theorem 1.3 there is a continuous path between (U_3,H_2) and (U_2,H_2) in the set $S_u(H_2)$. So (U_1,H_1) and (U_2,H_2) belong to the same connected component of the set S_u.

Now let (U_1,H_1) and (U_2,H_2) be in S_u. If $\text{sig}\, H_1 \neq \text{sig}\, H_2$, then clearly (U_1,H_1) and (U_2,H_2) belong to different connected components of S_u. Suppose $\text{sig}\, H_1 = \text{sig}\, H_2$, but the indices $H_1\text{-ind}_u(U_1)$ and $H_2\text{-ind}_u(U_2)$ are not equal. Assuming that there exists a continuous path between (U_1,H_1) and (U_2,H_2) we arrive at a contradiction, as in the proof of Theorem 1.1. □

1.3 E-Orthogonal Stably u-Diagonable Matrices

Let E be an invertible $n \times n$ real skew-symmetric $(E^* = -E)$ matrix. Recall that a real $n \times n$ matrix A is called E-orthogonal if $A^*EA = E$; such a matrix A is called stably u-diagonable if any E'-orthogonal matrix A' is similar to a diagonal matrix with (not necessarily real) unimodular eigenvalues, provided $||A-A'|| + ||E-E'||$ is small enough.

Denote by $S(E)$ the set of all E-orthogonal stably u-diagonable matrices. Theorem III.1.10 shows that for every $A \in S(E)$ the quadratic form (iEx,x) is either positive definite or negative definite on $\mathrm{Ker}(A-\lambda_0 I)$. In these two cases we say that the sign of $\lambda_0 \in \sigma(A)$ is positive, or negative, respectively. Note that since A is real, $\bar{\lambda}_0$ is an eigenvalue of A together with λ_0. It turns out that the signs of $\lambda_0 \in \sigma(A)$ and $\bar{\lambda}_0 \in \sigma(A)$ are opposite. In this connection we recall that by Lemma III.1.15 $\lambda \neq \bar{\lambda}_0$. Indeed,

$$\mathrm{Ker}(A-\bar{\lambda}_0 I) = \{\bar{x} \in \mathbb{C}^n \mid x \in \mathrm{Ker}(A-\lambda_0 I)\} ,$$

and for $x \in \mathrm{Ker}(A-\lambda_0 I)$ we have

$$(iE\bar{x},\bar{x}) = -(\overline{(iEx)},\bar{x}) = -(x,iEx) = -(iEx,x) .$$

Now let $A \in S(E)$, and let $(\alpha_0,\alpha_1),(\alpha_1,\alpha_2),\cdots,(\alpha_{p-1},\alpha_p)$, $\alpha_0 = 1$, $\alpha_p = -1$ be consecutive arcs on the upper half of the unit circle $\{e^{i\theta} \mid 0 < \theta < \pi\}$ such that every arc (α_i,α_{i+1}) contains the largest possible number of eigenvalues of A of the same sign (so adjacent arcs contain eigenvalues of opposite signs). Let $v_1,\cdots,v_p$ be the sum of multiplicities of the eigenvalues of A lying in $(\alpha_0,\alpha_1),\cdots,(\alpha_{p-1},\alpha_p)$ respectively, and multiplied by (-1) if the sign of these eigenvalues is negative. The sequence $\{v_1,\cdots,v_p\}$ will be called the index of A and denoted $\text{E-ind}(A)$. Clearly, $\text{E-ind}(A)$ does not depend on the choice of $\alpha_1,\cdots,\alpha_{p-1}$. The following properties of $\text{E-ind}(A)$ are easily verified:

$$\text{if } p > 1, \text{ then } v_i v_{i+1} < 0 , \quad i = 1,\cdots,p-1 ; \tag{1.8}$$

$$\sum_{i=1}^{p} |v_i| = \frac{n}{2} . \tag{1.9}$$

THEOREM 1.5 *Given integers* $v_1,\cdots,v_p$ *with the properties* (1.8) *and* (1.9), *the set of all* $A \in S(E)$ *such that* $\text{E-ind}(A)$ *is equal to* $\{v_1,\cdots,v_p\}$ *is a (non-empty) connected component in* $S(E)$. *Every connected component in* $S(E)$ *has this form.*

PROOF. If $\text{E-ind}\, A_1 \neq \text{E-ind}\, A_2$, where $A_1,A_2 \in S(E)$, then A_1 and A_2 belong to different connected components of $S(E)$ in view of Theorem 1.3 (the complex case).

Assume now that the indices E-ind A_1 and E-ind A_2 coincide for $A_1, A_2 \in S(E)$. Using the description of stably u-diagonable E-orthogonal matrices given in Theorem III.1.10, and using the connectedness of the group of E-orthogonal matrices (Theorem II.1.7), one proves that A_1 and A_2 are connected in $S(E)$. □

CHAPTER 2

DIFFERENTIAL AND DIFFERENCE EQUATIONS WITH CONSTANT COEFFICIENTS

Using the theory developed in Chapter 1 the problem of finding the connected components of higher order differential and difference equations with hermitian constant coefficients is to be examined. The problem will, in fact, only be solved in some special cases and the structure of the connected components in general remains an open question.

2.1 Connected Components of Differential Equations With Hermitian Coefficients and Stably Bounded Solutions

Consider the differential equations

$$L[i\frac{d}{dt}]x = \sum_{j=0}^{\ell} i^j A_j \frac{d^j x}{dt^j} = 0 ; \quad t \in \mathbb{R} \tag{2.1}$$

where $L(\lambda) = \sum_{i=0}^{\ell} A_i\lambda^i$ is a polynomial with $n \times n$ hermitian coefficients ($A_i = A_i^*$) and invertible leading coefficient A_ℓ. We shall assume that (2.1) has stably bounded solutions. By Theorem III.2.2, $\sigma(L)$ is real and the quadratic form $(L'(\lambda_0)x,x)$ is definite on the subspace $\operatorname{Ker} L(\lambda_0)$ for every $\lambda_0 \in \sigma(L)$. So we can assign a sign $\varepsilon(\lambda_0)$ to every eigenvalue of L; namely, $\varepsilon(\lambda_0) = +1$ (resp. $\varepsilon(\lambda_0) = -1$) if $(L'(\lambda_0)x,x)$ is positive (resp. negative) definite on the subspace $\operatorname{Ker} L(\lambda_0)$.

Now construct the index $\operatorname{Ind}(L) = \{r;n_1,\cdots,n_p\}$ associated with the polynomial L as follows (cf. the definition of the index of an H-self-adjoint stably r-diagonable matrix). Put $r = \operatorname{sig} A_\ell$. Further, define $\alpha_0 = -\infty$, $\alpha_p = \infty$ and let $(\alpha_0,\alpha_1),(\alpha_1,\alpha_2),\cdots,(\alpha_{p-1},\alpha_p)$ be consecutive intervals on the real line such that every interval (α_i,α_{i+1}) contains the maximal number of eigenvalues of $L(\lambda)$ of the same sign. Let $n_1,\cdots,n_p$

be the sum of multiplicities of the eigenvalues of $L(\lambda)$ lying in (α_{i-1},α_i), multiplied by (-1) if the sign of these eigenvalues in the sign characteristic is negative.

For brevity, a matrix polynomial L with hermitian coefficients and invertible leading coefficient, and such that the differential equation (2.1) has stably bounded solutions, will be called an SB *polynomial*.

THEOREM 2.1 *Let* L_1 *and* L_2 *be* SB *polynomials. If* $\mathrm{Ind}\, L_1 \neq \mathrm{Ind}\, L_2$, *then* L_1 *and* L_2 *belong to different connected components in the set of all* SB *polynomials.*

The topology in the set Ω of all SB polynomials is introduced naturally as follows: Ω is a disconnected union of the sets Ω_ℓ, $\ell = 1,2,\cdots$, where Ω_ℓ is the set of all SB polynomials of fixed degree ℓ. In turn, Ω_ℓ is given the topology as a subset of $\mathbb{C}^{n\times n} \times \cdots \times \mathbb{C}^{n\times n}$ ($\ell+1$ times), where $\mathbb{C}^{n\times n}$ is the set of all $n \times n$ complex matrices with the standard topology.

PROOF. Write $L_i(\lambda) = \sum_{j=0}^{\ell} \lambda^j A_{ij}$, $i = 1,2$, and let

$$\mathrm{Ind}(L_i) = \{r^{(i)};n_1^{(i)},\cdots,n_{p_i}^{(i)}\} .$$

We can assume that L_1 and L_2 have the same degree ℓ (otherwise Theorem 2.1 is trivial). Clearly, if the leading coefficients of L_1 and L_2 have different signatures (i.e. if $r^{(1)} \neq r^{(2)}$), then L_1 and L_2 belong to different connected components in the set Ω_ℓ.

Let C_{L_1} and C_{L_2} be the companion matrices of L_1 and L_2, respectively, and introduce the matrices

$$B_{L_i} = \begin{bmatrix} A_{i0} & A_{i1} & \cdots & A_{i\ell} \\ A_{i1} & & \cdot & \\ \cdot & \cdot & & \\ \cdot & \cdot & & \\ \cdot & A_{i\ell} & & 0 \\ A_{i\ell} & & & \end{bmatrix}, \quad i = 1,2 .$$

By Theorem III.1.4 (using Proposition II.2.2), for $i = 1,2$ the companion matrix C_{L_i} is B_{L_i}-selfadjoint and also B_{L_i}-stably r-diagonable. Also for $i = 1,2$

$$B_{L_i}\text{-ind}_r(C_{L_i}) = \{n_1^{(i)},\cdots,n_{p_i}^{(i)}\} .$$

By Theorem 1.2, if $\{n_1^{(1)},\cdots,n_p^{(1)}\} \neq \{n_1^{(2)},\cdots,n_p^{(2)}\}$, then (C_{L_1},B_{L_1}) and (C_{L_2},B_{L_2}) belong to different connected components in the set S_r. So in this case the polynomials L_1 and L_2 also belong to different connected components in Ω_ℓ. □

In conclusion we state the following conjecture.

CONJECTURE 2.2 *If* L_1 *and* L_2 *are* SB *polynomials of the same degree and with equal indices, then* L_1 *and* L_2 *belong to the same connected component of the set of all* SB *polynomials.*

Theorem 1.2 verifies this conjecture for polynomials of first degree. On the other hand, in the case of scalar polynomials $(n = 1)$, the conjecture can be checked easily. Indeed, a scalar polynomial $L(\lambda) = \sum_{j=0}^{\ell} \lambda^j A_j$, $A_\ell \neq 0$ with real coefficients of degree ℓ is SB if and only if it has ℓ distinct real zeros. Denote them by $x_1 < x_2 < \cdots < x_\ell$. Clearly, $L'(x_i)\cdot L'(x_{i+1}) < 0$ for $i = 1,\cdots,\ell-1$. Further, $L'(x_1)$ is positive if either ℓ is odd, $A_\ell > 0$ or ℓ is even, $A_\ell < 0$, and $L'(x_1)$ is negative if either ℓ is odd, $A_\ell < 0$ or ℓ is even, $A_\ell > 0$. So the index of $L(\lambda)$ is:

$\{1;1,-1,1,-1,\cdots,1\}$	if ℓ is odd and $A_\ell > 0$;
$\{-1;1,-1,1,-1,\cdots,-1\}$	if ℓ is even and $A_\ell < 0$;
$\{-1;-1,1,-1,\cdots,-1\}$	if ℓ is odd and $A_\ell < 0$;
$\{-1;1,-1,1,\cdots,-1\}$	if ℓ is even and $A_\ell > 0$.

On the other hand, it is easily verified that the set of all real polynomials of degree ℓ with positive (resp. negative) leading coefficient, and ℓ distinct real zeros, is connected. So Conjecture 2.2 does indeed hold for the scalar case.

2.2 A Special Case

We shall verify Conjecture 2.2 also for the class A of SB polynomial matrices of second degree of size 2×2 with positive definite leading coefficient. So every $L(\lambda) \in A$ can be written in the form

$$L(\lambda) = \lambda^2 A_2 + \lambda A_1 + A_0 , \tag{2.2}$$

where A_i are 2×2 hermitian matrices $(i = 0,1,2)$, and A_2 is positive definite.

We shall need a preliminary result.

LEMMA 2.3 *Let* $\underset{\sim}{\lambda}$ *(resp.* $\widetilde{\lambda}$*) be the smallest (resp. the greatest) eigenvalue of* $L(\lambda) \in A$. *Then the form* $(L'(\underset{\sim}{\lambda})x,x)$ *is negative definite on the space* $\operatorname{Ker} L(\underset{\sim}{\lambda})$, *and the form* $(L'(\widetilde{\lambda})x,x)$ *is positive definite on the space* $\operatorname{Ker} L(\widetilde{\lambda})$.

PROOF. Let $L(\lambda)$ be given by (2.2). Let us show first that the hermitian matrix $L(\lambda)$ is positive definite for $\lambda < \underset{\sim}{\lambda}$. Indeed, since A_2 is positive definite, $L(\lambda) = \lambda^2(A_2 + \frac{1}{\lambda} A_1 + \frac{1}{\lambda^2} A_0)$ is certainly positive definite for λ real and $|\lambda|$ large enough. Let $\mu = \inf\{\lambda \in \mathbb{R} \mid L(\lambda)$ is not positive definite$\}$. Then $L(\lambda)$ is positive definite for $\lambda < \mu$ and, by continuity, $L(\mu)$ is nonnegative definite. But $L(\mu)$ cannot be positive definite (otherwise $L(\lambda)$ would be positive definite also for $\mu < \lambda < \mu+\varepsilon$, for small $\varepsilon > 0$; a contradiction with the definition of μ). This means that $L(\mu)x = 0$ for some $x \neq 0$, i.e. μ is an eigenvalue of $L(\lambda)$, and therefore $\mu \geq \underset{\sim}{\lambda}$. So $L(\lambda)$ is positive definite for $\lambda < \underset{\sim}{\lambda}$.

Now pick $x \in \operatorname{Ker} L(\underset{\sim}{\lambda}) \setminus \{0\}$ and put $f(\lambda) = (L(\lambda)x,x)$. Then $f(\lambda) < 0$ for $\lambda < \underset{\sim}{\lambda}$ and $f(\underset{\sim}{\lambda}) = 0$. Consequently, $f'(\underset{\sim}{\lambda}) = (L'(\underset{\sim}{\lambda})x,x) \lesssim 0$. But the equality $(L'(\underset{\sim}{\lambda})x,x) = 0$ is impossible because $L(\lambda)$ is SB.

Similarly, one checks Lemma 2.3 for the largest eigenvalue $\widetilde{\lambda}$. □

Lemma 2.3 shows that the index for $L \in A$ can be only one of the following two: $\text{Ind}_1 = \{2;-1,1,-1,1\}$; $\text{Ind}_2 = \{2;-2,2\}$.

Let A_1 (resp. A_2) be the set of all polynomials $L \in A$ whose index is Ind_1 (resp. Ind_2). It is easily seen that both sets A_1 and A_2 are non-empty (for instance, $\operatorname{diag}[\lambda(\lambda-1),(\lambda-2)(\lambda-3)] \in A_1$ and $\operatorname{diag}[\lambda(\lambda-2),(\lambda-1)(\lambda-3)] \in A_2$).

Theorem 2.1 shows that if $L_1 \in A_1$, $L_2 \in A_2$, then L_1 and L_2 belong to different connected components in A. It will be proved that the sets A_1 and A_2 are connected, thereby verifying the conjecture for SB polynomials of size 2×2, degree 2 and with positive definite leading coefficients.

To this end consider the set A_0 of all polynomials $L \in A$ which have four distinct eigenvalues (necessarily real).

LEMMA 2.4 *The set* A_0 *consists of exactly two connected components.*

PROOF. Pick

$$L(\lambda) = \lambda^2 A_2 + \lambda A_1 + A_0 \in A_0 . \tag{2.3}$$

Since A_2 is positive definite, write $A_2 = S^* S$ for some invertible 2×2 matrix S. Let $S(t)$, $t \in [0,1]$ be a continuous path of invertible 2×2 matrices such that $S(0) = S$; $S(1) = I$. Then putting

$$L(\lambda,t) = (S(t))^* (\lambda^2 I + S^{*-1} A_1 S + S^{*-1} A_0 S) S(t) , \quad t \in [0,1]$$

we connect L with a polynomial $L_1(\lambda) = L(\lambda,1)$ with leading coefficient I, and the connecting path lies entirely in A_0. Further, let λ be the smallest eigenvalue of L_1. Putting

$$L_1(\lambda,t) = L_1(\lambda + t\underset{\sim}{\lambda}) , \quad t \in [0,1]$$

we obtain a continuous path in A_0 connecting $L_1(\lambda) = L_1(\lambda,0)$ with $L_2(\lambda) \overset{\text{def}}{=} L_1(\lambda + \underset{\sim}{\lambda})$, and the latter polynomial has leading coefficient I and its smallest eigenvalue is zero. Finally, by putting

$$L_2(\lambda,t) = U(t)^* L_2(\lambda) U(t) , \quad t \in [0,1]$$

where $U(t)$ is a suitable continuous path in the connected group of 2×2 unitary matrices, we connect $L_2(\lambda)$ in A_0 with a polynomial $L_3(\lambda) \in A_0$ of the following form:

$$L_3(\lambda) = \begin{bmatrix} \lambda^2 + \lambda a & \lambda(b+ic) \\ \lambda(b-ic) & \lambda^2 + \lambda d + x \end{bmatrix} , \tag{2.4}$$

where a, b, c, d, x are real numbers with $x \neq 0$, and $\det L_3(\lambda)$ has three different positive zeros.

Let $\widetilde{A}_0$ be the set of all polynomials of type (2.4). Let $0 < p < q < r$ be the eigenvalues of $L_3(\lambda) \in \widetilde{A}_0$, which is given by (2.4). Then the following relations hold:

$$ax = -pqr \; ; \quad d + a = -p - q - r \; ; \quad x + ad - b^2 - c^2 = pq + pr + qr , \tag{2.5}$$

which, together with inequalities $0 < p < q < r$, are necessary and sufficient in order that $L_3 \in \widetilde{A}_0$. It follows from (2.5) that, given $x \neq 0$ and p, q, r such that $0 < p < q < r$, one can solve (2.5) uniquely for a, d and $b^2 + c^2$ if and only if the number

$$x + ad - pq - pr - qr = \frac{1}{x^2}[x^3 - x^2(pq+pr+qr) + x(p+q+r)pqr - (pqr)^2]$$

is nonnegative. As the polynomial $x^3 - x^2(pq+pr+qr) + x(p+q+r)pqr - (pqr)^2$ has 3 different positive real zeros $pq < pr < qr$, we find that $\widetilde{A}_0$ has exactly two connected components, which can be identified in terms of the numbers p, q, r, x, b, c introduced above in the following way. One component is given by

$$\{(p,q,r,x,b,c) \in \mathbb{R}^6 \mid 0 < p < q < r ; \quad pq \lesssim x \lesssim pr ;$$
$$b^2 + c^2 = x^3 - x^2(pq+pq+qr) + x(p+q+r)pqr - (pqr)^2\} ;$$

the second is

$$\{(p,q,r,x,b,c) \in \mathbb{R}^6 \mid 0 < p < q < r ; \quad x \gtrsim qr ;$$
$$b^2 + c^2 = x^3 - x^2(pq+pr+qr) + x(p+q+r)pqr - (pqr)^2\} .$$

Hence the set A_0 also has exactly two connected components. □

Now one can easily show that the sets A_1 and A_2 are connected. Indeed, since $\widetilde{A}$ is dense in A, the set A consists of at most two connected components, and because of Theorem 2.1, A has exactly two connected components. Now $A = A_1 \cup A_2$ and $A_1 \cap A_2 = \emptyset$; so in view of the same Theorem 2.1, it is clear that A_1 and A_2 are the connected components of A.

2.3 Connected Components of Difference Equations

Consider again the class of difference equations studied in Section II.2.4. A typical equation has the form

$$A_0 x_i + A_1 x_{i+1} + \cdots + A_\ell x_{i+\ell} = 0 , \quad i = 0,1,\cdots , \tag{2.6}$$

where $\{x_i\}_{i=0}^{\infty}$ is a sequence of n-dimensional complex vectors to be found, and $A_0,\cdots,A_\ell$ are given complex $n \times n$ matrices. As before, we shall assume that ℓ is even,

$$A_j^* = A_{\ell-j}, \quad j = 0,\cdots,\ell$$

and A_ℓ is invertible (in particular, $A_{\frac{1}{2}\ell}$ is hermitian and A_0 is invertible). A system (2.6) has *stably bounded* solutions if there exists an $\varepsilon > 0$ such that all solutions of every equation $\sum_{m=0}^{\ell} A_m' x_{i+m} = 0$, $i = 0,1,\cdots$ with $A_j'^* = A_{\ell-j}'$, $j = 0,\cdots,\ell$ and $\sum_{m=0}^{\ell} ||A_m'-A_m|| < \varepsilon$ are bounded. A description of the difference equations with stably bounded solutions was given in Theorem III.2.6. Namely, the solutions of (2.6) are stably bounded if and only if the spectrum of the matrix polynomial $L(\lambda) = \sum_{j=0}^{\ell} \lambda^j A_j$ lies on the unit circle and the quadratic form $(x, i\lambda_0(\lambda^{-\frac{1}{2}\ell}L(\lambda))^{(1)}(\lambda_0)x)$, $x \in \operatorname{Ker} L(\lambda_0)$ is either positive definite or negative definite, for every $\lambda_0 \in \sigma(L)$.

There exists a natural topology on the set Ω_u of all equations (2.6) (with ℓ even, $A_{\ell-j}^* = A_j$, $j = 0,\cdots,\ell$ and A_ℓ invertible) with stably bounded solutions. Observe that Ω_u is a disconnected union of the sets $\Omega_{u,\ell}$, $\ell = 2,4,\cdots$, where $\Omega_{u,\ell}$ is the set of equations (2.6) with stably bounded solutions of degree ℓ; in turn, $\Omega_{u,\ell}$ is given a topology as a subset of $\mathbb{C}^{n\times n} \times \cdots \times \mathbb{C}^{n\times n}$ ($\ell+1$ times). So it makes sense to speak about the connected components of Ω_u.

In the sequel we shall identify the system (2.6) with the matrix polynomial $L(\lambda)$ formed by its coefficients.

To study the connected components of Ω_u we shall introduce the notion of an index for $L(\lambda)$: $\operatorname{Ind}_u(L) = \{n_1,\cdots,n_p\}$.

As we noticed above, every $\lambda_0 \in \sigma(L)$ is unimodular; we say that the sign of λ_0 is negative (resp. positive) if the quadratic form $(x, i\lambda_0(\lambda^{-\frac{1}{2}\ell}L(\lambda))^{(1)}(\lambda_0)x)$, $x \in \operatorname{Ker} L(\lambda_0)$ is negative (resp. positive) definite. Let $[\alpha_0,\alpha_1),(\alpha_1,\alpha_2),\cdots,(\alpha_{p-1},\alpha_p)$, $\alpha_0 = 0$, $\alpha_p = 2\pi$, be consecutive intervals such that every arc $[e^{i\alpha_0},e^{i\alpha_1}),\cdots,(e^{i\alpha_{p-1}},e^{i\alpha_p})$ contains the largest possible number of eigenvalues of $L(\lambda)$ having the same sign. Let n_i $(i = 1,\cdots,p)$ be the sum of multiplicities of the eigenvalues of $L(\lambda)$ lying on the i-th arc and multiplied by (-1) if the sign of these eigenvalues is negative.

THEOREM 2.5 *Let*

$$A_{01}x_i + A_{11}x_{i+1} + \cdots + A_{\ell 1}x_{i+\ell} = 0 , \quad i = 0,1,\cdots \tag{2.7}$$

and

$$A_{02}x_i + A_{12}x_{i+1} + \cdots + A_{\ell 2}x_{i+\ell} = 0 , \quad i = 0,1,\cdots \tag{2.8}$$

be two difference equations of type (2.6) $(A_{ji}^* = A_{\ell-j,i},\ j = 0,\cdots,\ell;$ $i = 1,2;$ ℓ *is even and* $A_{\ell 1}, A_{\ell 2}$ *are invertible*). *Suppose that both equations* (2.7) *and* (2.8) *have stably bounded solutions, and the indices* $\text{Ind}_u(L_1)$ *and* $\text{Ind}_u(L_2)$ *of* $L_1(\lambda) = \sum_{j=0}^{\ell} \lambda^j A_{j1}$ *and* $L_2(\lambda) = \sum_{j=0}^{\ell} \lambda^j A_{j2}$, *respectively, cannot be obtained from each other by a cyclic permutation. Then the equations* (2.7) *and* (2.8) *belong to different connected components in the set of all equations of type* (2.6) *with stably bounded solutions.*

PROOF. Let

$$C_j = \begin{bmatrix} 0 & I & \cdots & & 0 \\ 0 & 0 & I & \cdots & 0 \\ \vdots & & & \ddots & \vdots \\ 0 & 0 & 0 & \cdots & I \\ -A_{\ell j}^{-1}A_{0j} & -A_{\ell j}^{-1}A_{1j} & \cdots & & -A_{\ell j}^{-1}A_{\ell-1,j} \end{bmatrix}, \quad j = 1,2$$

$$\hat{B}_j = i \left[\begin{array}{ccc|ccc} & & & A_{\ell j} & \cdots & 0 \\ & 0 & & \vdots & \ddots & \vdots \\ & & & A_{k+1,j} & \cdots & A_{\ell j} \\ \hline -A_{0j} & \cdots & -A_{k-1,j} & & & \\ \vdots & \ddots & \vdots & & 0 & \\ 0 & \cdots & -A_{0j} & & & \end{array}\right], \quad j = 1,2 ,$$

where $k = \frac{\ell}{2}$. As we know from Section II.2.4, C_j is $\hat{B}_j$-unitary and, moreover, C_j is stably u-diagonable. Further, $\mathrm{Ind}_u(L_j)$ coincides with the index $\hat{B}_j$-$\mathrm{ind}_u(C_j)$ (Equation (II.2.20)). Now Theorem 1.4 shows that $(C_1,\hat{B}_1)$ and $(C_2,\hat{B}_2)$ belong to different connected components in the set of all pairs (U,H) where U is H-unitary and stably u-diagonable, and the conclusion of Theorem 2.5 follows. □

As in the case of SB polynomials, we propose the following conjecture.

CONJECTURE 2.6 *Let*

$$A_{0j}x_i + A_{1j}x_{i+1} + \cdots + A_{\ell j}x_{i+\ell} = 0\ , \quad i = 1,2,\cdots\ ; \quad j = 1,2$$

be difference equations with stably bounded solutions ($A^*_{ij} = A_{\ell-i,j}$, $i = 0,\cdots,\ell$; $j = 1,2$; ℓ *is even;* $A_{\ell j}$ *invertible,* $j = 1,2$). *Then the equations belong to the same connected component in the set of all difference equations of this type if and only if the indices* $\mathrm{Ind}_u(L_j)$, $j = 1,2$, *where* $L_j(\lambda) = \sum_{k=0}^{\ell} \lambda^k A_{kj}$ *are obtained from each other by a cyclic permutation.*

The "only if" part is the content of Theorem 2.5.

CHAPTER 3

CONNECTED COMPONENTS OF HAMILTONIAN EQUATIONS

Linear Hamiltonian differential equations with periodic coefficients and stably bounded solutions are studied in the chapter. The results obtained are based on a deeper study of the group of H-unitary matrices. It will be shown that the sign characteristic of the monodromy matrix plays an important role but does not give a complete picture of the structure of the connected components of these equations. To obtain the full description it is necessary to introduce an additional characteristic analogous to the winding number.

3.1 Definition and Explanation of the Problem

Consider the following system of differential equations

$$E\frac{dx}{dt} = iH(t)x \ , \quad t \in \mathbb{R} \tag{3.1}$$

where $H(t)$ is a complex hermitian piecewise continuous $n \times n$ matrix function with period $\omega \neq 0$, and $E = E^*$ is an invertible $n \times n$ complex matrix.

In Part III we characterized the system (3.1) whose solutions are stably bounded. This means that there exists an $\varepsilon > 0$ such that for all hermitian piecewise continuous $n \times n$ matrix functions $\widetilde{H}(t)$ with period ω such that $\max\limits_{0 \leq t \leq \omega} ||H(t)-\widetilde{H}(t)|| < \varepsilon$, the solutions of the system

$$E\frac{dx}{dt} = i\widetilde{H}(t)x \ , \quad t \in \mathbb{R}$$

are bounded on the whole real line. The characterization referred to is given in terms of the monodromy matrix X of (3.1); namely, X should be diagonable with unimodular eigenvalues, and for every $\lambda_0 \in \sigma(X)$ the quadratic form (Ex,x) is either positive definite or negative definite on $\mathrm{Ker}(\lambda_0 I-X)$

(in short, X is a stably u-diagonable E-unitary matrix).

In this chapter we study the connected components of all systems (3.1) with stably bounded solutions. Let us first explain the problem in more detail.

Denote by H_s the set of all complex hermitian piecewise continuous $n \times n$ matrix functions $H(t)$ with period ω (ω being fixed), for which the system (3.1) has stably bounded solutions. We say that two systems

$$E\,\frac{dx}{dt} = iH_1(t)x$$

and

$$E\,\frac{dx}{dt} = iH_2(t)x$$

with $H_1 \in H_s$, $H_2 \in H_s$ belong to the same connected component H_s if there exists a continuous H_s-valued function $H(\cdot;\nu)$, $\nu \in [0,1]$ such that $H(\cdot;0) = H_1$; $H(\cdot;1) = H_2$. Continuity is understood in the sense of the norm $|||H(\cdot;\nu)|||$, i.e. for every convergent sequence $\nu_m \to \nu \in [0,1]$; $\nu_m \in [0,1]$ we have

$$\lim_{m\to\infty} \max_{t\in[0,\omega]} ||H(t,\nu_m)-H(t,\nu)|| = 0 .$$

As we know (Theorem II.1.1), a system (3.1) can be identified with a piecewise continuously differentiable curve $X(t)$, $t \in [0,\omega]$ in the group $\mathbb{U}(E)$ of E-unitary matrices such that $X(0) = I$. The systems with $H(t) \in H_s$ are described by the curves $X(t)$ with the additional property that $X(\omega)$ is stably u-diagonable. So the description of the connected components of systems (3.1) with stably bounded solutions can be reduced to the description of connected components in the set Y of piecewise continuously differentiable curves $X(t)$, $t \in [0,\omega]$, for which $X(t) \in \mathbb{U}(E)$, $X(0) = I$, and $X(\omega)$ is stably u-diagonable. A necessary condition for two such curves $X_1(t)$ and $X_2(t)$ to belong to the same connected component in Y is easily given: namely, their ends $X_1(\omega)$ and $X_2(\omega)$ should be in the same connected component of the set of stably u-diagonable E-unitary matrices. However, this obvious necessary condition is not sufficient. To understand the situation better, we need more information about the group $\mathbb{U}(E)$ and about the topological structure of connected components of the set of stably u-diagonable E-unitary matrices. This information will be provided in a broader context (which is independently interesting) in the next sections.

3.2 Group of H-Unitary Matrices

Let H be an $n \times n$ hermitian invertible matrix. We have already proved some of the properties of the group $\mathbb{U}(H)$ of all H-unitary matrices. For instance, Lemma I.3.8 shows that $\mathbb{U}(H)$ is connected. However, we need more knowledge of the topological structure of $\mathbb{U}(H)$. The following result will be useful.

THEOREM 3.1 *Let* p, q *be the number of positive and negative eigenvalues of* H *respectively (counting multiplicities). Assume* $p > 0$ *and* $q > 0$. *Then the group* $\mathbb{U}(H)$ *is homeomorphic to the product*

$$\mathbb{SU}(p) \times \mathbb{SU}(q) \times \mathbb{T} \times \mathbb{T} \times \mathbb{R}^{2pq} , \tag{3.2}$$

where $\mathbb{T}$ *is the unit circle and* $\mathbb{SU}(k)$ *is the group of all unitary* $k \times k$ *matrices with determinant* 1. *If* H *is either positive definite or negative definite (i.e. if either* $p = 0$ *or* $q = 0$*) then* $\mathbb{U}(H)$ *is homeomorphic to* $\mathbb{SU}(n) \times \mathbb{T}$.

PROOF. The case when either $p = 0$ or $q = 0$ presents no difficulty. So consider the case $p,q > 0$. Without loss of generality we can assume

$$H = \begin{bmatrix} I_p & 0 \\ 0 & -I_q \end{bmatrix} . \tag{3.3}$$

Given H-unitary X, consider the polar representation $X = PU$, where P is positive definite and U is unitary. Then P and U are also H-unitary, because equality $HXH^{-1} = X^{*-1}$ implies

$$HPUH^{-1} = (U^*P^*)^{-1} = (U^{-1}P)^{-1} = P^{-1}U ,$$

or

$$PU = H^{-1}P^{-1}UH = H^{-1}P^{-1}H \cdot H^{-1}UH .$$

In view of the uniqueness of the polar representation of X, we have $P = H^{-1}P^{-1}H$, $U = H^{-1}UH$, so P and U are indeed H-unitary. Since the polar representation depends continuously on X (because $P = (XX^*)^{\frac{1}{2}}$), we find that $\mathbb{U}(H)$ is homeomorphic to the product

$$\{P \in \mathbb{U}(H) \mid P \text{ is positive definite}\} \times \{U \in \mathbb{U}(H) \mid UU^* = I\} .$$

Further, note that for every positive definite P there exists a unique hermitian Q such that $P = e^Q$, and this correspondence is a homeomorphism between the set of all $n \times n$ positive definite matrices and the set of all $n \times n$ hermitian matrices. If, in addition P is H-unitary, then uniqueness of the hermitian logarithm of P and (3.3) imply that

$$Q = -H^{-1}QH . \tag{3.4}$$

Write

$$Q = \begin{bmatrix} Q_{11} & Q_{12} \\ Q_{12}^* & Q_{22} \end{bmatrix}, \quad Q_{11} = Q_{11}^* , \quad Q_{22} = Q_{22}^* ,$$

where the size of Q_{11} (Q_{22}) is $p \times p$ $(q \times q)$. Taking into account the form (3.3) of H, we find that (3.4) holds if and only if $Q_{11} = 0$ and $Q_{22} = 0$. In conclusion, the set of $P \in \mathbb{U}(H)$ for which P is positive definite is homeomorphic to the set of all $p \times q$ complex matrices Q_{12}, i.e. to $\mathbb{R}^{2pq}$.

An H-unitary matrix U is unitary if and only if U commutes with H, i.e. if and only if U has the form $U = U_1 \oplus U_2$, where U_1 (resp. U_2) is a unitary $p \times p$ matrix (resp. unitary $q \times q$ matrix). So the set of $U \in \mathbb{U}(H)$ for which $UU^* = I$ is homeomorphic to the product $\mathbb{U}(p) \times \mathbb{U}(q)$, where $\mathbb{U}(k)$ is the group of unitary $k \times k$ matrices. But then $\mathbb{U}(k)$ is homeomorphic to the product $\mathbb{T} \times S\mathbb{U}(k)$, since

$$X = \operatorname{diag}[\det X, 1, \cdots, 1]Y ,$$

where $X \in \mathbb{U}(k)$ and $Y \in S\mathbb{U}(k)$. So the theorem follows. □

In this chapter the notion of homotopy will be important. Let X be an arcwise connected topological space, and let $x_0 \in X$. Two continuous loops $a\colon [0,1] \to X$ and $b\colon [0,1] \to X$ such that $a(0) = a(1) = b(0) = b(1) = x_0$ are called *homotopic* (with fixed ends) if there exists a continuous function $h : [0,1] \times [0,1] \to X$ such that $h(\tau,0) = a(\tau)$; $h(\tau,1) = b(\tau)$ and $h(0,t) = h(1,t) = x_0$, for $0 \le \tau,t \le 1$ in each case. The topological space X is called *simply connected* if any continuous loop $a : [0,1] \to X$ with $a(0) = a(1) = x_0$ is homotopic to a constant loop x_0 (because of the arcwise connectedness of X this property does not depend on the choice of $x_0 \in X$). For example, $\mathbb{R}^n$ is simply connected, but the unit circle $\mathbb{T}$ is not.

It will be important for future reference to note that the group $\mathbb{SU}(k)$ is simply connected for all integers k. A standard proof of this fact uses the concept of covering spaces for topological groups and is therefore omitted (see p. 60 of [9], for example).

Later, we shall need an explicit description of homotopy in the topological space $\mathbb{U}(H)$ (which follows from the proof of Theorem 3.1). Namely, assume H is not definite (positive or negative), and let f_1, f_2 be eigenvectors of H corresponding to a positive and to a negative eigenvalue of H. Then for each continuous loop $a : [0,1] \to \mathbb{U}(H)$ with $a(0) = a(1) = I$ there exist unique integers k, ℓ such that $a(\tau)$ is homotopic to the loop $\pi_{k,\ell} : [0,1] \to \mathbb{U}(H)$ defined as follows

$$\pi_{k,\ell}(\tau)f_1 = e^{2\pi i(k+\ell)\tau}f_1 \; ; \quad \pi_{k,\ell}(\tau)f_2 = e^{-2\pi ik\tau}f_2 \; ; \quad \pi_{k,\ell}(\tau)x = x \qquad (3.5)$$

for every $x \in \mathbb{C}^n$ orthogonal to f_1 and f_2. For those readers who are familiar with the notion of fundamental group we remark that $\pi_{1,0}$ and $\pi_{0,1}$ are generators in the fundamental group (isomorphic to $\mathbb{Z} \times \mathbb{Z}$) of $\mathbb{U}(H)$.

Consider now the group $\mathbb{SU}(H)$ of all H-unitary matrices with determinant 1. As in the proof of Theorem 3.1, one shows, assuming H is non-definite, that $\mathbb{SU}(H)$ is homeomorphic to

$$\mathbb{SU}(p) \times \mathbb{SU}(q) \times \mathbb{T} \times \mathbb{R}^{2pq} .$$

For each continuous loop $a : [0,1] \to \mathbb{SU}(H)$ with $a(0) = a(1) = I$ there is a unique integer k such that a is homotopic to $\pi_{k,0}$ (in the notation introduced in the preceding paragraph). In other words, $\pi_{1,0}$ is a generator in the fundamental group (isomorphic to $\mathbb{Z}$) of $\mathbb{SU}(H)$.

We note also that Theorem 3.1 (as well as Theorem I.5.8) can be easily deduced using a general result on pseudoalgebraic groups, which we shall state now. For every invertible complex $n \times n$ matrix A let $z_{ij}(A) \in \mathbb{C}$ be its (i,j)-entry, and let $x_{ij}(A)$ and $y_{ij}(A)$ be the real and the imaginary part of $z_{ij}(A)$, respectively. A group G of complex $n \times n$ matrices (with respect to the matrix multiplication) is called *pseudoalgebraic* if there exist polynomials $P_1(\cdots,x_{ij},y_{ij},\cdots), P_2(\cdots,x_{ij},y_{ij},\cdots),\cdots, P_r(\cdots,x_{ij},y_{ij},\cdots)$ of $2n^2$ real variables x_{ij}, y_{ij} $(i,j = 1,\cdots,n)$ with complex coefficients such that $A \in G$ if and only if

$$P_1(\cdots,x_{ij}(A),y_{ij}(A),\cdots) = \cdots = P_r(\cdots,x_{ij}(A),y_{ij}(A),\cdots) = 0 .$$

The following are examples of pseudoalgebraic groups: (i) all unitary matrices; (ii) all real orthogonal matrices; (iii) all matrices with determinant 1; (iv) all upper triangular matrices with 1 on the main diagonal.

THEOREM 3.2 *Let* G *be a pseudoalgebraic group of* $n \times n$ *complex matrices such that* $A \in G$ *implies* $A^* \in G$. *Then* G *is homeomorphic to the product of* $G \cap \mathbb{U}(n)$, *where* $\mathbb{U}(n)$ *is the set of all unitary matrices in* $\mathbb{C}^n$, *and* $\mathbb{R}^d$ *for some integer* $d \geq 0$. *In particular,* G *and* $G \cap \mathbb{U}(n)$ *have the same number of connected components.*

The proof of this theorem is beyond the scope of this book (we refer the reader to Chapter IV.4 of [21]).

3.3 Simple Connectedness of Unitary Similarity Classes

Let A be H-selfadjoint. We already have seen in Chapter I.3 that the set of all H-selfadjoint matrices B which are H-unitarily similar to A is connected. Using Theorem 3.1, we shall prove now that this set is also simply connected, provided A has certain spectral properties.

We start with an example showing that not every H-unitary similarity class is simply connected.

EXAMPLE 3.1 Let

$$A = \begin{bmatrix} 1 & -1 \\ 1 & -1 \end{bmatrix} ; \quad H = \begin{bmatrix} 1 & 0 \\ 0 & -1 \end{bmatrix} .$$

One easily checks that $HA = A^*H$, i.e. A is H-selfadjoint. We shall prove that the set of all 2×2 matrices A_1 of the form $A_1 = S^{-1}AS$, where $S^*HS = H$, is not simply connected. It will turn out that this property is a consequence of the fact that the unit circle is not simply connected.

Note that $\sigma(A) = \{0\}$. Let

$$A_1 = \begin{bmatrix} \alpha & \beta \\ \gamma & \delta \end{bmatrix}$$

be any matrix which is H-unitarily similar to A. Then

$$\alpha + \delta = 0 , \quad \alpha\delta - \gamma\beta = 0 , \quad \alpha \neq 0 . \tag{3.6}$$

Indeed, the first two properties are just consequences of the fact that $\mathrm{tr}\, A_1 = \mathrm{tr}\, A$ and $\det A_1 = \det A$. To verify the last property, suppose the contrary, i.e. that $\alpha = 0$. Then either $\gamma = 0, \beta \neq 0$ or $\gamma \neq 0, \beta = 0$. Assume, for instance, that $\gamma = 0$ and $\beta \neq 0$. Write

$$S \begin{bmatrix} 0 & \beta \\ 0 & 0 \end{bmatrix} = \begin{bmatrix} 1 & -1 \\ 1 & -1 \end{bmatrix} S , \tag{3.7}$$

where $S = \begin{bmatrix} s_1 & s_2 \\ s_3 & s_4 \end{bmatrix}$ is H-unitary. A computation in (3.7) implies $s_1 = s_3$. On the other hand, the equality $S^*HS = H$ implies $|s_1|^2 - |s_3|^2 = 1$, a contradiction.

Consider now the following loop $A(t)$ in the H-unitary similarity class $\mathcal{UE}(A) \ni A$:

$$A(\tau) = \begin{bmatrix} 1 & -e^{-2\pi i\tau} \\ e^{2\pi i\tau} & -1 \end{bmatrix}, \quad \tau \in [0,1] .$$

Since

$$A(\tau) = \begin{bmatrix} e^{-2\pi i\tau} & 0 \\ 0 & 1 \end{bmatrix} A \begin{bmatrix} e^{2\pi i\tau} & 0 \\ 0 & 1 \end{bmatrix}$$

and the matrix $\mathrm{diag}[e^{2\pi i\tau},1]$ is H-unitary, the loop $A(\tau)$ is indeed in $\mathcal{UE}(A)$. We claim that $A(\tau)$ is not homotopic to a constant loop A. Assume the contrary. Then there exists a continuous function $Q : [0,1] \times [0,1] \to$ $\to \mathcal{UE}(A)$ such that $Q(0,\tau) = A(\tau)$ and $Q(1,\tau) = A$ for $0 \leq \tau \leq 1$; and $Q(s,0) = Q(s,1) = A$ $(0 \leq s \leq 1)$. Write

$$Q(s,\tau) = \begin{bmatrix} q_1(s,\tau) & q_2(s,\tau) \\ q_3(s,\tau) & q_4(s,\tau) \end{bmatrix} .$$

By the above observation, $q_1(s,\tau) \neq 0$, and consequently (in view of (3.6)) $q_3(s,\tau) \neq 0$. Denote $x(s,\tau) = (1-q_1(s,\tau))q_3(s,\tau)^{-1}$ and put

$$R_1(s,\tau) = \begin{bmatrix} 1 & x(s,\tau) \\ 0 & 1 \end{bmatrix} Q(s,\tau) \begin{bmatrix} 1 & -x(s,\tau) \\ 0 & 1 \end{bmatrix} .$$

Then $R_1(s,\tau)$ is continuous,

$$\left.\begin{aligned} &R_1(0,\tau) = A(\tau) \; ; \quad R_1(1,\tau) = A \; ; \\ &R_1(s,0) = R_1(s,1) = A \end{aligned}\right\} \tag{3.8}$$

and for every $(s,\tau) \in [0,1] \times [0,1]$ the matrix $R_1(s,\tau)$ has the form

$$R_1(s,\tau) = \begin{bmatrix} 1 & -\frac{1}{r_1(s,\tau)} \\ r_1(s,\tau) & -1 \end{bmatrix}$$

where $r_1(s,\tau)$ is a non-zero number. Further, put

$$R_2(s,\tau) = \begin{bmatrix} \frac{1}{|r_1(s,\tau)|} & 0 \\ 0 & |r_1(s,\tau)| \end{bmatrix} \begin{bmatrix} 1 & -\frac{1}{r_1(s,\tau)} \\ r_1(s,\tau) & -1 \end{bmatrix} \begin{bmatrix} |r_1(s,\tau)| & 0 \\ 0 & \frac{1}{|r_1(s,\tau)|} \end{bmatrix} =$$

$$= \begin{bmatrix} 1 & -\frac{1}{r_2(s,\tau)} \\ r_2(s,\tau) & -1 \end{bmatrix},$$

where $|r_2(s,\tau)| = 1$ for $0 \leq s,\tau \leq 1$. Then the equalities (3.8) still hold when replacing R_1 by R_2; in other words, $r_2(s,\tau)$ is a homotopy between the loop $e^{2\pi i\tau}$ and the constant loop 1 on the unit circle. However, such a homotopy does not exist. □

The following result provides sufficient conditions for simple connectedness of H-unitary similarity classes.

THEOREM 3.3 *Let* A *be* H-*selfadjoint, where* H *is neither positive nor negative definite, and assume that* A *is similar to a real diagonal matrix. Then the* H-*unitary similarity class of* H-*selfadjoint matrices which contains* A, *is simply connected.*

PROOF. We shall assume $(A,H) = (J,P_{\varepsilon,J})$ is in the canonical form. Write $P = P_{\varepsilon,J}$ for short.

Let $F(\tau)$, $\tau \in [0,1]$ be a continuous loop in the H-unitary similarity class $\mathcal{UE}(J)$ containing J such that $F(0) = F(1) = J$ and $F(\tau_1) \neq F(\tau_2)$ for $0 < |\tau_1-\tau_2| < 1$ $(\tau_1,\tau_2 \in [0,1])$. We will show that there exists a continuous $\mathcal{UE}(J)$-valued function $G(\tau,s)$; $\tau,s \in [0,1]$ such that

$$G(\tau,0) = F(\tau) \; ; \quad G(0,s) = G(1,s) = J \; , \tag{3.9}$$

$$G(\tau,1) = J \quad \text{for all} \quad \tau \in [0,1] \; , \tag{3.10}$$

which will establish the simple connectedness of $\mathcal{UE}(J)$. By Corollary III.5.13 there exists a continuous invertible matrix function $S(\tau)$, $\tau \in [0,1]$ such that $F(\tau) = S(\tau)^{-1}JS(\tau)$, $P = S(\tau)^*PS(\tau)$, and $S(0) = I$. Two cases can occur: (1) $S(1) \neq I$; (2) $S(1) = I$.

We shall show first that the case (1) can be reduced to the case (2). Consider the set C of all matrices S such that $SJ = JS$ and $S^* P_{\varepsilon,J} S = P_{\varepsilon,J}$. The set C is connected. Indeed, write

$$J = \alpha_1 I_{m_1} \oplus \cdots \oplus \alpha_k I_{m_k} \;; \qquad P_{\varepsilon,J} = Q_1 \oplus \cdots \oplus Q_k$$

where $\alpha_1,\cdots,\alpha_k$ are different real numbers and Q_j is an invertible hermitian $m_j \times m_j$ matrix. Clearly, every $S \in C$ has the form $S = S_1 \oplus \cdots \oplus S_k$, where S_j is a Q_j-unitary matrix, $j = 1,\cdots,k$ and the connectedness of C follows from the connectedness of the group of all H-unitary matrices.

Assume now case (1) holds. By the connectedness of C, there exists a continuous path $T(\tau)$ such that $T(0) = S(1)$; $T(1) = I$; $T(\tau)J = JT(\tau)$ and $T(\tau)^* T(\tau) = P$. Then consider

$$F'(\tau) = \begin{cases} F(2\tau) & , \quad 0 \leq \tau \leq \frac{1}{2} \\ T(2\tau-1)^{-1} J T(2\tau-1) = J \;, & \frac{1}{2} \leq \tau \leq 1 \end{cases}$$

in place of $F(\tau)$ and

$$S'(\tau) = \begin{cases} S(2\tau) \;, & 0 \leq \tau \leq \frac{1}{2} \\ T(2\tau-1) \;, & \frac{1}{2} \leq \tau \leq 1 \end{cases}$$

in place of $S(\tau)$.

So consider the case (2). As we have seen in Section 3.1, there exist unique integers p and q such that the loop $S(\tau)$, $\tau \in [0,1]$ is homotopic to the loop $\pi_{p,q} : [0,1] \to \mathbb{U}(H)$ defined by the equalities

$$\pi_{p,q}(\tau)f_1 = e^{2\pi i(p+q)\tau} f_1 \;; \qquad \pi_{p,q}(\tau)f_2 = e^{-2\pi i p\tau} f_2 \;; \qquad \pi_{p,q}(\tau)x = x$$

for every $x \perp \mathrm{Span}\{f_1,f_2\}$, where f_1 (resp. f_2) is an eigenvector of P corresponding to a positive (resp. negative) eigenvalue. So we can assume $S(\tau) = \pi_{p,q}(\tau)$.

Now we shall choose f_1 and f_2 in a special way. Let

$$J = J_1 \oplus J_2 \oplus J_3 \;, \qquad P = P_1 \oplus P_2 \oplus P_3 \;, \tag{3.11}$$

where J_1, J_2 are the Jordan blocks of size 1 with real eigenvalues, and $\mathrm{sig}\, P_1 = 1$, $\mathrm{sig}\, P_2 = -1$. (Of course, the partition of J and P in (3.11) are consistent). Put $f_1 = \langle 1\ 0 \cdots 0\rangle$, $f_2 = \langle 0\ 1\ 0 \cdots 0\rangle$. Then

$(\pi_{pq}(\tau))^{-1}J\pi_{pq}(\tau) = J$ for all $\tau \in [0,1]$, so in fact $F(\tau) = S(\tau)^{-1}JS(\tau) = J$ for all $\tau \in [0,1]$, and one can choose $G(\tau,s) = F(\tau)$. □

Passing to H-unitary matrices by means of the Cayley transformation, we obtain the following statement from Theorem 3.3.

THEOREM 3.4 *Let* A *be an* H-*unitary matrix which is similar to a diagonable matrix with unimodular eigenvalues. Assume* H *is not definite. Then the* H-*unitary similarity class which contains* A *is simply connected.*

It turns out that for any H-selfadjoint matrix A the set of all matrices S such that SA = AS and $S^*HS = H$, is connected. Using this fact one can prove the simple connectedness of the H-unitary similarity class which contains A, provided the canonical form of (A,H) has two Jordan blocks of size 1 each with opposite signs in the sign characteristic. The proof is similar to the proof of Theorem 3.3.

3.4 Homotopy in Connected Components of u-Stably Diagonable Matrices

Let $H = H^*$ be an invertible $n \times n$ matrix. Recall that an H-unitary matrix A is called *stably* u-*diagonable* if all eigenvalues of A are unimodular, there exists a basis of eigenvectors of A in $\mathbb{C}^n$, and these properties are valid for every H-unitary matrix B sufficiently close to A.

Denote by $S_u(H)$ the set of all stably u-diagonable H-unitary matrices (for fixed H). By Theorem 1.3 the set $S_u(H)$ is not connected (unless H is positive definite or negative definite); this theorem also provides a description of the connected components in $S_u(H)$. We will be interested in homotopy in these connected components.

It will be convenient to introduce the following definition. Let $a : [0,1] \to X$, $a(0) = a(1) \in X$ be a continuous loop, where X is a topological space, and let k be an integer. The k-*th multiple* of $a(\tau)$ is the loop $a_k(\tau)$ defined as follows:

$$a_k(\tau) = a(k(\tau - \tfrac{i}{k})) \quad \text{for} \quad \tfrac{i}{k} \leq \tau \leq \tfrac{i+1}{k}, \quad i = 0,\cdots,k-1$$

if k is positive;

$$a_k(\tau) = a(|k|(\tfrac{i+1}{|k|} - \tau)) \quad \text{for} \quad \tfrac{i}{|k|} \leq \tau \leq \tfrac{i+1}{|k|}, \quad i = 0,\cdots,|k|-1$$

if k is negative; and

$$a_0(\tau) \equiv a(0) , \quad 0 \leq \tau \leq 1 .$$

THEOREM 3.5 *Let* $S_u^{(c)}(H)$ *be a connected component in* $S_u(H)$, *and let* $X_0 \in S_u^{(c)}(H)$. *Then there is a continuous loop* $a_0 : [0,1] \to S_u^{(c)}(H)$, $a_0(0) = a_0(1) = X_0$ *such that for any continuous loop* $a : [0,1] \to S_u^{(c)}(H)$ *with* $a(0) = a(1) = X_0$ *there exists a unique integer* k *with the property that* $a(\tau)$ *is homotopic to the* k-*th multiple of* $a_0(\tau)$.

Note that the loop $a_0(\tau)$ in Theorem 3.5 is obviously not unique.

If H is a diagonal matrix with ± 1's on the main diagonal and X_0 is a diagonal matrix with unimodular eigenvalues, the loop $a_0(\tau)$ can be conveniently described by

$$a_0(\tau) = e^{2\pi i\tau} X_0 , \quad 0 \leq \tau \leq 1 .$$

PROOF. Let $X(\tau)$, $\tau \in [0,1]$ be a continuous loop in $S_u(H)$ such that $X(0) = X(1) = X_0$. Without loss of generality assume (cf. the proof of Theorem 1.3 that

$$X_0 = \mathrm{diag}[e^{i\theta_1}, e^{i\theta_2}, \cdots, e^{i\theta_n}] ; \quad H = \mathrm{diag}[\zeta_1, \zeta_2, \cdots, \zeta_n] ; \quad \zeta_i = \pm 1 .$$

Here $\theta_1 \leq \theta_2 \leq \cdots \leq \theta_n$; $\theta_n - \theta_1 < 2\pi$. Assume: $\zeta_1 = \zeta_2 = \cdots = \zeta_{k_1} = 1$; $\zeta_{k_1+1} = \zeta_{k_1+2} = \cdots = \zeta_{k_2} = -1$; $\cdots$; $\zeta_{k_{p-1}+1} = \zeta_{k_{p-1}+2} = \cdots = \zeta_{k_p} = (-1)^{p-1}$; $k_p = n$. Then $\theta_{k_i} < \theta_{k_i+1}$; $i = 1,\cdots,p-1$. Let $e^{\theta_i(\tau)}$, $\tau \in [0,1]$ be a continuous eigenvalue of $X(\tau)$ such that $\theta_i(0) = \theta_i$ $(i = 1,\cdots,n)$. Then $\theta_i(1) = \theta_i + 2k\pi$, $i = 1,\cdots,n$ for some integer k (the integer k is the same for all i in view of the condition $X(\tau) \in S_u(H)$.

Define $\widetilde{X}(\tau)$, $\tau \in [0,1]$ as follows. For $i = 1,\cdots,p$, let $M_i(\tau)$ be the sum of root subspaces for $X(\tau)$ corresponding to the eigenvalues $e^{\theta_{k_{i-1}+1}(\tau)}, \cdots, e^{\theta_{k_i}(\tau)}$ (by definition $k_0 = 0$). Then $M_i(t)$ is an invariant subspace for $\widetilde{X}(t)$ and $\widetilde{X}(t)|_{M_i(t)} = e^{\theta_{k_i}(t)} I$.

The loop $\widetilde{X}(\tau)$, $\tau \in [0,1]$ is homotopic to $X(\tau)$. Indeed, if $\widetilde{M}$ is the root subspace for $X(\tau)$ corresponding to the set of equal eigenvalues

$$e^{\theta_j(\tau)} = e^{\theta_{j+1}(\tau)} = \cdots = e^{\theta_u(\tau)} ,$$

then put

$$G(\tau,s)|_{\widetilde{M}} = \exp[(1-s)\theta(\tau) + s\theta_{k_i}(\tau)]I|_{\widetilde{M}} ,$$

where $\theta(\tau) = \theta_j(\tau) = \cdots = \theta_u(\tau)$ and k_i is such that $k_{i-1} < j < u \leq k_i$. In this way a continuous $S_u(H)$-valued function $G(\tau,s)$, $0 \leq \tau,s \leq 1$ is defined, which provides the homotopy between $\widetilde{X}(\tau)$ and $X(\tau)$.

We shall prove that the loop $\widetilde{X}(\tau)$ homotopic (with fixed ends) to the loop

$$Y(\tau) = \mathrm{diag}[e^{i\eta_1(\tau)} I_{k_1}, e^{i\eta_2(\tau)} I_{k_2-k_1}, \cdots, e^{i\eta_p(\tau)} I_{k_p-k_{p-1}}],\ \tau \in [0,1],$$

where $\eta_i(\tau) = \theta_{k_i}(\tau)$, $i = 1,\cdots,p$. In other words, we shall prove that there exists a continuous $S_u(H)$-valued function $G_0(\tau,s)$ such that $G_0(\tau,0) = \widetilde{X}(\tau)$; $G_0(\tau,1) = Y(\tau)$; $G_0(0,s) = G_0(1,s) = \widetilde{X}(0)$ $(= Y(0) = \widetilde{X}(1) = = Y(1))$. It is sufficient to check that the loop $Z(\tau)$, $0 \leq \tau \leq 1$ defined by the equalities $Z(\tau) = \widetilde{X}(2\tau)$ for $0 \leq \tau \leq \frac{1}{2}$ and $Z(\tau) = Y(-2\tau+2)$ for $\frac{1}{2} \leq \tau \leq 1$, is homotopic to the loop which takes the same value X_0 for all $\tau \in [0,1]$.

By Corollary III.5.13, there exists a continuous matrix function $S(\tau)$, $\tau \in [0,1]$ such that $S(1) = I$ and for all τ the matrix $S(\tau)$ is H-unitary and $\widetilde{X}(\tau) = S(\tau)^{-1}Y(\tau)S(\tau)$. Define a continuous $S_u(H)$-valued function $G_1(\tau,s)$ as follows:

$$G_1(\tau,s) = \begin{cases} \widetilde{X}(2\tau) & , \quad 0 \leq \tau \leq \frac{1}{2}(1-s) ; \\ \widetilde{X}(1-s) & , \quad \frac{1}{2}(1-s) \leq \tau \leq \frac{1}{2} ; \\ S(2\tau-s)Y(1-s)(S(2\tau-s))^{-1} & , \quad \frac{1}{2} \leq \tau \leq \frac{1}{2}(1+s) ; \\ Y(-2\tau+2) & , \quad \frac{1}{2}(1+s) \leq \tau \leq 1 . \end{cases}$$

So $G_1(\tau,s)$ is a homotopy such that

$$G_1(\tau,0) = Z(\tau) ; \quad G_1(\tau,1) = \begin{cases} X_0 & , \ 0 \leq \tau \leq \frac{1}{2} ; \\ S(2\tau-1)Y(0)(S(2\tau-1))^{-1} & , \ \frac{1}{2} \leq \tau \leq 1 ; \end{cases}$$

$$G_1(0,s) = G_1(1,s) = X_0 .$$

Further, the loop $G_1(\tau,1)$ lies entirely in the H-unitary similarity class of H-unitary matrices. By Theorem 3.3, $G_1(\tau,1)$ is homotopic (with fixed ends) to a constant loop X_0. So the same is true for $Z(\tau)$.

We have proved that $\tilde{X}(\tau)$ is homotopic (with fixed ends) to $Y(\tau)$. From the construction of $\tilde{X}(\tau)$ it follows that the same is true for $X(\tau)$. Now $Y(\tau)$ is clearly homotopic (with fixed ends) to the loop

$$Y_1(\tau) = e^{2\pi ik\tau}X_0 , \quad \tau \in [0,1] .$$

So we have proved that an arbitrary loop $X(\tau)$, $\tau \in [0,1]$, $X(0) = X(1) = X_0$ is homotopic to some integer multiple of the loop $\varphi(\tau) = e^{2\pi i\tau}X_0$, $0 \leq \tau \leq 1$.

To prove the uniqueness of k, it remains to show that for $k \neq 0$, the loop $\varphi_k(\tau) = e^{2\pi ik\tau}X_0$ is not homotopic to a constant loop X_0. But this is trivial, since the winding number of $\det(\varphi_k)$ is $nk \neq 0$. □

3.5 Homotopy Indices of H-Unitary Matrices

Let H be a hermitian and invertible $n \times n$ matrix, and let X_+ (resp. X_-) be the subspace in $\mathbb{C}^n$ spanned by eigenvectors of H corresponding to positive (resp. negative) eigenvalues of H. For every H-unitary matrix A write

$$A = \begin{bmatrix} A_1 & A_2 \\ A_3 & A_4 \end{bmatrix} , \tag{3.12}$$

where $A_1 : X_+ \to X_+$; $A_2 : X_- \to X_+$; $A_3 : X_+ \to X_-$; $A_4 : X_- \to X_-$.

The linear transformations A_1 and A_4 are invertible. Indeed, the condition $A^*HA = H$ gives, in particular:

$$A_1^*H_1A_1 = H_1 - A_3^*H_2A_3 ; \tag{3.13}$$

$$A_4^*H_2A_4 = H_2 - A_2^*H_1A_2 , \tag{3.14}$$

where $H_1 = H|_{X_+}$, $H_2 = H|_{X_-}$. The right-hand side of (3.13) is positive definite, which implies invertibility of A_1. Similarly, (3.14) yields invertibility of A_4.

Now let $A(t)$, $t \in [0,1]$ be a continuous curve in the group $\mathbb{U}(H)$ of all H-unitary matrices. Partition $A(t)$ as in (3.12) and define

$$n_+ = \frac{1}{2\pi} \{\mathrm{Arg}(\det A_1(1)) - \mathrm{Arg}(\det A_1(0))\}$$

$$n_- = \frac{1}{2\pi}\{\mathrm{Arg}(\det A_4(1)) - \mathrm{Arg}(\det A_4(0))\} .$$

(Of course, the branch of $\mathrm{Arg}(\det A_1(t))$ and $\mathrm{Arg}(\det A_4(t))$ is supposed to be continuous on t). In other words, assuming the end $A(1)$ of the curve coincides with the beginning $A(0)$, n_+ (resp. n_-) is the winding number of the curve $\det A_1(t)$ (resp. $\det A_4(t)$), $0 \leq t \leq 1$. The numbers $n_\pm$ are called the *homotopy indices* of the continuous curve $A(t)$, $t \in [0,1]$. The number n_+ (resp. n_-) will be denoted $\mathrm{Ind}_+\{A(t)\}$ (resp. $\mathrm{Ind}_-\{A(t)\}$). If H is definite, then we have only one homotopy index of $A(t)$, denoted $\mathrm{Ind}\{A(t)\}$.

Let $A(t)$ be a continuous loop in $\mathbb{U}(H)$, so $A(0) = A(1)$. As the proof of Theorem 3.1 shows, the loop $A(t)$ is homotopic to the loop π_{-n_-,n_++n_-}, where $n_\pm = \mathrm{Ind}_\pm\{A(t)\}$, and $\pi_{k\ell}$ is defined as in (3.5).

We shall be interested in homotopy of continuous curves in $\mathbb{U}(H)$.

THEOREM 3.6 *Let* $A_1(t), A_2(t) \in \mathbb{U}(H)$, $t \in [0,1]$ *be continuous curves with common ends:* $A_1(0) = A_2(0)$; $A_1(1) = A_2(1)$. *Then* $A_1(t)$ *and* $A_2(t)$ *are homotopic with fixed ends (i.e. there exists a continuous function* $A : [0,1] \times [0,1] \to \mathbb{U}(H)$ *such that* $A(t,0) = A_1(t)$, $A(t,1) = A_2(t)$; $t \in [0,1]$; $A(0,s) = A_1(0)$; $A(1,s) = A_1(1)$; $s \in [0,1]$) *if and only if* $\mathrm{Ind}_+\{A_1(t)\} = \mathrm{Ind}_+\{A_2(t)\}$ *and* $\mathrm{Ind}_-\{A_1(t)\} = \mathrm{Ind}_-\{A_2(t)\}$.

PROOF. It is easily seen that $A_1(t)$ and $A_2(t)$ are homotopic (with fixed ends) if and only if the loop $B(t)$ defined by the equalities $B(t) = A_1(2t)$ for $0 \leq t \leq \frac{1}{2}$, $B(t) = A_2(-2t+2)$ for $\frac{1}{2} \leq t \leq 1$, is homotopic to a constant loop. The result now follows from the remark preceding Theorem 3.6. □

Let $S_u^{(c)}(H)$ be a connected component in the set $S_u(H)$ of all stably u-diagonable H-unitary matrices. Let $X_1(t)$, $X_2(t)$, $t \in [0,1]$ be continuous curves of H-unitary matrices such that $X_1(0) = X_2(0) = I$ and $X_1(1), X_2(1) \in S_u^{(c)}(H)$. We say that $X_1(t)$ and $X_2(t)$ are *equivalent* if there exists a continuous function $X : [0,1] \times [0,1] \to \mathbb{U}(H)$ such that $X(t,0) = X_1(t)$ and $X(t,1) = X_2(t)$ for $t \in [0,1]$; $X(0,s) = I$ and $X(1,s) \in S_u^{(c)}(H)$ for $s \in [0,1]$. Clearly, this equivalence is a true equivalence relation, i.e. it is symmetric, reflexive and transitive. If $\mathrm{Ind}_\pm\{X_1(t)\} = \mathrm{Ind}_\pm\{X_2(t)\}$, then Theorem 3.6 implies that $X_1(t)$ and $X_2(t)$ are equivalent. Indeed, $X_1(t)$ is clearly equivalent to the curve

$$X_3(t) = \begin{cases} X_1(2t) , & 0 \leq t \leq \frac{1}{2} \\ Y(2t-1) , & \frac{1}{2} \leq t \leq 1 , \end{cases}$$

where $Y(t)$ is a continuous curve in $S_u^{(c)}(H)$ connecting $X_1(1)$ and $X_2(1)$; the equivalence between $X_1(t)$ and $X_3(t)$ is given by the formula $X(t,s) = X_1((1 - \frac{1}{2} s)^{-1}t))$ for $0 \leq t \leq 1 - \frac{1}{2} s$, $0 \leq s \leq 1$; $X(t,s) = Y(2t+2-s)$ for $1 - \frac{1}{2} s \leq t \leq 1$, $0 \leq s \leq 1$.

In the next theorem we describe the equivalence classes.

THEOREM 3.7 *Let* $H = H^*$ *be an invertible* $n \times n$ *matrix, and let* $S_u^{(c)}(H)$ *be a connected component in* $S_u(H)$. *Let* p, q *be the number of positive and negative eigenvalues of* H, *respectively (so that* $p + q = n$). *Let* $X_1(t)$, $X_2(t)$ *be continuous paths in* $\mathbb{U}(H)$ *such that* $X_1(0) = X_2(0) = I$ *and* $X_1(1), X_2(1) \in S_u^{(c)}(H)$.

a) *if* $p = 0$, *or* $q = 0$, *then* $X_1(t)$ *and* $X_2(t)$ *are equivalent if and only if there exists a continuous path* $Y_1(t)$, $t \in [0,1]$ *with values in* $S_u^{(c)}(H)$ *and* $Y_1(0) = X_1(1)$, $Y_1(1) = X_2(1)$ *such that the integer* $\frac{1}{2\pi} \operatorname{Arg}[\det X_2(t)Y_2^{-1}(t)]_{t=0}^{t=1}$ *is divisible by* n; *here the continuous path* $Y_2(t)$ *is defined as follows:*

$$Y_2(t) = \begin{cases} X_1(2t) , & 0 \leq t \leq \frac{1}{2} ; \\ Y_1(2t-1) , & \frac{1}{2} \leq t \leq 1 . \end{cases}$$

b) *if* p *and* q *are both positive, then* $X_1(t)$ *and* $X_2(t)$ *are equivalent if and only if there exists a continuous* $S_u^{(c)}(H)$*-valued path* $Y_1(t)$, $t \in [0,1]$ *connecting* $X_1(1)$ *and* $X_2(t)$ *such that the numbers*

$$\frac{1}{q} (\mathrm{Ind}_-\{X_2(t)\} - \mathrm{Ind}_-\{Y_2(t)\})$$

and

$$\frac{1}{p} (\mathrm{Ind}_+\{X_2(t)\} - \mathrm{Ind}_-\{Y_2(t)\})$$

are integers and coincide, where $Y_2(t)$ *is defined as in* a).

PROOF. It is easily seen that the paths $X_1(t)$ and $Y_2(t)$ are equivalent, so we shall consider $Y_2(t)$ in place of $X_1(t)$. Without loss of generality we may assume that $H = \operatorname{diag}[I_p, -I_q]$.

Assume first that $p,q > 0$. Let $X_0 \in S_u^{(c)}(H)$ be a diagonal matrix with unimodular entries on the main diagonal. Theorem 3.1 shows that for any continuous loop in $\mathbb{U}(H)$ which starts and ends at X_0 there exist unique integers n_1 and n_2 such that this loop is homotopic to the loop

$$X_0 \operatorname{diag}[e^{2\pi i(n_1+n_2)\tau},1,\cdots,1,e^{-2\pi i n_2\tau},1,\cdots,1] , \quad 0 \leq \tau \leq 1$$

($e^{-2\pi i n_2\tau}$ appears in the (p+1)-th place). Here $X_0 \in S_u^{(c)}(H)$ is fixed. In particular, let n_1 and n_2 be such integers for the loop $Y(\tau)$ defined by the equalities $Y(\tau) = X_0X_2(2\tau)$ for $0 \leq \tau \leq \frac{1}{2}$; $Y(\tau) = X_0Y_2(-2\tau+2)$ for $\frac{1}{2} \leq \tau \leq 1$.

On the other hand, Theorem 3.5 shows (see also the remark after this theorem) that any loop in $S_u^{(c)}(H)$ which starts and ends at X_0 is homotopic to a multiple of the loop $e^{2\pi i\tau}X_0$, $\tau \in [0,1]$. So $X_2(t)$ and $Y_2(t)$ are equivalent if and only if for some integer n_3 the loop

$$X_0\operatorname{diag}[e^{2\pi i(n_1+n_2-n_3)\tau},e^{-2\pi i n_3\tau},\cdots,e^{-2\pi i n_3\tau},e^{-2\pi i(n_2+n_3)\tau},e^{-2\pi i n_3\tau},\cdots,$$
$$,e^{-2\pi i n_3\tau}] , \quad \tau \in [0,1] \tag{3.15}$$

is homotopic to the constant loop X_0. Using the fact that the group $S\mathbb{U}(k)$ is simply connected for $k = 1,2,\cdots$ (cf. the remark after the proof of Theorem 3.1) it is easily seen that the loop (3.15) is homotopic to the loop

$$X_0\operatorname{diag}[e^{2\pi i(n_1+n_2-pn_3)\tau},1,\cdots,1,e^{-2\pi i(n_2+qn_3)\tau},1,\cdots,1] , \tau \in [0,1].$$

This loop is homotopic to the constant loop X_0 if and only if $n_1 + n_2 - pn_3 = 0$ and $n_2 + qn_3 = 0$, which means $n_1 = (p+q)n_3$, $n_2 = -qn_3$ for some integer n_3. Observing that

$$n_1 = \operatorname{Ind}_+\{X_2(t)\} - \operatorname{Ind}_+\{Y_2(t)\} + \operatorname{Ind}_-\{X_2(t)\} - \operatorname{Ind}_-\{Y_2(t)\}$$

and

$$n_2 = -[\operatorname{Ind}_-\{X_2(t)\} - \operatorname{Ind}_-\{Y_2(t)\}]$$

we finish the proof of Theorem 3.7 in case (b).

The proof in case (a) is similar. □

3.6 Connected Components of Hamiltonian Systems With Stably Bounded Solutions

Return now to systems of the form (3.1), which are the main objects of concern in this chapter, and recall the definition of the set H_s made in Section 3.1. Now we will describe the structure of the connected components in H_s.

Recall that the $n \times n$ matrix solution $X(t)$ of the initial value problem

$$E \frac{dX}{dt} = iH(t)X \ ; \quad X(0) = I$$

is called the matrizant of (3.1) and is a piecewise differentiable curve in the group $\mathbb{U}(E)$, and the matrix $X(\omega)$ is the monodromy matrix of (3.1). Also recall from Chapter II.1 that the monodromy matrix of system (3.1) is E-unitary, and the system (3.1) has stably bounded solutions if and only if the monodromy matrix is stably u-diagonable (as an E-unitary matrix), see Theorem III.2.7. So (assuming $H \in H_s$) the homotopy indices $\mathrm{Ind}_{\pm}\{X(t)\}$ ($\mathrm{Ind}\{X(t)\}$ in case E is either positive definite or negative definite), as well as the index $E\text{-}\mathrm{ind}_u(X(\omega))$, make sense (we shall assume without loss of generality that $\omega = 1$).

THEOREM 3.8 *Assume* E *is neither positive definite nor negative definite, and let* p *(resp.* q*) be the number of positive (resp. negative) eigenvalues of* E *counting multiplicities. Let*

$$E \frac{dx}{dt} = iH_1(t)x \ , \quad t \in \mathbb{R} \tag{3.16}$$

and

$$E \frac{dx}{dt} = iH_2(t)x \ , \quad t \in \mathbb{R} \tag{3.17}$$

be differential equations of type (3.1) *with stably bounded solutions and corresponding matrizants* $X_1(t)$ *and* $X_2(t)$. *The equations* (3.16) *and* (3.17) *belong to the same connected component in the set of all differential equations of type* (3.1) *with stably bounded solutions if and only if the following two conditions hold:*

1) $E\text{-}\mathrm{ind}_u(X_1(1))$ *and* $E\text{-}\mathrm{ind}_u(X_2(1))$ *coincide up to a cyclic permutation;*

2) *there exists a continuous path* $Y_1(t)$, $t \in [0,1]$ *with E-unitary stably u-diagonable values and* $Y_1(0) = X_1(1)$, $Y_1(1) = X_2(1)$ *such that the numbers*

$p^{-1}(\mathrm{Ind}_+\{X_2(t)\} - \mathrm{Ind}_+\{Y_2(t)\})$ *and* $q^{-1}(\mathrm{Ind}_-\{X_2(t)\} - \mathrm{Ind}_-\{Y_2(t)\})$ *are integers and coincide; here*

$$Y_2(t) = \begin{cases} X_1(2t) , & 0 \leq t \leq \frac{1}{2} ; \\ Y_1(2t-1) , & \frac{1}{2} \leq t \leq 1 . \end{cases}$$

If E *is either positive definite or negative definite, then* 1) *and* 2) *are replaced by the following: there exists a continuous* E*-unitary stably* u-*diagonable path* $Y_1(t)$, $t \in [0,1]$ *connecting* $X_1(1)$ *and* $X_2(1)$ *such that the integer*

$$\frac{1}{2\pi} \mathrm{Arg}[\det X_2(t)Y_2^{-1}(t)]$$

is divisible by n, *where* $Y_2(t)$ *is defined as in* 2).

PROOF. Assume H_1 and H_2 belong to the same connected component in $\mathcal{H}_s$, so there exists a continuous curve $H(\zeta) \in \mathcal{H}_s$, $\zeta \in [0,1]$ such that $H(0) = H_1$, $H(1) = H_2$. By Theorem II.1.2 the matrizants $X(t;\zeta)$ of $H(\zeta)$ also form a continuous curve. In particular, the monodromy matrices $X(1;\zeta)$ form a continuous curve in the set $\mathcal{S}_u(E)$. So, in view of Theorem 1.3, $E\text{-ind}_u(X_1(1))$ and $E\text{-ind}_u(X_2(1))$ are obtained from each other by a cyclic permutation. The matrizants $X(t,\zeta)$ also provide an equivalence between the curves $X(t,0)$ and $X(t,1)$, hence the condition 2) holds by Theorem 3.7.

Now assume that 1) and 2) hold: by Theorem 1.3, the E-unitary matrices $X_1(1)$ and $X_2(1)$ belong to the same connected component $\mathcal{S}_u^{(c)}(H)$ in the set of all E-unitary stably u-diagonable matrices. In view of Theorem 3.7, the curves $X_1(t)$ and $X_2(t)$, $t \in [0,1]$ are equivalent, which means that there exists a continuous function $X : [0,1] \times [0,1] \to \mathbb{U}(E)$ such that $X(t,0) = X_1(t)$; $X(t,1) = X_2(t)$ for $t \in [0,1]$; $X(0,s) = I$; $X(1,s) \in \mathcal{S}_u^{(c)}(H)$ for $s \in [0,1]$. It is intuitively clear (and can be proved formally) that the function $X(t,s)$ can also be chosen in such a way that $X(t,s)$ is piecewise continuously differentiable in t, for every fixed $s \in [0,1]$, and $\frac{\partial X(t,s)}{\partial t}$ is a continuous function of s. Then (for every fixed s_0) $X(t,s_0)$ is the matrizant of a system (3.1) for some complex hermitian piecewise continuous function $H(t,s_0)$ with period ω (Theorem II.1.1), and, moreover, $H(t,s_0)$ depends continuously on s_0 (in the sense of the norm $\max_{t\in[0,\omega]} ||H(t,s_0)||$), see the proof of Theorem II.1.1). □

Let $\mathcal{V}$ be the set of sequences of integers $V = \{v_1,\cdots,v_r\}$ (where

$r \geq 1$ may depend on V) and the integers satisfy: $v_i v_{i+1} < 0$, $i = 1,\cdots,r$ ($v_{r+1} = v_1$ by definition); $\sum_{i=1}^{r} |v_i| = n$; $\sum_{i=1}^{r} v_i = \text{sig } E$.

Fix $V \in \mathcal{V}$ and fix a piecewise continuously differentiable curve $X_0(t)$, $t \in [0,\omega]$ with $X_0(0) = I$ and $E\text{-ind}_u(X_0(\omega)) = V$. It is easily seen that for every pair of integers (n_+,n_-) there exists a differential equation of type (3.1) with stably bounded solutions such that its matrizant $X(t)$ has the properties that $X(\omega) = X_0(\omega)$ and

$$\text{Ind}_{\pm}\{X(t)\} - \text{Ind}_{\pm}\{X_0(t)\} = n_{\pm}$$

(cf. Theorem 3.8); it is assumed that E is not definite. Consequently, Theorem 3.8 shows, in particular, that the number of connected components in H_s is countably infinite, except for the case when E is either positive definite or negative definite (in this case there are exactly n connected components in H_s, where n is the size of E).

The description of the connected components of H_s given in Theorem 3.8 can also be obtained using the notion of fundamental group. Let $S_u^{(c)}(E)$ be a connected component in the set of all E-unitary stably u-diagonable matrices. It turns out that the fundamental group of the factor topological space $\mathbb{U}(E)/S_u^{(c)}(E)$ is isomorphic to $\mathbb{Z}_n$ if the $n \times n$ matrix E is either positive definite or negative definite, and isomorphic to $\mathbb{Z} \times \mathbb{Z}_m$ if E is not definite. Here m is the greatest common divisor of the number of positive eigenvalues and the number q of negative eigenvalues of E (counting multiplicities), and $\mathbb{Z}_m$ is the commutative group of integers modulo m. (In particular, the fundamental group of $\mathbb{U}(E)/S_u^{(c)}(E)$ is isomorphic to $\mathbb{Z}$ if p and q are relatively prime.) Moreover, the connected components of the set H_s are in natural one-to-one correspondence with the set of all pairs (V,g), where $V \in \mathcal{V}$ and g belongs to the fundamental group of $\mathbb{U}(E)/S_u^{(c)}(E)$.

3.7 Simple Connectedness of Connected Components of Real E-Orthogonal Matrices

In the next section we shall describe connected components in the set of all Hamiltonian equations with *real* periodic coefficients whose solutions are stably bounded (a real analogue of Theorem 3.8). To this end we prove in this section the simple connectedness of the connected components of real E-orthogonal matrices. The presentation of the real case will be less detailed than the corresponding presentation in the complex case (Sections 3.1-3.6).

Let E be a real $n \times n$ skew-symmetric invertible matrix, and let $S(E)$ be the set of all (real) E-orthogonal stably u-diagonable matrices. The structure of the connected components in $S(E)$ was described in Section 1.3.

We shall need the notion of an argument of a continuous curve of E-orthogonal matrices, which will now be introduced. As we know (Theorem II.1.7) the group $\mathbb{U}_R(E) \cap \mathbb{O}(n)$ of all $n \times n$ matrices which are both E-orthogonal and orthogonal, is homeomorphic to the group $\mathbb{U}(\frac{n}{2})$ of all $\frac{n}{2} \times \frac{n}{2}$ unitary matrices. If $E = \begin{bmatrix} 0 & I \\ -I & 0 \end{bmatrix}$, this homeomorphism $\varphi : \mathbb{U}_r(E) \cap \mathbb{O}(n) \to \mathbb{U}(\frac{n}{2})$ can be chosen to be particularly simple: write $X = \begin{bmatrix} A & B \\ C & D \end{bmatrix}$, where X is orthogonal and E-orthogonal, and A, B, C, D are $\frac{n}{2} \times \frac{n}{2}$ matrices; then $A = D$, $C = -B$ and $\varphi(X) = A + iB$. In the sequel we shall denote by φ a fixed homeomorphism of $\mathbb{U}_R(E) \cap \mathbb{O}(n)$ onto $\mathbb{U}(\frac{n}{2})$.

Let $Y(t)$, $t \in [0,\omega]$ be a continuous curve in the group $\mathbb{U}_r(E)$ of (real) E-orthogonal matrices. For each $Y(t)$, let $U(t) = (Y(t)Y(t)^T)^{-\frac{1}{2}}Y(t)$; then (Section II.1.5) $U(t)$ is E-orthogonal and orthogonal. Now put

$$k = \frac{1}{2\pi} [\mathrm{Arg}(\det \varphi(U(t)))]_{t=0}^{t=\omega} ;$$

where $\mathrm{Arg}\, z$ is a continuously changing value of the complex number $z \neq 0$. The number k will be called the *argument* of $Y(t)$. The importance of this notion lies in the following fact.

THEOREM 3.9 *Two continuous curves in* $\mathbb{U}_R(E)$ *with common beginning and common ends are homotopic in* $\mathbb{U}_R(E)$ *(with homotopy preserving the common beginnings and ends) if and only if their arguments are equal.*

PROOF. In view of Theorem II.1.7 we have to check that two continuous curves $Y_1(t)$, $Y_2(t)$, $t \in [0,\omega]$ in the group $\mathbb{U}(m)$ $(m = \frac{n}{2})$, and such that $Y_1(0) = Y_2(0)$, $Y_1(\omega) = Y_2(\omega)$, are homotopic in $\mathbb{U}(m)$ (with homotopy preserving the beginnings $Y_1(0)$ and the ends $Y_1(\omega)$) if and only if

$$[\mathrm{Arg}(\det Y_1(t))]_{t=0}^{t=\omega} = [\mathrm{Arg}(\det Y_2(t))]_{t=0}^{t=\omega} . \tag{3.18}$$

Now the group $\mathbb{U}(m)$ is homeomorphic to the topological product of the simply connected group $\mathbb{SU}(m) = \{X \in \mathbb{U}(m) \mid \det X = 1\}$ and the unit circle $\mathbb{T}$ (cf. end of the proof of Theorem 3.1). So $Y_1(t)$ and $Y_2(t)$ are homotopic if and only if their projections $\det Y_1(t)$ and $\det Y_2(t)$ on $\mathbb{T}$ are homotopic. As it is well-known, the condition for homotopy of continuous $\mathbb{T}$-valued curves $\det Y_1(t)$ and $\det Y_2(t)$ is exactly (3.18). □

THEOREM 3.10 *Each connected component of the set* $S(E)$ *of all stably* u-*diagonable* E-*orthogonal matrices is simply connected.*

PROOF. By Corollary III.1.12 we can assume $E = \begin{bmatrix} 0 & I \\ -I & 0 \end{bmatrix}$. Let $S(E;v_1,\cdots,v_p)$ be a connected component of $S(E)$ determined by the sequence of integers $v_1,\cdots,v_p$ (see Theorem 1.5). Let $A_1(t)$, $A_2(t)$, $t \in [0,1]$ be two continuous curves in $S(E;v_1,\cdots,v_p)$ with $A_1(0) = A_2(0)$, $A_1(1) = A_2(1)$. We shall prove that $A_1(t)$ and $A_2(t)$ are homotopic (with homotopy preserving $A_1(0)$ and $A_1(1)$). By Theorem III.1.10(iv), $A_k(t)$, $k = 1,2$ admits the representation

$$A_k(t) = G_k(t) \begin{bmatrix} \cos\theta_k(t) & \sin\theta_k(t) \\ -\sin\theta_k(t) & \cos\theta_k(t) \end{bmatrix} G_k(t)^{-1}, \quad t \in [0,1]$$

where $G_k(t)$ is E-orthogonal, $\theta_k(t) = \mathrm{diag}[\theta_{1k}(t),\cdots,\theta_{\frac{n}{2},k}(t)]$ with $-\pi < \theta_{ik}(t) < \pi$, $\theta_{ik}(t) + \theta_{jk}(t) \neq 0$ for $1 \leq i,j \leq \frac{n}{2}$. It turns out that $G_k(t)$ and $\theta_k(t)$ can be chosen to depend continuously on t. Indeed, this follows from the proof of (iii)→(iv) in Theorem III.1.10, taking into account Theorem A.10 in the Appendix to Part III. (One applies Theorem A.10 with H=iE, and for fixed $k = 1,2$; $q = 1,\cdots,p$ the subspace $M(t)$, $t \in [0,1]$ is equal to the sum of root subspaces of $A_k(t)$ corresponding to the eigenvalues in the q-th consecutive arc $(\alpha_{q-1},\alpha_q) = (\alpha_{q-1}(t),\alpha_q(t))$ on the upper half of the unit circle taken from the definition of E-ind$(A_k(t))$). Moreover, by the same Theorem A.10 we can arrange that

$$\theta_1(0) = \theta_2(0) ; \quad \theta_1(1) = \theta_2(1) ;$$

$$G_1(0) = G_1(1) ; \quad G_1(1) = G_2(1) .$$

Since $A_k(t) \in S(E;v_1,\cdots,v_p)$; $k = 1,2$; $t \in [0,1]$, it is easily seen that the curves $\theta_1(t)$ and $\theta_2(t)$ are homotopic (with fixed ends) in the set of all matrices of type $\mathrm{diag}[\mu_1,\cdots,\mu_{\frac{n}{2}}]$, $-\pi < \mu_i < \pi$, $\mu_i + \mu_j \neq 0$ for $1 \leq i,j \leq \frac{n}{2}$. If the arguments of $G_1(t)$ and $G_2(t)$ are equal, then $G_1(t)$ and $G_2(t)$ are homotopic (see Theorem 3.9), and we are done. Suppose the argument of $G_1(t)$ is greater than that of $G_2(t)$ by an integer $m \neq 0$. Put $\hat{G}_2(t) = G_2(t)N(t)$, where

$$N(t) = \begin{bmatrix} \mathrm{diag}[\cos 2\pi tm, 1, \cdots, 1] & \mathrm{diag}[\sin 2\pi tm, 1, \cdots, 1] \\ \mathrm{diag}[-\sin 2\pi tm, 1, \cdots, 1] & \mathrm{diag}[\cos 2\pi tm, 1, \cdots, 1] \end{bmatrix}$$

is an $n \times n$ matrix represented as a 2×2 block matrix, each block being of size $\frac{n}{2} \times \frac{n}{2}$. It is easily seen that $N(t)$ commutes with $R_2(t) = \begin{bmatrix} \cos \theta_2(t) & \sin \theta_2(t) \\ -\sin \theta_2(t) & \cos \theta_2(t) \end{bmatrix}$; so $A_2(t) = \hat{G}_2(t)R_2(t)(\hat{G}_2(t))^{-1}$. Clearly, $\hat{G}_2(t)$ is E-orthogonal. Hence we can replace $G_2(t)$ by $\hat{G}_2(t)$ from the very beginning. But now the argument of $\hat{G}_2(t)$ is greater than the argument of $G_2(t)$ by m, as one sees from the definition of the argument, using $\varphi\begin{bmatrix} A & B \\ -B & A \end{bmatrix} = A + iB$. Thus, the arguments of $G_1(t)$ and $G_2(t)$ are equal, and the proof can be finished as indicated above. □

Now we introduce an analogue for the notion of equivalence of $\mathbb{U}(H)$-valued curves defined in Section 3.5. Let $S^{(c)}(E)$ be a connected component in the set $S(E)$ of all E-orthogonal stably u-diagonable matrices. Two continuous curves $X_i(t)$, $t \in [0,1]$, $i = 1,2$ of E-orthogonal matrices such that $X_i(0) = I$ and $X_i(1) \in S^{(c)}(E)$, $i = 1,2$ are called *real equivalent* if there exists a continuous function $X : [0,1] \times [0,1] \to \mathbb{U}_r(E)$ (the group of all E-orthogonal matrices) such that $X(t,0) = X_1(t)$, $X(t,1) = X_2(t)$ for $t \in [0,1]$; $X(0,s) = I$, $X(1,s) \in S^{(c)}(E)$ for $s \in [0,1]$.

The notion of equivalence can be described easily in terms of the argument of a continuous $\mathbb{U}_r(E)$-valued curve, as follows. The continuous curves $X_1(t)$ and $X_2(t)$ as above are real equivalent if and only if there exists a $S^{(c)}(E)$-valued continuous path $Y_0(t)$, $t \in [0,1]$ with $Y_0(0) = X_1(1)$, $Y_0(1) = X_2(1)$ such that the curve $Z(t)$ defined by

$$Z(t) = \begin{cases} X_1(3t) , & 0 \leq t \leq \frac{1}{3} \\ Y_0(3t-1) , & \frac{1}{3} \leq t \leq \frac{2}{3} \\ X_2(-3t+3), & \frac{2}{3} \leq t \leq 1 \end{cases}$$

has zero argument. In view of Theorem 3.10 if some $S^{(c)}(E)$-valued continuous path $Y_0(t)$ has this property, then the same is true for any $S^{(c)}(E)$-valued continuous path $Y(t)$, $t \in [0,1]$ with $Y(0) = X_1(1)$, $Y(1) = X_2(1)$.

3.8 Connected Components of Real Hamiltonian Systems With Stably Bounded Solutions

Now we return to Hamiltonian systems of the form

$$E\frac{dx}{dt} = H(t)x\ , \quad t \in \mathbb{R} \tag{3.19}$$

where $E = -E^T$ is a real and invertible $n \times n$ matrix, and the piecewise continuous $n \times n$ matrix function $H(t)$ is real, symmetric and periodic with period ω. The system (3.19) is said to have *stably bounded solutions* (in the class of systems with real coefficients) if there exists $\varepsilon > 0$ such that for every real symmetric piecewise continuous matrix function $\tilde{H}(t)$ with period ω such that $||H(t)-\tilde{H}(t)|| < \varepsilon$, $t \in [0,\omega]$, all solutions of the system

$$E\frac{dx}{dt} = \tilde{H}(t)x\ , \quad t \in \mathbb{R}$$

are bounded. In particular, in this case all solutions of (3.19) are bounded. As in the case of Hamiltonian systems with complex coefficients, the set of systems of type (3.19) can be identified with the set C of all piecewise continuously differentiable curves $X(t)$, $t \in [0,\omega]$ with the properties that $X(0) = I$ and $X(t)$ is E-orthogonal for all $t \in [0,\omega]$ (see Section II.1.4). In fact, $X(t)$ is the matrizant of (3.19). Moreover, the set of systems (3.19) with stably bounded solutions is identified with the set of all curves $X(t) \in C$ such that $X(\omega)$ is a stably u-diagonable matrix (see Theorems II.1.8 and II.1.9).

THEOREM 3.11 *Let*

$$E\frac{dx}{dt} = H_j(t)x\ , \quad t \in \mathbb{R}\ , \quad j = 1,2 \tag{3.20}$$

be differential equations of type (3.19) *with stably bounded solutions and corresponding matrizants* $X_j(t)$, $j = 1,2$. *The equations* (3.20) *belong to the same connected component in the set of all differential equations of type* (3.19) *with stably bounded solutions if and only if the following conditions hold:*

1) $\text{E-ind}(X_1(\omega))$ *coincides with* $\text{E-ind}(X_2(\omega))$;

2) *there exists a continuous path* $Y_0(t)$, $t \in [0,\omega]$ *with* E-*orthogonal stably* u-*diagonable values satisfying* $Y_0(0) = X_1(\omega)$, $Y_0(\omega) = X_2(\omega)$ *such that the*

argument of the curve $Z(t)$, $t \in [0,\omega]$ *defined by*

$$Z(t) = \begin{cases} X_1(3t) \quad , & 0 \leq t \leq \frac{\omega}{3} \\ Y_0(3t-\omega) \quad , & \frac{\omega}{3} \leq t \leq \frac{2\omega}{3} \\ X_2(-3t+3\omega) \ , & \frac{2\omega}{3} \leq t \leq \omega \end{cases}$$

is zero.

To prove Theorem 3.11 observe the identification between Hamiltonian equation (3.19) and certain piecewise differentiable curves with E-orthogonal values explained at the beginning of this section. Now use Theorems 3.9 and 3.10, as well as Theorem 1.5.

If condition 2) in Theorem 3.11 holds for some $Y_0(t)$, then it will hold for any continuous path $Y(t)$, $t \in [0,\omega]$ with E-orthogonal stably u-diagonable values satisfying $Y(0) = X_1(\omega)$, $Y(\omega) = X_2(\omega)$.

In particular, Theorem 3.11 shows that the set of all differential equations of type (3.19) with stably bounded solutions has countably many connected components.

Again, the result of Theorem 3.11 can be stated in terms of the fundamental group. It turns out that the connected components of the set of all equations (3.19) with stably bounded solutions are in natural one-to-one correspondence with the fundamental group of the factor topological space $\mathbf{U}_r(E)/S^{(c)}(E)$, where $S^{(c)}(E)$ is a connected component in $S(E)$. This fundamental group is isomorphic to $\mathbb{Z}$.

NOTES TO PART IV

The contents of Chapters 1 and 2 (except for Section 1.3) are taken from [19c]. Section 1.3 is probably new.

The main results of Chapter 3 in the real case are due to Gelfand and Lidskii [15] and in the complex case to Coppel and Howe [11]. Section 3.5 is based on [50]. The proofs of the main results in Chapter 3 seem to be new. The auxiliary results obtained (e.g., Theorems 3.3 and 3.4; see also the remark after Theorem 3.4) are of independent interest, and probably also new.

REFERENCES

1. B.D.O. Anderson, S. Vongpanitlerd: Network analysis and synthesis. Prentice Hall, Englewood Cliffs, 1973.

2. T. Ya. Azizov, I.S. Iohvidov: Linear operators in spaces with an indefinite metric and their applications. *Itogi Nauki i Techniki, Matem. Analiz.* Vol. 17, 113-205 (1979) (Russian).

3. J.A. Ball, J.W. Helton: Factorization results related to shifts in an indefinite metric. *Integral Equations and Operator Theory,* Vol. 5, 632-658 (1982).

4. H. Bart, I. Gohberg, M.A. Kaashoek: Minimal factorization of matrix and operator functions. *Operator Theory: Advances and Applications,* Vol. 1. Birkhauser, Basel, 1979.

5. H. Baumgartel: Endlich-dimensionale Analytische Störungstheorie. Akademie-Verlag, Berlin, 1972.

6. J. Bognar: Indefinite inner product spaces. Springer-Verlag, Berlin and New York, 1974.

7. R. Brockett: Finite dimensional linear systems. John Wiley, New York, London, Sydney, Toronto, 1970.

8. N. Burgoyne, R. Cushman: Normal forms for real linear Hamiltonian systems. Ames Research Center (NASA) conference on geometric control theory, Moffett Field, Calif., 483-529 (1976).

9. C. Chevalley: Theory of Lie groups, I. Princeton, 1946.

10. W.A. Coppel: Matrix quadratic equations. *Bull. Austral. Math. Soc.*, Vol. 10, 377-401 (1974).

11. W.A. Coppel, A. Howe: On the stability of linear canonical systems with periodic coefficients. *J. Australian Math. Soc.*, Vol. 5, 169-195 (1965).

12. Ju. L. Daleckii, M.G. Krein: Stability of solutions of differential equations in Banach space. *Amer. Math. Soc. Transl. Math. Monographs,* Vol. 43, Providence, Rhode Island, 1974.

13a. P.A. Fuhrmann: On symmetric rational matrix functions. *Linear Alg. Applic.*, to appear.

13b. P.A. Fuhrmann: On Hamiltonian rational transfer functions, preprint.

14. F.R. Gantmacher. The theory of matrices. Vols. I and II, Chelsea, New York, 1959.

15. I.M. Gelfand, V.B. Lidskii: On the structure of the domains of stability of linear canonical systems of differential equations with periodic coefficients (Russian). *Uspehi Mat. Nauk*, Vol. 10, 3-40 (1955); *Amer. Math. Soc. Transl.*, Vol. 8, 143-181 (1958).

16. I.M. Glazman, Ju. I. Ljubic: Finite-dimensional linear analysis. MIT Press, Cambridge and London, 1974.

17a. I.C. Gohberg, M.G. Krein: The basic propositions on defect numbers, root numbers and indices of linear operators. *Uspehi Mat. Nauk.* Vol. 12, 43-118 (1957); *Russian Math. Surveys*, Vol. 13, 185-264 (1960).

17b. I.C. Gohberg, M.G. Krein: Theory of Volterra operators in Hilbert space and its applications. *Amer. Math. Soc. Transl. Math. Monographs,* Vol. 24, Providence, Rhode Island, 1970.

18. I. Gohberg, N. Krupnik: Einführung in die Theorie der eindimensionalen singulären Integraloperatoren. Birkhauser, Basel, 1979.

19a. I. Gohberg, P. Lancaster, L. Rodman: Spectral analysis of selfadjoint matrix polynomials. *Ann. of Math.*, Vol. 112, 34-71 (1980).

19b. I. Gohberg, P. Lancaster, L. Rodman: Matrix polynomials. Academic Press, 1982.

19c. I. Gohberg, P. Lancaster, L. Rodman: Perturbations of H-selfadjoint matrices, with applications to differential equations. *Integral Equations and Operator Theory*, Vol.5, 718-757 (1982).

19d. I. Gohberg, P. Lancaster, L. Rodman: A sign characteristic for selfadjoint meromorphic matrix functions. *Applicable Analysis*, to appear.

20. I. Gohberg, E.I. Sigal: On operator generalizations of the logarithmic residue theorem and the theorem of Rouché. *Math. USSR-Sb.* Vol. 13, 603-625 (1971).

21. S. Helgason: Differential geometry and symmetric spaces. Academic Press, New York, 1962.

22. E. Hille: Lectures on ordinary differential equations. Addison-Wesley, 1969.

23. I.S. Iohvidov, M.G. Krein, H. Langer: Introduction to spectral theory of operators in spaces with an indefinite metric. Math. Research, Vol. 9, Akademie-Verlag, Berlin, 1982.

24. C.R. Johnson, C.R. De Prima: The range of $A^{-1}A^*$ in GL(n,C). *Linear Alg. Applic.* Vol. 9, 209-222 (1974).

25. T. Kato: Perturbation theory for linear operators. Springer-Verlag, Berlin and New York, 1966.

26a. M.G. Krein: The basic propositions in the theory of λ-zones of stability of a canonical system of linear differential equations with periodic coefficients (Russian). Pamyati A.A. Andronova, *Izdat. Akad. Nauk. SSR*, Moscow, 413-498 (1955). English transl. in: M.G. Krein: Topics in differential and integral equations and operator theory. *Operator Theory: Advances and Applications,* Vol. 7, Birkhauser, Basel, 1983.

26b. M.G. Krein: Introduction to the geometry of indefinite J-spaces and to the theory of operators in those spaces. *AMS Transl.* (2) 93, 103-176 (1970).

27. M.G. Krein, H. Langer: On some mathematical principles in the linear theory of damped oscillations of continua, I, II. *Integral Equations and Operator Theory*, Vol. 1, 364-399, 539-566 (Translation) (1978).

28. L. Kronecker: Algebraishe Reduktion der Scharen bilinearer Formen. S.-B. Akad. Berlin, 763-776 (1890).

29. H. Kwakernaak, R. Sivan: Linear optimal control systems. Wiley-Interscience, New York, London, Sydney, Toronto, 1972.

30a. P. Lancaster: Lambda-matrices and vibrating systems. Pergamon, Oxford, 1966.

30b. P. Lancaster: Theory of matrices. Academic Press, New York, 1969.

31. P. Lancaster, L. Rodman: Existence and uniqueness theorems for the algebraic Riccati equation. *Internat. J. Control*, Vol. 32, 285-309 (1980).

32. N. Levinson: The stability of linear, real, periodic self-adjoint systems of differential equations. *J. of Math. Anal. Appl.*, Vol. 6, 473-482 (1963).

33. V.B. Lidskii, P.A. Frolov: The structure of the domain of stability of selfadjoint systems of differential equations with periodic coefficients (Russian). *Mat. Sbornik*, Vol. 71 (113), 48-64 (1966).

34. Loo-Keng, Hua: On the theory of automorphic functions of a matrix variable, II. The classification of hypercircles under the symplectic group. *Amer. J. Math.*, Vol. 66, 531-563 (1944).

35. A.I. Mal'cev: Foundations of Linear Algebra. W.H. Freeman, San Francisco and London, 1963.

36. A.S. Marcus, V.I. Matsaev, G.I. Russu: On some generalizations of the theory of strongly damped bundles to the case of bundles of arbitrary order (Russian). *Acta Sci. Math.* Szeged 34, 245-271 (1973).

37. M.A. Naimark: Linear differential operators, Parts I, II. George G. Harrap & Co., London, 1967.

38. A.C.M. Ran: Minimal factorization of selfadjoint rational matrix functions. *Integral Equations and Operator Theory*, Vol. 5, 850-869 (1982).

39. A.C.M. Ran, L. Rodman: Stability of neutral invariant subspaces in indefinite inner products and stable symmetric factorizations. *Integral Equations and Operator Theory*, to appear.

40a. L. Rodman: On extremal solutions of the algebraic Riccati equation. Algebraic and geometric methods in linear systems theory, *AMS Lectures in Applied Mathematics*, Vol. 18, 311-327 (1980).

40b. L. Rodman: Maximal invariant neutral subspaces and application to the algebraic Riccati equation. Submitted to *Manuscripta Matematica*.

41. M.A. Shayman: Geometry of the algebraic Riccati equations, Parts I and II. *SIAM J. Control*, to appear.

42. G.E. Shilov: Mathematical analysis. Finite dimensional linear spaces. *Nauka*, Moscow, 1969 (Russian). English transl.: *Linear Algebra*, Prentice-Hall, 1971.

43. Ju. L. Shmuljan: Finite dimensional operators depending analytically on a parameter. *Ukrainian Math. J.* IX, 195-204 (1957)(Russian).

44. G.R. Trott: On the canonical form of a non-singular pencil of hermitian matrices. *Amer. J. Math.*, Vol. 56, 359-371 (1934).

45a. F. Uhlig: A canonical form for a pair of real symmetric matrices that generate a nonsingular pencil. *Lin. Alg. Applic.*, Vol. 14, 189-209 (1976).

45b. F. Uhlig: Inertia and eigenvalue relations between symmetrized and symmetrizing matrices for the real and the general field case. *Lin. Alg. Applic.*, Vol. 35, 203-226 (1981).

46. K. Weierstrass: Zur Theorie der quadratischen und bilinearen Formen. Monatsber. Akad. Wiss., Berlin, 310-338 (1868).

47. J.C. Willems: Least squares stationary optimal control and the algebraic Riccati equation. *IEEE Trans. on Aut. Cont. AC*-16, 621-634 (1971).

48a. J. Williamson: The equivalence of non-singular pencils of hermitian matrices in an arbitrary field. *Amer. J. Math.* Vol. 57, 475-490 (1935).

48b. J. Williamson: On the algebraic problem concerning the normal forms of linear dynamical systems. *Amer. J. Math.*, Vol. 58, 141-163 (1936).

49. H. Wimmer: On the algebraic Riccati equation. *Bull. Austr. Math. Soc.*, Vol. 14, 457-461 (1976).

50. V.A. Yakubovic: Arguments on the group of symplectic matrices. *Matem. Sbornik*, Vol. 55, 255-280 (1961) (Russian).

51. V.A. Yakubovic, V.M. Starzhinskii: Linear differential equations with periodic coefficients, Vols. I, II. John Wiley, New York, Toronto, 1975.

LIST OF NOTATIONS

$\bar{Z}$ matrix whose entries are complex conjugate to those of the matrix Z

Z^T transposed matrix

Z^* conjugate transposed matrix

I_n $n \times n$ identity matrix

Im Z image (or range) of the matrix Z (subspace spanned by its columns)

Ker Z kernel of Z

$\sigma(Z) = \{\lambda \in ¢ \mid \det(\lambda I - Z) = 0\}$; spectrum of a square matrix Z

$E_Z(\lambda_0) = \mathrm{Ker}(\lambda_0 I - Z)^n$ root subspace of the $n \times n$ matrix Z corresponding to $\lambda_0 \in \sigma(Z)$

$Z|_M$ restriction of a matrix Z to an invariant subspace M

$\mathrm{diag}[Z_1,\cdots,Z_k]$, or $Z_1 \oplus \cdots \oplus Z_k$, block diagonal matrix whose diagonal blocks are $Z_1,\cdots,Z_k$

$\mathrm{Span}\{x_1,\cdots,x_k\}$ subspace spanned by the vectors $x_1,\cdots,x_k$

sig H signature of a hermitian matrix H (difference between the number of positive eigenvalues and the number of negative eigenvalues, counting multiplicities)

$Re\ \lambda$, $Im\ \lambda$ real part, imaginary part of $\lambda \in ¢$

sgn λ sign of a non-zero real number λ

$¢^n$ the complex linear space of n-dimensional column vectors with complex coordinates

$\langle x_1,\cdots,x_n \rangle$ typical vector from $¢^n$

$||x|| = (\sum_{i=1}^{n} |x_i|^2)^{\frac{1}{2}}$, $x = \langle x_1,\cdots,x_n \rangle \in ¢^n$

$¢^{n \times n}$ linear space of all $n \times n$ (complex) matrices

$||Z||$ euclidean norm of the matrix Z

$\mathbb{R}$ real numbers

$\mathbb{Z}$	integers
$\mathbb{T}$	unit circle
$\mathbb{U}(n)$	the group of $n \times n$ unitary matrices

$\mathbb{SU}(n) = \{Z \in \mathbb{U}(n) \mid \det Z = 1\}$

$\mathbb{O}(n)$	the group of $n \times n$ real orthogonal matrices
$\mathbb{U}(H)$	the group of H-unitary matrices
$\mathbb{U}_r(H)$	the group of real H-unitary matrices ($H = H^*$ real and invertible)
$\mathbb{U}_r(E)$	the group of (real) E-orthogonal matrices ($E = -E^*$ real and invertible)
$GL_r(n)$	the group of $n \times n$ real invertible matrices

SUBJECT INDEX